W9-APZ-748

Tensor Analysis on Manifolds

Richard L. Bishop
University of Illinois

Samuel I. Goldberg
University of Illinois

Dover Publications, Inc.
New York

Copyright © 1968, 1980 by Richard L. Bishop and Samuel
I. Goldberg.
All rights reserved under Pan American and International
Copyright Conventions.

Published in Canada by General Publishing Company, Ltd.,
30 Lesmill Road, Don Mills, Toronto, Ontario.

This Dover edition, first published in 1980, is an unabridged
and corrected republication of the work originally published
by The Macmillan Company in 1968.

International Standard Book Number: 0-486-64039-6
Library of Congress Catalog Card Number: 80-66959
Manufactured in the United States of America
Dover Publications, Inc.
180 Varick Street
New York, N.Y. 10014

Preface

*"Sie bedeutet einen wahren Triumph der durch Gauss,
Riemann, Christoffel, Ricci . . . begründeten Methoden des
allgemeinen Differentialcalculus."*   ALBERT EINSTEIN, 1915

SINCE ITS DEVELOPMENT BY RICCI between 1887 and 1896, tensor analysis has had a rather restricted outlook despite its striking success as a mathematical tool in the general theory of relativity and its adaptability to a wide range of problems in differential equations, geometry, and physics. The emphasis has been on notation and manipulation of indices. This book is an attempt to broaden this point of view at the stage where the student first encounters the subject. We have treated tensor analysis as a continuation of advanced calculus, and our standards of rigor and logical completeness compare favorably with parallel courses in the curriculum such as complex variable theory and linear algebra.

For students in the physical sciences, who acquire mathematical knowledge on a "need-to-know" basis, this book provides organization. On the other hand, it can be used by mathematics students as a meaningful introduction to differential geometry.

A broad range of notations is explained and interrelated, so the student will be able to continue his studies among either the classical references, those in the style of E. Cartan, or the current abstractions.

The material has been organized according to the dictates of mathematical structure, proceeding from the general to the special. The initial chapter has been numbered 0 because it logically precedes the main topics. Thus Chapter 0 establishes notation and gives an outline of a body of theory required to put the remaining chapters on a sound and logical footing. It is intended to be a handy reference but not for systematic study in a course. Chapters 1 and 2 are independent of each other, representing a division of tensor analysis into its function-theoretical and algebraic aspects, respectively. This material is combined and developed in several ways in Chapters 3 and 4, without specialization of mathematical structure. In the last two chapters (5 and 6) several important special structures are studied, those in Chapter 6 illustrating how the previous material can be adapted to clarify the ideas of classical mechanics.

Advanced calculus and elementary differential equations are the minimum background necessary for the study of this book. The topics in advanced calculus which are essential are the theory of functions of several variables, the implicit function theorem, and (for Chapter 4) multiple integrals. An understanding of what it means for solutions of systems of differential equations to exist and be unique is more important than an ability to crank out general solutions. Thus we would not expect that a student in the physical sciences would be ready for a course based on this book until his senior year. Mathematics students intent on graduate study might use this material as early as their junior year, but we suggest that they would find it more fruitful and make faster progress if they wait until they have had a course in linear algebra and matrix theory. Other courses helpful in speeding the digestion of this material are those in real variable theory and topology.

The problems are frequently important to the development of the text. Other problems are devices to enforce the understanding of a definition or a theorem. They also have been used to insert additional topics not discussed in the text.

We advocate eliminating many of the parentheses customarily used in denoting function values. That is, we often write fx instead of $f(x)$.

The end of a proof will be denoted by the symbol █.

We wish to thank Professor Louis N. Howard of MIT for his critical reading and many helpful suggestions; W. C. Weber for critical reading, useful suggestions, and other editorial assistance; E. M. Moskal and D. E. Blair for proofreading parts of the manuscript; and the editors of The Macmillan Company for their cooperation and patience.

Suggestions for the Reader

The bulk of this material can be covered in a two-semester (or three-quarter) course. Thus one could omit Chapter 0 and several sections of the later chapters, as follows: 2.14, 2.22, 2.23, 3.8, 3.10, 3.11, 3.12, the Appendix in Chapter 3, 4.4, 4.5, 4.10, 5.6, and all of Chapter 6. If it is desired to cover Chapter 6, Sections 2.23 and 4.4 and Appendix 3A should be studied. For a one-semester course one should try to get through most of Chapters 1 and 2 and half of Chapter 3. A thorough study of Chapter 2 would make a reasonable course in linear algebra, so that for students who have had linear algebra the time on Chapter 2 could be considerably shortened. In a slightly longer course, say two quarters, it is desirable to cover Chapter 3, Sections 4.1, 4.2, and 4.3, and most of the rest of Chapter 4 or all of Chapter 5. The choice of either is possible because Chapter 5 does not depend on Sections 4.4 through 4.10. The parts in smaller print are more difficult or tangential, so they may be considered as supplemental reading.

R. L. B.
S. I. G.

Contents

Chapter 2/Tensor Algebra   *59*

Chapter 3/Vector Analysis on Manifolds   *116*

CHAPTER **0**

Set Theory and Topology

0.1. SET THEORY

Since we cannot hope to convey the significance of set theory, it is mostly for the sake of logical completeness and to fix our notation that we give the definitions and deduce the facts that follow.

0.1.1. Sets

Set theory is concerned with abstract objects and their relation to various collections which contain them. We do not define what a set is but accept it as a primitive notion. We gain an intuitive feeling for the meaning of sets and, consequently, an idea of their usage from merely listing some of the synonyms: class, collection, conglomeration, bunch, aggregate. Similarly, the notion of an object is primitive, with synonyms element and point. Finally, the relation between elements and sets, the idea of an element being in a set, is primitive. We use a special symbol to indicate this relation, $\in$, which is read "is an element of." The negation is written $\notin$, read "is not an element of."

As with all modern mathematics, once the primitive terms have been specified, axioms regarding their usage can be specified, so the set theory can be developed as a sequence of theorems and definitions. (For example, this is done in an appendix to J. Kelly, *General Topology*, Van Nostrand, Princeton, N.J., 1955.) However, the axioms are either very transparent intuitively or highly technical, so we shall use the naïve approach of dependence on intuition, since it is quite natural (deceptively so) and customary.

We do not exclude the possibility that sets are elements of other sets. Thus we may have $x \in A$ and $A \in \tau$, which we interpret as saying that A and τ are sets, x and A are elements, and that x belongs to A and A belongs to τ. It may also be that x belongs to the set B, that x is itself a set, and that τ is an element

1

of some set. In fact, in formal set theory no distinction is made between sets and elements.

We specify a set by placing all its elements or a typical element and the condition which defines "typical" within braces, { }. In the latter case we separate the typical element from the condition by a vertical |. For example, the set having the first three odd natural numbers as its only elements is $\{1, 3, 5\}$. If Z is the set of all integers, then the set of odd integers is $\{x \mid$ there is $n \in Z$ such that $x = 2n + 1\}$, or, more simply, $\{x \mid x = 2n + 1, \ n \in Z\}$ or $\{2n + 1 \mid n \in Z\}$.

Set A is a *subset* of set B if every element of A is also an element of B. The relation is written $A \subset B$ or $B \supset A$, which can also be read "A is contained in B" or "B contains A." Although the word "contain" is used for both "$\in$" and "$\subset$," the meaning is different in each case, and which is meant can be determined from the context. To make matters worse, frequently an element x and the single-element set $\{x\}$ (called *singleton* x) are not distinguished, which destroys the distinction (notationally) between "$x \subset x$," which is always true, and "$x \in x$," which is usually false.

The sets A and B are *equal*, written $A = B$, if and only if $A \subset B$ and $B \subset A$. We shall abbreviate the phrase "if and only if" as "iff."

0.1.2. Set Operations

For two sets A and B, the *intersection* of A and B, $A \cap B$, read "A intersect B," is the set consisting of those elements which belong to both A and B. The *union* of A and B, $A \cup B$, consists of those elements which belong to A or B (or both). The operations of union and intersection are easily described in terms of the notation given above:

$$A \cap B = \{x \mid x \in A \text{ and } x \in B\},$$
$$A \cup B = \{x \mid x \in A \text{ or } x \in B\}.$$

Note that the use of "or" in mathematics is invariably inclusive, so that "or both" is not needed.

It is sometimes convenient to use the generalization of the operations of union and intersection to more than two sets. To include the infinite cases we start with a collection of sets which are labeled with subscripts ("indexed") from an index set J. Thus the collection of sets which we wish to unite or intersect has the form $\{A_\alpha \mid \alpha \in J\}$. The two acceptable notations in each case, with the first the more usual, are

$$\bigcap_{\alpha \in J} A_\alpha = \bigcap \{A_\alpha \mid \alpha \in J\}, \text{ the general intersection,}$$
$$\bigcup_{\alpha \in J} A_\alpha = \bigcup \{A_\alpha \mid \alpha \in J\}, \text{ the general union.}$$

Frequently J will be finite, for example, the first n positive integers, in which case we shall use one of the following forms:

$$\bigcap_{i=1}^{n} A_i = A_1 \cap A_2 \cap \cdots \cap A_n, \cdot$$

and similarly for union,

$$\bigcup_{i=1}^{n} A_i = A_1 \cup A_2 \cup \cdots \cup A_n.$$

In order that the intersection of sets be a set even when they have no common elements, we introduce the *empty set* $\varnothing$, the set which has no elements. For this and other reasons, appearing below, $\varnothing$ is a useful gadget. The empty set is a subset of every set.

The set-theoretic *difference* between two sets A and B is defined by $A - B = \{x \mid x \in A \text{ and } x \notin B\}$. We do not require that B be a subset of A in order for this difference to be formed. If $A \supset B$, then $A - B$ is called the *complement* of B with respect to A. Frequently we are concerned primarily with a fixed set A and its subsets, in which case we shall speak of the complement of a subset, omitting the phrase "with respect to A."

Problem 0.1.2.1.   The *disjunctive union* or *symmetric difference* of two sets A and B is $A \triangle B = A \cup B - A \cap B = (A - B) \cup (B - A)$. Observe that $A \triangle B = B \triangle A$. Prove the last equality. A distributive law is true for these set operations: $(A \triangle B) \cap C = A \cap C \triangle B \cap C$. However, $A \triangle A = \varnothing$ for every A.

0.1.3.   Cartesian Products

An *ordered pair* is an object which consists of a pair of elements distinguished as a *first element* and a *second element* of the ordered pair. The ordered pair whose first element is $a \in A$ and second element is $b \in B$ is denoted (a, b). In contrast we may also consider nonordered pairs, sets having two elements, say a and b, which would be denoted $\{a, b\}$ in accordance with what we said above. To be called a pair we should have $a \neq b$, and in any case $\{a, b\} = \{b, a\}$. On the other hand, we do consider ordered pairs of the form (a, a), and if $a \neq b$, then $(a, b) \neq (b, a)$. Indeed, $(a, b) = (c, d)$ iff $a = c$ and $b = d$.

The set of ordered pairs of elements from A and B, denoted $A \times B$,

$$A \times B = \{(a, b) \mid a \in A, b \in B\},$$

is called the *cartesian product* of A and B.

Problem 0.1.3.1.   Is $A \times B = B \times A$?

The operation of taking cartesian products may be iterated, in which case certain obvious identifications are made. For example, $A \times (B \times C)$ and $(A \times B) \times C$ are both considered the same as the *triple cartesian product*, which is defined to be the set of triplets (3-tuples)

$$A \times B \times C = \{(a, b, c) \mid a \in A, b \in B, c \in C\}.$$

Thus no distinction is made between $((a, b), c)$, $(a, (b, c))$, and (a, b, c). More generally, we only use one *n*-fold cartesian product $A_1 \times A_2 \times \cdots \times A_n$ rather than the many different ones which could be obtained by distributing parentheses so as to take the products two at a time. If the same set is used repeatedly, we generally use exponential notation, so $A \times A \times A$ is denoted A^3, etc.

A subset S of $A \times B$ is called a *relation* on A to B. An alternative notation for $(a, b) \in S$ is aSb, which can be read "*a* is *S*-related to *b*," although in many common examples it is read as it stands. For example, if $A = B = R$ we have the relation $<$, called "is less than," which formally consists of all those ordered pairs of real numbers (x, y) such that x is less than y. A function (see Section 0.1.4) is a special kind of relation.

Of particular importance in analysis and its special topic, tensor analysis, is the *real cartesian n-space* R^n, where R is the set of real numbers. In the case when $n = 2$ or 3 this is not quite the same as the analytic euclidean plane or analytic euclidean space in that the word "euclidean" indicates that the additional structure derived from a particular definition of distance is being considered. Moreover, in euclidean space no single point or line has preference over any other, whereas in R^3 the point $(0, 0, 0)$ and the coordinate axes are obviously distinguishable from other points and lines in R^3.

0.1.4. Functions

A *function from A into B*, denoted $f: A \to B$, is a rule which assigns to each $a \in A$ an element $fa = b \in B$. The idea of a "rule" is apparently a primitive notion in this definition, but need not be, since it can be defined in terms of the other notions previously given—"element of" and "cartesian product." This is done by means of the *graph of a function*—the subset

$$\{(a, fa) \mid a \in A\} \text{ of } A \times B.$$

The properties of a subset of $A \times B$ which are necessary and sufficient for the subset to be the graph of a function can be given in purely set-theoretic terms and the function itself can likewise be recaptured from its graph. In fact, it is customary to say that the function *is* its graph, but we shall use the distinction indicated by our phrasing of the definition given above.

Synonyms for "function" are "transformation," "map," "mapping," and "operator." Some authors use the convention that "function" is to be used for real-valued transformations.

We shall avoid the customary parentheses unless they are required to resolve ambiguity. Thus it is customary to write $f(a)$ instead of fa, which we used above. Parentheses must be used where a is itself composite; for example, $f(a + b)$ is not the same as $fa + b$. In fact, the latter is meaningless, except that we take it conventionally to be $(fa) + b$, the general rule being that operations such as addition are to be performed after evaluation of functions in the operands.

The *domain* of a function $f: A \to B$ is A. The *range* (image, target) of f is $fA = \{fa \mid a \in A\} \subset B$. The set B is called the *range set* of f. An element of the range, $b = fa$, is called a *value* of f, or the *image of a under f*.

If $fA = B$, then we say that f is *onto*, or that f *maps A onto B* (in contrast to "into" above).

If for every $b \in fA$ there is just one $a \in A$ such that $b = fa$, then f is said to be *one-to-one*, abbreviated 1–1. In this case we can define the *inverse of f*, $f^{-1}: fA \to A$, by setting $f^{-1}fa = a$.

If $f: A \to B$ and $C \subset A$, then the *restriction of f to C* is denoted $f|_C: C \to B$. It is frequently unnecessary to distinguish between f and $f|_C$, since they have the same rule, but merely apply to different sets.

If $C \subset A$, then the *inclusion map* $i_C: C \to A$ is defined simply by $i_C c = c$. If $C = A$, then i_C is called the *identity map on C*.

If $f: A \to B$ and $g: C \to D$, then the *composition* of g and f, denoted $g \circ f$, is the function obtained by following f by g, applied to every $a \in A$ for which this makes sense: $(g \circ f)a = g(fa)$. The domain of $g \circ f$ is thus $E = \{a \mid a \in A \text{ and } fa \in C\}$. (If $C \cap B = \varnothing$, then $g \circ f$ is the *empty function* $\varnothing : \varnothing \to D$.) If g and f are defined by formulas, or sets of formulas, the formula(s) for $g \circ f$ is obtained by *substituting* the formula(s) for f into the formula(s) for g.

For any functions, f, g, h, composition is associative; that is, $(f \circ g) \circ h = f \circ (g \circ h)$.

Problem 0.1.4.1. Let $f: A \to B$. Suppose there is $g: B \to A$ such that $f \circ g = i_B$. Then f is onto, g is 1–1, $h = f|_{gB}$ is 1–1 onto, and $g = i_{gB} \circ h^{-1}$. Show by an example that f need not be 1–1.

Problem 0.1.4.2. $f: A \to B$ is 1–1 onto iff there is $g: B \to A$ such that $g \circ f = i_A$ and $f \circ g = i_B$. This characterizes $g = f^{-1}$.

Examples. (a) If N is the set consisting of the first n natural numbers, $N = \{z \mid z \in Z, 0 < z < n + 1\}$, then R^n may be considered to be the set of all functions, $f: N \to R$. For such a function we obtain the n-tuple $(f1, f2, \ldots, fn)$, and from this it is obvious how, conversely, we get a function from an n-tuple.

(b) The ith *coordinate function* $u^i: R^n \to R$, also called the *projection into the ith factor*, or *cartesian coordinate function*, is defined by $u^i(x^1, \ldots, x^n) = x^i$. If we think of R^n as being functions $f: N \to R$, then we would define $u^i f = fi$.

(c) Using the idea of Example **(a)**, infinite cartesian products may be defined: If $\{A_\alpha \mid \alpha \in J\}$ is a collection of sets, then their cartesian product is

$$\prod_{\alpha \in J} A_\alpha = \{f \mid f: J \to \bigcup_{\alpha \in J} A_\alpha \text{ and } f\alpha \in A_\alpha \text{ for every } \alpha\}.$$

The *projections* or coordinate functions, $u_\alpha: \prod_{\beta \in J} A_\beta \to A_\alpha$, are defined as in Example **(b)**, by setting $u_\alpha f = f\alpha$. Projections are always onto.

0.1.5. Functions and Set Operations

If A is a set, we denote by $\mathscr{P}A$ the collection of all subsets of A, $\mathscr{P}A = \{C \mid C \subset A\}$. $\mathscr{P}A$ is called the *power set* of A.

If $f: A \to B$, then we define the *power map* of f, $f: \mathscr{P}A \to \mathscr{P}B$ by $fC = \{fc \mid c \in C\}$ for every $C \in \mathscr{P}A$. In particular, the range of f may still be denoted fA.

If $f: A \to B$, we also define the *complete inverse image map* of f, $f^{-1}: \mathscr{P}B \to \mathscr{P}A$, by $f^{-1}D = \{a \mid fa \in D\}$, for every $D \in \mathscr{P}B$. If f is 1–1 and onto, then the set map f^{-1} agrees with the power map of the inverse of f.

The facts to be established in the following problems show, generally, that the inverse image map is better behaved than the power map with respect to set operations.

Problem 0.1.5.1. The map f is onto iff the inverse image map f^{-1} is 1–1.

Problem 0.1.5.2. (a) $f^{-1}(D_1 \cap D_2) = (f^{-1}D_1) \cap (f^{-1}D_2)$.
(b) $f^{-1}(D_1 \cup D_2) = (f^{-1}D_1) \cup (f^{-1}D_2)$.
(c) $f(C_1 \cap C_2) \subset (fC_1) \cap (fC_2)$.
(d) $f(C_1 \cup C_2) = (fC_1) \cup (fC_2)$.

Problem 0.1.5.3. Find an example of f, C_1, C_2 such that $(fC_1) \cap (fC_2) \neq f(C_1 \cap C_2)$.

Problem 0.1.5.4. If $C \subset A$, we define the *characteristic function* $\Phi_C: A \to \{0,1\}$ by $\Phi_C a = 0$ if $a \in A - C$ and $\Phi_C a = 1$ if $a \in C$. Denote the set of all functions $f: A \to \{0,1\}$ by 2^A. Show that the function $\Phi: \mathscr{P}A \to 2^A$ given by $\Phi C = \Phi_C$ is 1–1 and onto, so that $\mathscr{P}A$ and 2^A are essentially the same.

Problem 0.1.5.5. If A is finite, show that 2^A is finite. How many elements does 2^A have?

Problem 0.1.5.6. If $F: A \to 2^A$, define $f \in 2^A$ by $fa \neq (Fa)a$ for every $a \in A$. This definition of f makes sense because there are only two possibilities for $(Fa)a$. Show that f is not in the range of F, so that F cannot be onto. In particular, there can be no 1–1 correspondence between A and 2^A. This is a precise statement of the intuitively clear contention that 2^A is "larger" than A.

A set is *countable* if it is either finite or its members can be arranged in an infinite sequence; or, what is the same, there is a 1–1 map from the set into the positive integers. The set of all integers, Z, is countable, as can be seen from the sequence $0, 1, -1, 2, -2, 3, -3, \ldots$. The cartesian product of the positive integers with itself is countable, as can be seen from the 1–1 map taking (m, n) into $2^m 3^n$. From this last statement it is easy to conclude that the union of a countable collection of countable sets is countable. It can be shown that the rational numbers are countable.

By Problem 0.1.5.6 we conclude that 2^Z is not countable. A similar trick using binary expansions of real numbers shows that the real numbers are not countable.

0.1.6. Equivalence Relations

An *equivalence relation* on a set P with elements $m, n, p, \ldots$, is a relation E which satisfies three properties:

(a) *Reflexivity*: For every m, mEm.

(b) *Symmetry*: If mEn, then nEm.

(c) *Transitivity*: If mEn and nEp, then mEp.

(mEn can be read "m is E-related to n.")

For every equivalence relation there is an exhaustive partition of P into disjoint subsets, the *equivalence classes of E*, for which the equivalence class to which an arbitrary m belongs is

$$[m] = \{n \mid nEm\}.$$

From (a), (b), and (c) we have

for every $m, m \in [m]$;
if $m \in [n]$, then $n \in [m]$;
if $m \in [n]$ and $n \in [p]$, then $m \in [p]$;

from which it follows that

$$[m] = [n] \qquad \text{iff } mEn.$$

Conversely, if we are given an exhaustive partition of P into disjoint subsets, we define two elements of P to be E-related if they are in the same subset, and thus obtain an equivalence relation E for which the subsets of the partition are the equivalence classes.

The set of equivalence classes, called the *quotient*, or *P divided by E*, is denoted

$$P/E = \{[m] \mid m \in P\}.$$

0.2.   TOPOLOGY

0.2.1.   Topologies

We cannot expect to convey here much of the significance of topological spaces. It is mostly for the sake of greater logical completeness that we give the definitions and theorems that follow. An initial study of tensor analysis can almost ignore the topological aspects since the topological assumptions are either very natural (continuity, the Hausdorff property) or highly technical (separability, paracompactness). However, a deeper analysis of many of the existence problems encountered in tensor analysis requires assumption of some of the more difficult-to-use topological properties, such as compactness and paracompactness. For example, the existence of complete integral curves of vector fields (Theorem 3.4.3) and existence of maxima and minima of continuous functions (Proposition 0.2.8.3) both require compactness; existence of riemannian metrics is proved using paracompactness (Section 5.2). Finally, we expect and hope that the extensive theory of algebraic topological invariants (Betti numbers, etc.) will be used a great deal more in applied mathematics and therefore we have included a few examples and remarks hinting of such uses (cf. Morse theory in Section 3.10 and de Rham's theorem in Section 4.5).

A *topology* on a set X is a subset T of $\mathscr{P}X$, $T \subset \mathscr{P}X$, such that

(a) If $G_1, G_2 \in T$, then $G_1 \cap G_2 \in T$.

(b) If $\{G_\alpha \mid \alpha \in J\} \subset T$, then $\bigcup_{\alpha \in J} G_\alpha \in T$.

(c) $\varnothing \in T$ and $X \in T$.

The combination (X, T) is called a *topological space*. The elements of T are called the *open sets* of the topological space. Frequently we shall have a specific topology in mind and then speak of the topological space X, with T being understood. The same space, however, can have many different topologies. In particular, there are always the *discrete topology* for which $T = \mathscr{P}X$ and the *concrete topology* for which $T = \{\varnothing, X\}$. These are so trivial as to be practically useless.

Problem 0.2.1.1. How many distinct topologies does a finite set having two or three points admit?

The *closed sets* of a topology T on X are the complements of the members of T, that is, the sets $X - G$ where $G \in T$. A topology could equally well be defined in terms of closed sets, with axioms corresponding to those above, which we state as theorems.

Proposition 0.2.1.1. (a) *A finite union of closed sets is a closed set.*

(b) *An arbitrary intersection of closed sets is closed.*

(c) $\varnothing$ *and* X *are closed sets.*

We emphasize that closedness and openness are not negations of each other or even contrary to each other; a set may be only closed, or only open, or both, or neither.

If $A \subset X$, X a topological space, then the union of all open sets contained in A is the *interior* of A, denoted A^0. Thus $A^0 = \bigcup \{B \mid B \subset A \text{ and } B \in T\}$. By (b), the interior of A is an open set itself, and is in fact one of the open sets of which we take the union in its definition. It is the largest open subset of A.

Just as "open" and "closed," "union" and "intersection" are "dual" notions, the dual notion to "interior" is "closure." The closure of $A \subset X$ is the intersection of all closed sets containing A and is denoted A^-. Thus $A^- = \bigcap \{B \mid A \subset B \text{ and } X - B \in T\}$ is closed by (b), and is the smallest closed set containing A. The following theorem shows that a complete knowledge of the operations of taking the interior or the closure is adequate to determine the topology.

Proposition 0.2.1.2. *A set is open iff the interior of the set equals the set. A set is closed iff the closure of the set equals the set.*

Axioms for the closure operation, which is really a function $^- : \mathscr{P}X \to \mathscr{P}X$, have been formulated by Kuratowski. When they are taken as axioms, Proposition 0.2.1.2 is essentially the definition of a closed set, and the axioms for closed sets, (a), (b), (c) of Proposition 0.2.1.1 are then theorems. In our scheme Kuratowski's axioms become theorems, as follows.

Proposition 0.2.1.3. *For all subsets A, B of X:*

(a) $(A \cup B)^- = A^- \cup B^-$.

(b) $A \subset A^-$.

(c) $(A^-)^- = A^-$.

(d) $\varnothing^- = \varnothing$.

Problem 0.2.1.2. Prove Proposition 0.2.1.3 and state and prove the dual proposition for the operation of taking the interior $^0 : \mathscr{P}X \to \mathscr{P}X$.

The *boundary* (also called the *frontier*, or the *derived set*) of a set $A \subset X$ is the set $\partial A = A^- - A^0$. The elements of ∂A are called *boundary points* of A. Again, it is possible to axiomatize topology by taking $\partial\colon \mathscr{P}X \to \mathscr{P}X$ as the fundamental concept. For example, if we know all about ∂, then open sets may be defined as those G for which $G \cap \partial G = \varnothing$.

A *neighborhood of* $x \in X$ is any $A \subset X$ such that $x \in A^0$. In particular, any open set containing x is a neighborhood of x. A *basis of neighborhoods at* x is a collection of neighborhoods of x such that every neighborhood of x contains one of the basis neighborhoods. In particular, the collection of all open sets containing x is a basis of neighborhoods at x, but generally there are many other possibilities for bases of neighborhoods. A *basis of neighborhoods of* X is a specification of a basis of neighborhoods for each $x \in X$.

Topologies are frequently defined by the specification of a basis of neighborhoods. The definitive procedure is as follows.

A neighborhood of x is any set which contains a basis neighborhood of x. An open set is then any set which is a neighborhood of every one of its points.

It is interesting that closed sets, closure, and boundary points can be defined directly in terms of basis neighborhoods. A set G is closed iff whenever every basis neighborhood of x intersects G, then $x \in G$. The closure of A consists of those x such that every basis neighborhood of x intersects A. The boundary of A consists of those points x such that every basis neighborhood of x intersects both A and $X - A$.

0.2.2.   Metric Spaces

Basis neighborhoods, and hence a topology, are frequently defined in turn by means of a *metric* or *distance function*, which is a function $d\colon X \times X \to R$ satisfying axioms as follows.

   (a) For all $x, y \in X$, $d(x, y) \geq 0$ (*positivity*).
   (b) If $d(x, y) = 0$, then $x = y$ (*nondegeneracy*).
   (c) For all $x, y \in X$, $d(x, y) = d(y, x)$ (*symmetry*).
   (d) For all $x, y, z \in X$, $d(x, y) + d(y, z) \geq d(x, z)$ (the *triangle inequality*).
There is no essential change if we also allow $+\infty$ as a value of d. A set with a metric function is called a *metric space*.

The *open ball* with center x and radius $r > 0$ with respect to d is defined as $B(x, r) = \{y \mid d(x, y) < r\}$. It can then be demonstrated that such open balls will serve as basis neighborhoods for a topology of X, the *metric topology* of d. Two metrics are *equivalent* if they give rise to the same topology.

Two metrics $d, d_1\colon X \times X \to R$ are *strongly equivalent* if there are positive constants c, c_1 such that for every $x, y \in X$, $d(x, y) \leq c_1 d_1(x, y)$ and $d_1(x, y) \leq cd(x, y)$. Strongly equivalent metrics are equivalent but not conversely. In fact, a metric d is always equivalent to $d_1 = d/(1 + d)$, but these two are strongly equivalent iff d is *bounded*; that is, there is a constant k such that $d(x, y) \leq k$ for all x, y. The metric $d_1 = d/(1 + d)$ is always bounded ($k = 1$) whether d is bounded or not, but a bounded metric cannot be strongly equivalent to an unbounded one.

0.2.3. Subspaces

If $A \subset X$ and X has a topology T, then we get the *relative*, or *induced*, topology T_A by defining

$$T_A = \{G \cap A \mid G \in T\}.$$

It is easy to verify that T_A actually is a topology on A. When A is given this topology, it is said to be a (topogical) *subspace* of X. The closed sets of a subspace A are the intersections of closed sets of X with A.

0.2.4. Product Topologies

If X and Y are topological spaces, then we define a topology on $X \times Y$ by specifying the basis neighborhoods of (x, y) to be $G \times H \subset X \times Y$, where G is a neighborhood of x and H is a neighborhood of y. The choices for G and H may be restricted to basis systems and there will be no difference in the resulting topology on $X \times Y$. When $X \times Y$ is provided with this topology it is called the *topological product* of X and Y.

If X and Y are metric spaces with metrics d_X, d_Y, then we define d_p, a metric on $X \times Y$, for every $p \geq 1$, by

$$d_p((x, y), (x_1, y_1)) = [d_X(x, x_1)^p + d_Y(y, y_1)^p]^{1/p}$$

The limiting case as $p \to \infty$ is the metric d_∞, given by

$$d_\infty((x, y), (x_1, y_1)) = \max[d_X(x, x_1), d_Y(y, y_1)].$$

Although these metrics are all different, they are all strongly equivalent, so give the same topology on $X \times Y$; in fact, this topology is the product topology. Indeed, the balls with respect to d_∞ are just the products of balls with respect to d_X and d_Y of the same radii.

The *standard topology* on R is that of the metric defined by absolute value of differences, $(x, y) \to |x - y|$. The *standard topology* on R^n is obtained by taking repeated products of the standard topology on R. It is thus the topology of any of the metrics, for $p \geq 1$, $x, y \in R^n$,

$$d_p(x, y) = \left[\sum_{i=1}^{n} |u^i x - u^i y|^p \right]^{1/p}$$

$$d_\infty(x, y) = \max[|u^i x - u^i y|, \; i = 1, 2, \ldots, n].$$

Of these, d_2 is the usual *euclidean* metric on R^n, but they are all strongly equivalent to each other. Unless otherwise specified, we shall assume that a topology on R^n is the standard one.

Problem 0.2.4.1. (a)   For fixed x,y show that $d_p(x, y)$ is a nonincreasing function of $p \geq 1$. (*Hint:* Show that the derivative is ≤ 0.)

(b) For every $x,y \in R^n$, $d_1(x, y) \leq nd_\infty(x, y)$ and $d_\infty(x, y) = \lim\limits_{p \to \infty} d_p(x, y)$.

(c) All d_p, $1 \leq p \leq \infty$, are strongly equivalent.

0.2.5.   Hausdorff Spaces

A topological space X is a *Hausdorff space* if for every $x,y \in X$, $x \neq y$, there are neighborhoods U, V or x, y, respectively, such that $U \cap V = \varnothing$.

In a Hausdorff space the singleton sets $\{x\}$ are closed sets.

A metric topology is always Hausdorff.

Problem 0.2.5.1.   The product of Hausdorff spaces is a Hausdorff space.

0.2.6.   Continuity

Let X, Y be topological spaces. A function $f: X \to Y$ is *continuous* if for every open set G in Y, $f^{-1}G$ is open in X. In other words, $f^{-1}: \mathscr{P}Y \to \mathscr{P}X$ maps open sets into open sets.

Since f^{-1} behaves well with respect to set operations and, in particular, preserves complementation, we have immediately that f is continuous iff f^{-1} maps closed sets into closed sets.

The above definition of continuity is the most convenient one for working abstractly with topological spaces. For example, it is trivial to prove

Proposition 0.2.6.1.   *The composition of continuous functions is continuous.*

However, we can recast this definition into forms which are more directly abstractions from the $\varepsilon = \delta$ definition of continuity of real-valued functions of a real variable. In that definition we first define continuity at x, and continuity itself is obtained by requiring it at every x. In the definition of continuity of $f: R \to R$ at x, where $y = fx$, the ε served to define a basis neighborhood of y, given a priori, and the requirement was that there be a basis neighborhood of x determined by δ, such that f map the δ-neighborhood into the ε-neighborhood. The student should be able to show that this description is the essential content of the customary definition: "For every $\varepsilon > 0$ there is a $\delta > 0$ such that for every x_1 for which $|x - x_1| < \delta$ it is true that $|fx_1 - y| < \varepsilon$."

Abstracting the description in terms of neighborhoods is not a great chore. If $f: X \to Y$ we say that f is *continuous at* $x \in X$ if for every neighborhood V ($\leftrightarrow \varepsilon$ neighborhood) of $y = fx$ there is a neighborhood U ($\leftrightarrow \delta$ neighborhood) of x such that $U \subset f^{-1}V$ (or $fU \subset V$).

The following theorem shows that our definition of a continuous function is a correct abstraction of the usual one.

Proposition 0.2.6.2. *A function $f: X \to Y$ is continuous iff f is continuous at every $x \in X$.*

Problem 0.2.6.1. Show that all functions $f: X \to X$ are continuous in the discrete topology and that the only continuous functions in the concrete topology are the constant functions.

The notion of a limit can also be abstracted. We define that $\lim_{x \to x_0} fx = y$ iff for every neighborhood V of y there is a neighborhood U of x_0 such that $(U - \{x_0\}) \subset f^{-1}V$. It follows, as usual, that f is continuous at x_0 iff **(a)** $\lim_{x \to x_0} fx = y$ and **(b)** $fx_0 = y$.

A *homeomorphism* $f: X \to Y$ is a 1–1 onto function such that f and $f^{-1}: Y \to X$ are both continuous. If $f: X \to Y$ is 1–1 but not onto, then f is said to be a *homeomorphism into* if f and $f^{-1}: (\text{range } f) \to X$ are both continuous, where range f is given the relative topology from Y. A homeomorphism f is also called a *topological equivalence* because $f |_{T_X}$ and $f^{-1} |_{T_Y}$ are then 1–1 onto; that is, they give a 1–1 correspondence between the topologies T_X and T_Y of X and Y. A property of a topological space is said to be a *topological property* if every homeomorphic space has the property. A *topological invariant* is a rule which associates to topological spaces an object which is a topological property of the space. The object usually consists of a number or some algebraic system.

Problem 0.2.6.2. $\tan: (-\pi/2, \pi/2) \to R$ is a homeomorphism, where $\tan = \sin/\cos$.

0.2.7. Connectedness

A topological space X is *connected* if the only subsets of X which are both open and closed are $\varnothing$ and X. Another formulation of the same concept, in terms of its negation, is

Proposition 0.2.7.1. *A topological space X is not connected iff there are nonempty open sets G, H such that $G \cap H = \varnothing$, $G \cup H = X$.*

A subset A of X is *connected* if A with the relative topology is connected. The following is not hard to prove.

Proposition 0.2.7.2. (Chaining Theorem). *If* $\{A_\alpha \mid \alpha \in J\}$ *is a family of connected subsets of* X *and* $\bigcap\limits_{\alpha \in J} A_\alpha \neq \varnothing$, *then* $\bigcup\limits_{\alpha \in J} A_\alpha$ *is connected.*

A harder theorem is the following.

Proposition 0.2.7.3. *If* A *is connected and* $A \subset B$, $B \subset A^-$, *then* B *is connected. In particular,* A^- *is connected.*

The situation for real numbers is particularly simple. An *interval* (general sense) is a subset of R of one of the forms

$$(a, b) = \{x \mid a < x < b\},$$
$$(a, b] = \{x \mid a < x \leq b\},$$
$$[a, b) = \{x \mid a \leq x < b\},$$
$$[a, b] = \{x \mid a \leq x \leq b\},$$

where we allow $a = -\infty$, $b = \infty$ at open ends, with obvious meanings. The connected sets in R are precisely these intervals. In particular, R itself is connected.

Problem 0.2.7.1. Connectedness is a topological property; that is, the image of a connected set under a homeomorphism is connected.

Proposition 0.2.7.4. *If* $f: X \to Y$ *is continuous, and* $A \subset X$ *is connected, then* fA *is connected. In particular, if* $Y = R$, *then* fA *is an interval. (This is a generalization of the intermediate-value theorem for continuous functions of a real variable defined on an interval.)*

In particular, if $f: [a, b] \to Y$ is continuous, the range of f is connected. Such an f is called a continuous curve in Y from $y_a = fa$ to $y_b = fb$.

A topological space Y is *arcwise connected* if for every $y_1, y_2 \in Y$ there is a continuous curve from y_1 to y_2. It follows from Propositions 0.2.7.2 and 0.2.7.4 that an arcwise connected space is connected.

The *connected component of* X *containing* x is the union of all the connected subsets of X which contain x. By Proposition 0.2.7.2 we see that the component containing x is itself connected and that the components containing two different points are either identical or do not meet. Thus X is split up into a disjoint union of connected sets, the *components* of X, each of which is *maximal-connected*, that is, is not contained in a larger connected set. It follows from Proposition 0.2.7.3 that the components of X are closed. The number of components is a topological invariant.

If we substitute "arcwise connected" for "connected" above, we arrive at the notion of the *arc components* of a topological space. The subdivision into arc components is generally finer than the subdivision into components, and the arc components are not necessarily closed. Both these facts are illustrated by the space A in the following example, since A is connected but has two arc components, only one of which is closed.

Example.   The subset of R^2,

$$A = \{(x, \sin 1/x) \mid 0 < x \le 1\}^-$$

is connected but not arcwise connected. That it is connected is easy by Proposition 0.2.7.3, since it is the closure of an arcwise connected set $B = \{(x, \sin 1/x) \mid 0 < x \le 1\}$. However, the points on the boundary $\partial B = \{(0, y) \mid -1 \le y \le 1\}$ cannot be joined to those in B by a continuous curve in A.

For the open subsets of R^n the notions of connectedness and arcwise connectedness coincide. Indeed, in a connected open set of R^n any two points can be joined by a *polygonal continuous curve*, that is, a continuous curve for which the range consists of a finite number of straight-line segments.

Problem 0.2.7.2.   (**a**) Show that if A is an open set in R^n and $a \in A$, then the set of points in A which can be joined to a by a polygonal continuous curve is an open subset of A.

   (**b**) Prove that A is polygonally connected if A is connected.

0.2.8.   Compactness

If $A \subset X$, a *covering* of A is a family $\{C_\alpha \mid \alpha \in J\}$ in $\mathscr{P}X$ such that $A \subset \bigcup\limits_{\alpha \in J} C_\alpha$. An *open covering* is one for which the family consists of open sets. A *subcovering* of a covering $\{C_\alpha \mid \alpha \in J\}$ is a covering $\{C_\alpha \mid \alpha \in K\}$, where $K \subset J$. A *finite* covering is one for which J is finite.

   A subset A of X is *compact* if every open covering of A has a finite subcovering.

Problem 0.2.8.1.   Compactness is a topological property.

   To illustrate how the definition of compactness operates, we prove that a compact subset A of R is bounded. Consider the open covering of R consisting of open intervals of length 2, $\{(n, n + 2) \mid n \in Z\}$. Since this also is an open covering of A, there must be a finite subcovering. Among the $(n, n + 2)$'s which occur in the finite subcovering we must have one for which n is greatest, $n = n_1$, and one for which n is least, $n = n_0$. Then clearly $A \subset [n_0, n_1 + 2]$. We shall see below that A is also closed (see Proposition 0.2.8.2).

Conversely, the closed bounded subsets of R are compact, since this is only a restatement of the *Heine-Borel covering theorem*. This result generalizes to R^n—the compact subsets of R^n are exactly those which are closed and bounded. A *bounded* set is one which is contained in some ball with respect to one, and hence all, of the metrics d_p given previously. This notion is *not* related to the notion of a boundary of a set.

A dual formulation of compactness is given in terms of closed sets and finite intersections. A family of sets $\{C_\alpha \mid \alpha \in J\}$ has the *finite intersection property* (abbreviated FIP) if for every finite subset K of J, $\bigcap_{\alpha \in K} C_\alpha \neq \varnothing$.

Proposition 0.2.8.1. *A subset A of X is compact iff for every family of relatively closed subsets $\{C_\alpha \mid \alpha \in J\}$ of A which has the FIP, $\bigcap_{\alpha \in J} C_\alpha \neq \varnothing$. (FIP means that no finite number of complements $A - C_\alpha$ cover A, whereas total intersection being nonempty means that all the complements do not cover A.)*

Proposition 0.2.8.2. **(a)** *A compact subset of a Hausdorff space is closed.*
(b) *A closed subset in a compact space is also compact.*

Proof. For part **(b)** note that the complement of the closed subset may be added to any open covering of the closed subset so as to obtain an open covering of the containing compact space. A finite subcovering of the whole space exists and the complement of the closed subset may be deleted if it is there, leaving a finite subcovering of the closed subset.

Suppose that A is a compact subset in a Hausdorff space X and $A \neq A^-$, so there is an $x \in A^- - A$. For every $a \in A$ there are open sets G_a, G_a^x such that $G_a \cap G_a^x = \varnothing$, $a \in G_a$, and $x \in G_a^x$, because X is Hausdorff. Then $\{G_a \mid a \in A\}$ is an open covering of A, so there is a finite subcovering $\{G_a \mid a \in J\}$, where J is a finite subset of A. But then $\bigcap_{\alpha \in J} G_a^x$ is a neighborhood of x which does not meet $\bigcup_{\alpha \in J} G_a \supset A$, so x cannot be in A^-, a contradiction. ∎

Proposition 0.2.8.3. *Let $f: X \to Y$ be continuous and A a compact subset of X. Then fA is compact. In particular, if $Y = R$, then f has a maximum and a minimum on A (since fA is closed and bounded its supremum exists and is in fA); that is, there is an $a_M \in A$ such that for every $a \in A$, $fa \leq fa_M$, and similarly for a minimum.*

The proof is automatic.

Proposition 0.2.8.4. *Let $f: X \to Y$ be continuous, 1-1, and onto, where X is compact and Y is Hausdorff. Then f is a homeomorphism. In particular, X is Hausdorff.*

Proof. The problem is to show that f^{-1} is continuous. We do this in the form f (closed set) is closed. But for F closed in X, F is compact [Proposition

0.2.8.2(b)], fF is compact (Proposition 0.2.8.3), so fF is closed [Proposition 0.2.8.2(a)]. The last step uses the Hausdorff property of Y. ∎

0.2.9.   Local Compactness

A topological space X is called *locally compact* if each point of X has a compact neighborhood. Thus a compact space is automatically locally compact.

Problem 0.2.9.1.   (a) A closed subspace of a locally compact space is locally compact.

(b) A discrete space is locally compact.

(c) R^n is locally compact.

0.2.10.   Separability

A topological space X is called *separable* if it has a countable basis of neighborhoods.

Problem 0.2.10.1.   (a) Suppose that the metric space X has a countable subset A such that $A^- = X$. Show that the open balls with centers at points of A and rational radii is a basis of neighborhoods for X, and hence that X is separable.

(b) R^n is separable.

Problem 0.2.10.2.   The product of two separable spaces is separable.

0.2.11.   Paracompactness

A family of sets U_α of a topological space X is said to be *locally finite* if every point of X has a neighborhood meeting only a finite number of the U_α. A covering V_β of X is called a *refinement* of a covering U_α of X if for every index β there is at least one set U_α such that $V_\beta \subset U_\alpha$. A topological space X is said to be *paracompact* if it is Hausdorff and if every open covering has an open refinement which is locally finite.

Proposition 0.2.11.1.   *If X is a locally compact separable Hausdorff space, then X is the union of countable family of compact subsets $\{A_i\}$. This sequence of compact subsets may be taken to be increasing; that is, $A_i \subset A_{i+1}$ for every i.*

Proof. Let $\{U_i\}, i = 1, 2, \ldots$ be an open countable basis for X. We claim that those U_i such that $U_i{}^-$ is compact are still a basis. It suffices to show that if a subset G is open, then for every $x \in G$ there is a $U_i \subset G$ such that $U_i{}^-$ is compact and $x \in U_i$. Since X is locally compact there is a compact neighborhood V of x. Then $V^0 \cap G$ is open, so there is $U_i \subset V^0 \cap G$ such that $x \in U_i$.

But then $U_i{}^- \subset V^{0-} \subset V$, since V is closed by Proposition 0.2.8.2, so U_i is compact by Proposition 0.2.8.2.

Discarding those U_i for which $U_i{}^-$ is not compact, we have a countable basis whose elements have compact closures, which we again denote $\{U_i\}$. We define a sequence of compact sets with the increasing property by letting $A_i = \bigcup_{j=1}^{i} U_j{}^-$. ∎

Lemma.   *If a locally compact Hausdorff space X is the union of a countable family of compact subsets, then it is the union of the interiors of such a family.*

Proof. Let $X = \bigcup_{i=1}^{\infty} A_i$, A_i compact. Each A_i can be covered by open neighborhoods having compact closures and hence by a finite number of such neighborhoods. The closures of these neighborhoods, a finite number for each A_i, comprise a countable family of compact sets whose interiors cover X.

Proposition 0.2.11.2.   *If a locally compact Hausdorff space X is the countable union of compact sets, then X is paracompact.*

Proof. By the lemma we may suppose that $X = \bigcup_{i=1}^{\infty} A_i^0$, where A_i is compact and $A_i \subset A_{i+1}^0$ for every i.

Now if $\{W_\alpha\}$ is an open covering of X, then for each i the sets $(A_{i+2}^0 - A_{i-1}) \cap W_\alpha$ comprise an open covering of $A_{i+1} - A_i^0$. Therefore, we can choose a finite subcovering $V_{i1}, \ldots, V_{ip_i}$. Since the sets $A_{i+1} - A_i^0$ cover X, the V_{ij}, $i, j = 1, 2 \ldots$, cover X. Moreover, $\{V_{ij}\}$ refines the covering $\{W_\alpha\}$. Now let $x \in A_k$; then A_{k+1}^0 is a neighborhood of x which does not intersect any V_{ij} for $i > k + 1$. Thus $\{V_{ij}\}$ is locally finite. ∎

Example.   In R^n the compact sets are the closed, bounded sets. If we let A_i be the closed ball with radius i, $i = 1, 2, \ldots$, and center a fixed $x \in R^n$, then R^n is the union of the increasing sequence of the interiors of the compact sets A_i.

Proposition 0.2.11.3.   *A locally compact separable Hausdorff space is paracompact.*

This follows immediately from the previous two propositions.

Problem 0.2.11.1.   If a Hausdorff space X is the countable union of subspaces homeomorphic to open subsets of R^n, then X is paracompact.

Problem 0.2.11.2.   The space of rational numbers, with the induced topology from the reals, is paracompact but not locally compact.

Remark. A continuous function has as its domain a topological space. To generalize the notion of a differentiable function on R^n we shall require the concept of a differentiable manifold on which it will make sense to speak of differentiable functions.

CHAPTER 1

Manifolds

1.1. Definition of a Manifold

A manifold, roughly, is a topological space in which some neighborhood of each point admits a coordinate system, consisting of real coordinate functions on the points of the neighborhood, which determine the position of points and the topology of that neighborhood; that is, the space is locally cartesian. Moreover, the passage from one coordinate system to another is smooth in the overlapping region, so that the meaning of "differentiable" curve, function, or map is consistent when referred to either system. A detailed definition will be given below.

The mathematical models for many physical systems have manifolds as the basic objects of study, upon which further *structure* may be defined to obtain whatever system is in question. The concept generalizes and includes the special cases of the cartesian line, plane, space, and the surfaces which are studied in advanced calculus. The theory of these spaces which generalizes to manifolds includes the ideas of differentiable functions, smooth curves, tangent vectors, and vector fields. However, the notions of distance between points and straight lines (or shortest paths) are not part of the idea of a manifold but arise as consequences of additional structure, which may or may not be assumed and in any case is not unique.

A manifold has a dimension. As a model for a physical system this is the number of degrees of freedom. We limit ourselves to the study of finite-dimensional manifolds.

Some preliminary definitions will facilitate the definition of a manifold. If X is a topological space, a *chart* at $p \in X$ is a function $\mu: U \to R^d$, where U is an open set containing p and μ is a homeomorphism onto an open subset of R^d. The *dimension* of the chart $\mu: U \to R^d$ is d. The *coordinate functions* of the chart are the real-valued functions on U given by the entries of values of μ;

that is, they are the functions $x^i = u^i \circ \mu : U \to R$, where $u^i : R^d \to R$ are the standard coordinates on R^d. [The u^i are defined by $u^i(a^1, \ldots, a^d) = a^i$. The superscripts are not powers, of course, but are merely the customary tensor indexing of coordinates. If powers are needed, extra parentheses may be used, $(x)^3$ instead of x^3 for the cube of x, but usually the context will contain enough distinction to make such parentheses unnecessary.] Thus for each $q \in U$, $\mu q = (x^1 q, \ldots, x^d q)$, so we shall also write $\mu = (x^1, \ldots, x^d)$. In other terminology we call μ a *coordinate map*, U the *coordinate neighborhood*, and the collection $(x^1, \ldots, x^d)$ *coordinates* or a *coordinate system at p*.

We shall restrict the symbols "u^i" to this usage as standard coordinates on R^d. For R^2 and R^3 we shall also use x, y, z as coordinates as is customary, except that we shall usually treat them as functions.

A real-valued function $f : V \to R$ is C^∞ (continuous to order ∞) if V is an open set in R^d and f has continuous partial derivatives of all orders and types (mixed and not). A function $\varphi : V \to R^e$ is a C^∞ *map* if the components $u^i \circ \varphi : V \to R$ are C^∞, $i = 1, \ldots, e$.

More generally φ is C^k, k a nonnegative integer, if all partial derivatives up to and including those of order k exist and are continuous. (C^0 means merely continuous.) A map φ is *analytic* if $u^i \circ \varphi$ are real-analytic, that is, may be expressed in a neighborhood of each point by means of a convergent power series in cartesian coordinates having their origin at the point. Analytic maps are C^∞ but not conversely.

Problem 1.1.1. (**a**) Define $f : R \to R$ by

$$fx = \begin{cases} 0 & \text{if } x \leq 0, \\ e^{-1/x} & \text{if } x > 0. \end{cases}$$

Show that f is C^∞ and that all the derivatives of f at 0 vanish; that is, $f^{(k)}0 = 0$ for every k.

(**b**) If $g : R \to R$ is analytic in a neighborhood of 0, then

$$gx = \sum_{k=0}^{\infty} (g^{(k)}0) x^k / k!$$

for all x in a symmetric interval with center 0. Thus f in part (**a**) cannot be analytic at 0.

Example. Letting $z = x + iy$, a complex variable, we define $u(x, y)$ by $u + iv = e^{-1/z^4}$, $u(0, 0) = 0$. Then u is not C^∞, and in fact not even continuous at $(0, 0)$, but the partial derivatives of u of all orders exist everywhere, including $(0, 0)$. Thus the requirements of continuity in the definition of C^∞ is not superfluous. For functions of one variable, it is of course true that differentiable functions are continuous.

Two charts $\mu: U \to R^d$ and $\tau: V \to R^e$ on a topological space X are C^∞-*related* if $d = e$ and either $U \cap V = \varnothing$ (the empty set) or $\mu \circ \tau^{-1}$ and $\tau \circ \mu^{-1}$ are C^∞ maps. The domain of $\mu \circ \tau^{-1}$ is $\tau(U \cap V)$, an open set in R^d (see Figure 1).

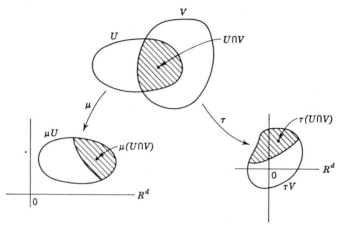

Figure 1

Other degrees of relatedness are defined by replacing "C^∞" by "C^k" or "analytic." Two charts of the same dimension are always C^0-related because coordinate maps are continuous.

A *topological* (C^0) *manifold* is a separable Hausdorff space such that there is a d-dimensional chart at every point. The *dimension* of the manifold is the same as the dimension of the charts. Thus there is a collection of charts $\{\mu_\alpha: U_\alpha \to R^d \mid \alpha \in I\}$ such that $\{U_\alpha \mid \alpha \in I\}$ is a covering of the space. Such a collection is called an *atlas*. A C^∞ *atlas* is one for which every pair of charts is C^∞-related. A chart is *admissible* to a C^∞ atlas if it is C^∞-related to every chart in the atlas. In particular the members of a C^∞ atlas are themselves admissible.

A C^∞ *manifold* is a topological manifold together with all the admissible charts of some C^∞ atlas. In this book the term "manifold," with no adjective, will always mean "C^∞ manifold." (The reason for including all admissible charts rather than merely those which are in some given atlas is to convey the idea that no particular coordinate systems are to be preferred over any others and also to resolve the logical problem of saying just what a manifold *is*. The source of this logical difficulty is the fact that two different atlases can have the same collection of admissible charts, in which case we should like to say we have only one manifold, not two different manifolds, one for each atlas. On the other hand, it is almost invariably the case that a manifold is specified by giving just one atlas, not the whole collection of admissible charts.)

The C^k *manifolds* and *real-analytic manifolds* are defined by replacing "C^∞" by "C^k" and "analytic," respectively, throughout the above chain of definitions. It should be clear that a C^∞ manifold becomes a C^k manifold simply by enlarging the collection of admissible charts to include all the C^k-related ones, and, similarly, a real-analytic manifold becomes a C^∞ manifold. Conversely, a C^1 manifold becomes a real-analytic (and hence C^∞) manifold, in many ways, by discarding a suitable collection of C^1 admissible charts so as to leave only charts which are mutually analytically related, but this result is not at all obvious, being a very difficult theorem of Whitney. That a C^0 manifold may fail to become a C^1 manifold is known, and even more difficult to prove.

Remark. In the definition of a coordinate system we have required that the coordinate neighborhood and the range in R^d be open sets. This is contrary to popular usage, or at least more specific than the usage of curvilinear coordinates in advanced calculus. For example, spherical coordinates are used even along points of the z axis where they are not even 1–1. The reasons for the restriction to open sets are that it forces a uniformity in the local structure which simplifies analysis on a manifold (there are no "edge points") and, even if local uniformity were forced in some other way, it avoids the problem of spelling out what we mean by differentiability at boundary points of the coordinate neighborhood; that is, one-sided derivatives need not be mentioned. On the other hand, in applications, boundary value problems frequently arise, the setting for which is a *manifold with boundary*. These spaces are more general than manifolds and the extra generality arises from allowing a *boundary manifold* of one dimension less. The points of the boundary manifold have a coordinate neighborhood in the boundary manifold which is attached to a coordinate neighborhood of the interior in much the same way as a face of a cube is attached to the interior. Just as the study of boundary value problems is more difficult than the study of spatial problems, the study of manifolds with boundary is more difficult than that of mere manifolds, so we shall limit ourselves to the latter.

1.2. Examples of Manifolds

(a) CARTESIAN SPACES. We define a manifold structure on R^d in the most obvious way by taking as atlas the single chart $I: R^d \to R^d$, the identity map. The coordinate functions of this chart are thus the standard (cartesian) coordinates u^i. When we speak of R^d as a manifold we shall intend this standard structure, unless otherwise stated.

A C^∞ admissible coordinate map on R^d is a 1–1 C^∞ map $\mu\colon U \to R^d$, where U is an open set and the jacobian determinant $|\, \partial x^i/\partial u^j\,| \neq 0$, where $x^i = u^i \circ \mu$ are the coordinate functions. Nonvanishing of the jacobian determinant is just another way of requiring the map μ^{-1} to be C^∞.

If f^i, $i = 1,\ldots, d$, are real-valued C^∞ functions on some open set of R^d and at some $p \in R^d$ we have $|\, \partial f^i/\partial u^j\,| \neq 0$, then the *inverse function* theorem states that there is a neighborhood U of p and a neighborhood V of $(f^1 p,\ldots,f^d p)$ such that the map $\mu = (f^1,\ldots,f^d)$ takes U onto V, is 1–1, and has a C^∞ inverse. This gives an effective means of obtaining admissible coordinates. In particular, polar coordinates, cylindrical coordinates, spherical coordinates, and the other customary curvilinear coordinates are admissible coordinates for R^2 and R^3 provided they are suitably restricted so as to be 1–1 and have nonzero jacobian determinant.

Example.   Let $\mu = (x^2 + 2y^2, 3xy)\colon R^2 \to R^2$, $u = x^2 + 2y^2$, $v = 3xy$. The jacobian determinant is $(\partial u/\partial x)(\partial v/\partial y) - (\partial u/\partial y)(\partial v/\partial x) = 6(x^2 - 2y^2)$, which is nonzero except on the two lines $y = x/\sqrt{2}$, $y = -x/\sqrt{2}$ of *singular points*. For every point except those on these lines there is some neighborhood on which μ is an admissible coordinate map. To find what these neighborhoods might be requires a more detailed analysis. By eliminating x and y from $u = x^2 + 2y^2$, $v = 3xy$, $y = x/\sqrt{2}$, we obtain $v = 3u/2\sqrt{2}$, and we note that $u \geq 0$. Thus the line of singular points $y = x/\sqrt{2}$ is mapped into the half-line $v = 3u/2\sqrt{2}$, $u \geq 0$; similarly, we find that $y = -x/\sqrt{2}$ is mapped into $v = -3u/2\sqrt{2}$, $u \geq 0$. Letting $x = c$ and eliminating y we get a parabola $u = c^2 + 2v^2/9c^2$ which is found to be tangent to the two half-lines just found and, except for the tangent points, lying in the open angle region V between the two half-lines (see Figure 2). Each of the four connected regions of non-singular points is mapped by μ 1–1 onto V, so for any nonsingular point p the one of these four regions which contains p, or any smaller neighborhood of p,

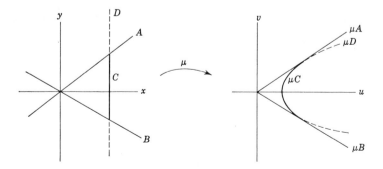

Figure 2

may be taken as the neighborhood U asserted to exist by the inverse function theorem. No neighborhood of a singular point is mapped 1–1 by μ; such neighborhoods are folded over onto themselves, and neighborhoods of $(0, 0)$ are folded twice, so that μ is generally 4–1 in neighborhoods of $(0, 0)$.

Problem 1.2.1. What restrictions on the domains and/or ranges of spherical and cylindrical coordinates can be imposed so as to make them admissible C^∞ coordinates for R^3? Show that all points (but not all simultaneously for one system), except those where the cylindrical radius $r = 0$, may be included in domains of systems of both types.

Problem 1.2.2. If $u: R \to R$ is the identity map, then its cube $u^3: R \to R$ is also 1–1, continuous, and has continuous inverse $u^{1/3}: R \to R$. If we take $\{u^3: R \to R\}$ as an atlas for R, this defines a manifold structure on R with a single chart. Show that this is not the standard manifold structure since $u^3: R \to R$ is not an admissible chart in the standard structure.

(b) OPEN SUBMANIFOLDS. If M is a manifold and N is any open subset of M, then N inherits a manifold structure by restricting the topology and coordinate maps of M to N. We call N an *open submanifold* of M. (A general submanifold may have a smaller dimension and will be defined in Section 1.4.) In particular, any open subset of R^d is a d-dimensional manifold.

Problem 1.2.3. Show that a manifold may be considered as an open submanifold of R^d iff the manifold has an atlas with only one chart.

(c) PRODUCT MANIFOLDS. If M and N are manifolds of dimensions d and e, respectively, then $M \times N$ is given a manifold structure by taking the product topology as its topology (basic neighborhoods are products of those in M and N) and as atlas the products of charts from atlases for M and N. If $\mu: U \to R^d$ is a chart on M, and $\varphi: V \to R^e$ is a chart on N, their product is $(\mu, \varphi): U \times V \to R^{d+e}$, which is defined by $(\mu, \varphi)(m, n) = (\mu m, \varphi n)$. If x^i are the coordinate functions of μ and y^j are the coordinate functions of φ, then the coordinates of (m, n) in the product chart are $(x^1 m, \ldots, x^d m, y^1 n, \ldots, y^e n)$. Thus if $p: M \times N \to M$ and $q: M \times N \to N$ are the projections, $p(m, n) = m$, $q(m, n) = n$, the coordinate functions on $U \times V$ are $z^1 = x^1 \circ p, \ldots, z^d = x^d \circ p, z^{d+1} = y^1 \circ q, \ldots, z^{d+e} = y^e \circ q$.

This product operation can obviously be iterated, and we may take different copies of the same manifold as factors. Thus even as a manifold $R^d = R \times R \times \cdots \times R$ (d factors). It is easy to see that a circle S^1 (the curve) is a one-dimensional manifold. Picturing S^1 as a part of R^2 we see that a cylinder (the surface) is the manifold $S^1 \times R$ and may be pictured in $R^3 = R^2 \times R$.

We may consider $S^1 \times S^1$ as a union, $\{\{p\} \times S^1 \mid p \in S^1\}$, of circles $\{p\} \times S^1$, one for each $p \in S^1$. Now if we picture the first factor as being in the xy plane of R^3, satisfying the equations $x^2 + y^2 = 1$, $z = 0$, and for each p in the first factor picture $\{p\} \times S^1$ as being a smaller circle with center p and diameters perpendicular to the first circle at p, then the union $S^1 \times S^1$ is the surface of revolution of the small circle about the z axis—a torus (see Figure 3). It is not difficult to see that the topology induced from R^3 on the torus is the product topology.

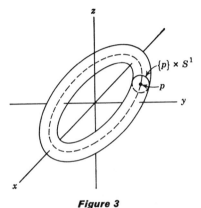

Figure 3

The torus is the underlying manifold which models the set of positions (the *configuration space*) of a double pendulum. We are thinking of a mechanical system consisting of two rods, the first of which is free to rotate in a plane about a fixed axis and the second of which rotates about an axis in a plane which is fixed relative to the first rod—usually, but not necessarily, the plane of the first rod. The angles these rods make with a coordinate axis in their planes may be matched with the angles u, v which occur in the parametrization of the torus given below, giving a 1–1 correspondence between the positions of the double pendulum and the torus. The linkage must be arranged so that each rod is free to make a complete circuit about its axis, or else only a part of the torus is the model. In fact, if the second rod is blocked by the axis of the first, so that v is restricted to $0 < \varepsilon < v < 2\pi$, then the model is a cylinder rather than a torus.

By adding more rods we obtain physical systems for which the model is the product of more copies of S^1. If the linkage is arranged so that the rod is free to move in space rather than in a plane, then some factors S^2 (see below) may be needed. Finally, if one end of the first rod is not fixed at all but is allowed to move freely in space (or a plane), then a factor R^3 (or R^2) may be needed.

More generally, if a physical system is a composite of two systems, each of which can assume all its positions independently of the other, then the composite system has as its manifold of positions the product of the manifolds of positions of the two component systems. This is so even though there is some dynamic linkage (e.g., gravitational or elastic) between the components.

Problem 1.2.4.   Consider a spring with a weight attached to each end which is allowed to move freely in space except that the length L of the spring is restricted to $L_1 < L < L_2$. Describe the configuration space as a triple product of R^3 and two other manifolds.

(d) LOW DIMENSIONS. A manifold of dimension 0 is a set of isolated points, that is, a set with discrete topology.

A manifold of dimension 1 which is connected is either R or S^1. (This is not obvious, but a proof will not be given here.) The other manifolds of dimension 1 consist of disjoint unions of copies of R and S^1. The number of copies of each must be finite or countably infinite in order that the manifold have a countable basis of neighborhoods.

Problem 1.2.5.   Let $M = S^1 = \{(a, b) \mid a^2 + b^2 = 1, a, b \in R\}$. As topology on M we take the induced topology from R^2. The following conditions define a unique $f: M \to R$: (α) For every $p \in S^1$, $0 \le fp < 2\pi$; and (β) if $p = (a, b) \in S^1$, then $a = \cos fp$, $b = \sin fp$.
   (a) Of the properties of a coordinate map listed, which does f satisfy?
      (1) A coordinate map has open domain.
      (2) A coordinate map is 1–1.
      (3) A coordinate map has open range.
      (4) A coordinate map is continuous.
      (5) The inverse of a coordinate map is continuous.
   (b) What is the largest set to which f can be restricted so as to be a coordinate map f^-?
   (c) Let g be defined in the same way as f^- except that the range of g is a different (and open) interval in R. For some specific choice of interval show that $\{f^-: U \to R, g: V \to R\}$ is an analytic atlas for S^1.

A manifold of dimension 2 may reasonably be called a surface, although there are such manifolds which cannot be placed in R^3. (See Problems 1.2.13 and 1.2.14.) Also, to make what are usually called surfaces in R^3 into manifolds, it is necessary to eliminate singular points, but these singular points cannot be handled by the usual methods of analysis from advanced calculus anyway. (For example, the tangent plane is customarily defined only at non-singular points.) To see that surfaces in R^3 are manifolds we examine how they usually arise.

(1) If the surface is the level surface of a C^∞ function $f: R^3 \to R$, then the singular points are those at which $df = 0$, that is, at which all three partial derivatives of f vanish. At a nonsingular point $p = (x_0, y_0, z_0)$, at which say $\partial f / \partial y(p) \neq 0$, there is an open neighborhood U of (x_0, z_0) in R^2 such that the equation $f(x, y, z) = c$ has a unique C^∞ solution $y = g(x, z)$ with $y_0 = g(x_0, z_0)$, where $c = fp$. This follows from the implicit function theorem. Then

$$V = \{(x, g(x, z), z) \mid (x, z) \in U\}$$

is an open subset of the surface with respect to the induced topology from R^3, and the projection from V to the xz plane,

$$\mu: V \to U, \text{ given by } \mu(x, g(x, z), z) = (x, z),$$

is a coordinate map on V.

We can form an atlas for the nonsingular part of the surface $f^{-1}c$ from such maps. If $\partial f / \partial z(q) \neq 0$ and

$$\varphi: W \to X \text{ is given by } \varphi(x, y, h(x, y)) = (x, y)$$

in a neighborhood W of q, where $z = h(x, y)$ is the C^∞ solution of $f(x, y, z) = c$ for z on X such that $q = (x_1, y_1, h(x_1, y_1))$, then on the overlap of W and V,

$$\mu \circ \varphi^{-1}(x, y) = \mu(x, y, h(x, y))$$
$$= (x, h(x, y)),$$

and similarly,

$$\varphi \circ \mu^{-1}(x, z) = (x, g(x, z)).$$

Since h and g are C^∞ functions, and x is a C^∞ function of either (x, y) or (x, z), the maps $\mu \circ \varphi^{-1}$ and $\varphi \circ \mu^{-1}$ are C^∞. This shows that the described atlas is C^∞-related and that the nonsingular points on the surface form a C^∞ manifold.

More specifically, if we take $f = x^2 + y^2 + z^2$ and $c = 1$, then the set of solutions to $f = 1$ is a sphere, S^2, and since $df = 0$ only at $(0, 0, 0)$, all points of the sphere are nonsingular.

The equation $x^2 + y^2 + z^2 = 1$ has two analytic solutions for z in the open disk $U_z = \{(x, y) \mid x^2 + y^2 < 1\}$, namely, $z = \sqrt{(1 - x^2 - y^2)}$ and $z = -\sqrt{(1 - x^2 - y^2)}$. The corresponding charts on S^2 are $\mu_z^+: U_z^+ \to U_z$, $\mu_z^-: U_z^- \to U_z$, where U_z^+ is the open upper hemisphere and U_z^- is the open lower hemisphere, and $\mu_z^+(x, y, z) = (x, y)$ for $(x, y, z) \in U_z^+$. The other map, μ_z^-, has the same formula as μ_z^+, but it is defined on U_z^-, where the third coordinate is negative rather than positive. In the same way we get charts $\mu_y^+ : U_y^+ \to U_y$, $\mu_y^- : U_y^- \to U_y$, $\mu_x^+ : U_x^+ \to U_x$, $\mu_x^- : U_x^- \to U_x$ on the left, right, front, and back open hemispheres. These six charts form an analytic atlas for S^2, so S^2 is an analytic manifold.

(2) Surfaces are sometimes given parametrically. That is, three C^∞ functions $x = f(u, v)$, $y = g(u, v)$, $z = h(u, v)$ are defined in some open region in the uv plane. The singular points are those for which the two triples of partial derivatives $(\partial f/\partial u, \partial g/\partial u, \partial h/\partial u)$ and $(\partial f/\partial v, \partial g/\partial v, \partial h/\partial v)$, are proportional (including one or both having all three entries $= 0$).

At nonsingular points these two triples will be direction numbers for two nonparallel lines which determine the tangent plane, but at singular points the two lines unite or are indeterminate (one or both all 0) and the tangent plane may not exist.

If (u_0, v_0) are the parameters of a nonsingular point, then there is an open neighborhood U of (u_0, v_0) in R^2 on which the parametrization is 1–1 onto an open set V in the surface. Indeed, nonsingularity implies that one of the jacobian determinants

$$\begin{vmatrix} \partial f/\partial u & \partial f/\partial v \\ \partial g/\partial u & \partial g/\partial v \end{vmatrix}, \quad \begin{vmatrix} \partial f/\partial u & \partial f/\partial v \\ \partial h/\partial u & \partial h/\partial v \end{vmatrix}, \quad \begin{vmatrix} \partial g/\partial u & \partial g/\partial v \\ \partial h/\partial u & \partial h/\partial v \end{vmatrix},$$

is nonzero, say the first one, in which case there is an open neighborhood U of (u_0, v_0) such that $(u, v) \to (f(u, v), g(u, v))$ is 1–1 with a C^∞ inverse on U, so certainly $(u, v) \to (f(u, v), g(u, v), h(u, v))$ is also 1–1 on U. The inverse of this map $U \to V$ is then a coordinate map $\mu : V \to U$, the parameters u, v themselves being the coordinate functions. The projection into $R^2, \varphi : V \to W$, $\varphi(x, y, z) = (x, y)$ is also 1–1 and C^∞-related to μ, so it can serve as an alternative coordinate map. However, the parametrization is usually 1–1 on a larger neighborhood than U on which one of the three jacobians is nonzero, so that μ may be extended to a more inclusive coordinate map and is thus usually to be preferred over φ.

The complete parametrization map $(u, v) \to (x, y, z)$ may not be 1–1 even on the nonsingular part, but may cover the same part of the surface with several different regions of the uv plane. Thus there can be nonidentical coordinate transformations from the uv plane into itself. These will be C^∞ at nonsingular points, so the set of nonsingular points forms a two-dimensional manifold.

In a neighborhood of a nonsingular point a normal vector can be chosen to vary as a C^∞ function of (u, v). Letting f be the directed distance to the surface, with the direction determined by the chosen normal field, we get the surface locally as the solutions of $f = 0$, where f is a C^∞ function. Thus nonsingular level surfaces are locally parametrized surfaces (the coordinates are parameters) and nonsingular parametrized surfaces are locally level surfaces. Methods (1) and (2) of specifying surfaces are locally equivalent. However, they are not globally equivalent, since nonsingular level surfaces are always orientable (two-sided, having a global continuous nonzero normal field), whereas nonsingular parametrized surfaces may be nonorientable (one-sided). In fact, the gradient of f is a normal field to the surface $f = c$, and it is not difficult to realize the Möbius band, which is nonorientable, as a parametrized surface.

The singularities of a parametrization may be either an unavoidable consequence of the shape of the surface (it may have a cusp or a corner at which no tangent space can be defined) or it may be an accident of the parametrization itself. An example of the latter is the standard spherical coordinate parametrization of the unit sphere,

$$x = \sin u \cos v,$$
$$y = \sin u \sin v,$$
$$z = \cos u,$$

for which the points $(0, 0, 1)$ and $(0, 0, -1)$ are singular points. For this parametrization the uv coordinate transformations assume one of two forms:

$$u_\alpha = u_\beta + 2p\pi,$$
$$v_\alpha = v_\beta + 2r\pi,$$

or

$$u_\alpha = u_\gamma + (2q + 1)\pi,$$
$$v_\alpha = -v_\gamma + 2s\pi,$$

where p, q, r, and s are integers and the three coordinate maps $\mu_\alpha = (u_\alpha, v_\alpha)$, $\mu_\beta = (u_\beta, v_\beta)$, $\mu_\gamma = (u_\gamma, v_\gamma)$ are related.

Problem 1.2.6. Show that S^2 has an atlas with two charts.

The torus in R^3 may be parametrized without singularities:

$$x = (a + b \sin v) \cos u,$$
$$y = (a + b \sin v) \sin u,$$
$$z = b \cos v,$$

where a is the radius of the first circle S^1 in the xy plane and b is the radius of the small second circles having their diameters perpendicular to the first circle, as in the above description of the torus as a product $S^1 \times S^1$. The parameters u and v measure the angles around the first and second circles.

The possible uv coordinate transformations are of the form

$$u_\alpha = u_\beta + 2p\pi,$$
$$v_\alpha = v_\beta + 2q\pi,$$

where p and q are integers.

Problem 1.2.7. Show that the parametrization of the torus given above may be inverted on three different domains so as to obtain an atlas of three charts for the torus.

(e) HYPERSURFACES. The idea of a surface may be generalized to higher dimensions. In a manner analogous to that for surfaces, we may show that the nonsingular points of a *level hypersurface*,

$$M = \{m \mid fm = c,\, df_m \neq 0\},$$

where $f: R^d \to R$ is a C^∞ function and c is a constant, form a manifold of dimension $d - 1$. Local coordinates are obtained by projections into the $(d - 1)$-dimensional coordinate hyperplanes and are shown to be C^∞-related by means of the implicit function theorem. Alternatively, we may consider parametric manifolds in R^d, with the number of parameters any number less than d, in particular, $d - 1$ parameters for a hypersurface. Nonsingularity is defined in terms of rank of jacobian matrices.

In particular, we define the *d-dimensional sphere* to be

$$S^d = \left\{ p \in R^{d+1} \mid \sum_{i=1}^{d+1} (u^i p)^2 = 1 \right\}.$$

In analogy with S^2, the projections which kill one component $u^i p$ give $2(d + 1)$ coordinate maps on the hemispheres for which a given $u^i p$ is constant in sign.

Problem 1.2.8. An open subset of R^d is not compact (cf. Section 0.2.8). Show that a compact manifold (e.g., S^d, which is a closed bounded subset of R^{d+1}) cannot have an atlas consisting of just one chart (cf. Problem 1.2.3).

Problem 1.2.9. Consider a rod of length L in space R^3. Letting the standard coordinates of one end be u^1, u^2, u^3 and of the other end be u^4, u^5, u^6, the collection of positions of this rod can be viewed as the hypersurface in R^6 given by the equation

$$(u^1 - u^4)^2 + (u^2 - u^5)^2 + (u^3 - u^6)^2 = L^2.$$

Show how this manifold is also the same as $R^3 \times S^2$.

The manner in which $S^1 \times S^1$ is placed in R^3 to get a torus may be generalized to an imbedding of $S^d \times S^e$ in R^{d+e+1} as a hypersurface; that is, a small copy of S^e is placed in an R^{e+1} perpendicular to S^d at each point of S^d as it is contained in $R^{d+1} = R^{d+1} \times \{0\} \subset R^{d+e+1}$.

(f) MANIFOLDS PATCHED TOGETHER. A manifold can be given by specifying the coordinate ranges of an atlas, the images in those coordinate ranges of the overlapping parts of the coordinate domains, and the coordinate transformations for each of those overlapping domains. When a manifold is specified in this way, a rather tricky condition on the specifications is needed to give the Hausdorff property, but otherwise the topology can be defined completely by simply requiring the coordinate maps to be homeomorphisms. Two examples follow.

(1) Let there be two charts $\mu\colon U \to S$, $\varphi\colon V \to S$ such that the range of each is the rectangular strip

$$S = \{(a, b) \mid -5 < a < 5, -1 < b < 1\}.$$

The overlapping domain $U \cap V$ corresponds to the union of two end rectangles under both μ and φ,

$$
\begin{aligned}
T &= \mu(U \cap V) \\
&= \varphi(U \cap V) \\
&= \{(a, b) \mid -5 < a < -4 \text{ or } 4 < a < 5 \text{ and } -1 < b < 1\}.
\end{aligned}
$$

It remains to define $\mu \circ \varphi^{-1}$ (or $\varphi \circ \mu^{-1}$) on T, which we do by the formula

$$\mu \circ \varphi^{-1}(a, b) = \begin{cases} (a + 9, b) & \text{if } -5 < a < -4, \\ (a - 9, -b) & \text{if } 4 < a < 5. \end{cases}$$

The reader should paste two strips of paper together in accordance with this formula (at least mentally) if he wishes to see what this manifold represents. Since the formula components represent rigid euclidean transformations, the paper need not be torn or stretched.

To obtain the manifold more specifically as a set of elements with topology, etc., we take disjoint copies of the ranges of the coordinate maps and "identify" points in these ranges which correspond under the overlap formulas. The precise meaning of "identification" comes from the idea of an "equivalence relation," which is a modification of the idea of equality in sets to mean something other than "identically the same." The idea is not new, since it is necessary to give precise meaning to such things as $4/6 = 6/9$.

In the case at hand the coordinate ranges are not already disjoint, so we manufacture disjoint copies of their common range S by tagging the elements of S with a 1 or a 2:

$$S_\alpha = \{(s, \alpha) \mid s \in S\},$$

where $\alpha = 1$ or 2 and let $P = S_1 \cup S_2$. We define an equivalence relation on P in accordance with a desire to identify a member of S_1 with a member of S_2

if they are connected by the coordinate transformation $F = \mu \circ \varphi^{-1}$, but otherwise to make no identifications between members of P: For all $s, t \in S$,

$$(s, 1)E(t, 1) \text{ iff } s = t,$$
$$(s, 2)E(t, 2) \text{ iff } s = t,$$
$$(s, 1)E(t, 2) \text{ iff } t \in T \text{ and } s = Ft,$$
$$(s, 2)E(t, 1) \text{ iff } s \in T \text{ and } t = Fs.$$

In this case the equivalence classes have only one or two elements: If $s \notin T$ and $t \in T$, then

$$[s, \alpha] = \{(s, \alpha)\}, \quad \text{where } \alpha = 1 \text{ or } 2,$$
$$[t, 1] = \{(t, 1), (F^{-1}t, 2)\},$$
$$[t, 2] = \{(Ft, 1), (t, 2)\}.$$

We unify the definitions of the coordinate maps μ and φ by calling them μ_α, $\alpha = 1$ or 2. Their domains are $U_\alpha = [S_\alpha]$, the collection of all equivalence classes of members of S_α. The maps are given simply by

$$\mu_\alpha[s, \alpha] = s.$$

Since these μ_α are to be the coordinate maps on $M = P/E$, the topology on M must be defined in such a way that they are homeomorphisms. Accordingly we define the open sets of M to be of three types: (a) A subset of U_1 is open iff it corresponds under μ_1 to an open set in S; (b) a subset of U_2 is open iff it corresponds under μ_2 to an open set of S; (c) a subset of M which is neither a subset of U_1 nor a subset of U_2 is open iff the intersections of the subset with U_1 and U_2 are both open according to (a) and (b).

Problem 1.2.10. Complete the demonstration that the M defined above is an analytic manifold, including the proof that it is a Hausdorff space.

Problem 1.2.11. By extending S and the formula to include points where $b = \pm 1$, a boundary manifold is attached to this M. What is this boundary manifold intrinsically?

(2) In this example there are three coordinate systems in the given atlas, all with R^2 as their range. Let them be $\mu_1 = (x^1, x^2)$, $\mu_2 = (y^1, y^2)$, $\mu_3 = (z^1, z^2)$. The overlapping domains correspond to as much of R^2 in each case as makes sense in the following formulas.

$$x^1 = 1/y^2, \quad x^2 = y^1/y^2,$$
$$y^1 = 1/z^2, \quad y^2 = z^1/z^2,$$
$$z^1 = 1/x^2, \quad z^2 = x^1/x^2.$$

We could proceed as in (1) to manufacture the manifold by taking three copies of R^2 and defining an equivalence relation corresponding to these formulas. The manifold defined by these coordinate transformations admits a more concrete interpretation. Let S^2 be the unit sphere in R^3 with center at the origin. We define two opposite points of S^2 to be equivalent and M to be the set of equivalence classes; thus an element of M is a nonordered pair $\{p, -p\}$, where $p \in S^2$. If $p = (a, b, c)$ we have written $-p$ for $(-a, -b, -c)$. We could also consider the elements of M to be the lines through the origin in R^3, where the line through p and $-p$ corresponds to $\{p, -p\}$. The name for M is the *analytic real projective plane*.

If x, y, z are the cartesian coordinates on R^3, then the ratios, x/y, x/z, y/z, etc., have the same values on p and $-p$, so they are well-defined functions on the subsets of M on which the denominators are nonzero. We obtain the coordinate maps on M from pairs of these ratios.

$$\mu_1 = (y/x, z/x) = (x^1, x^2),$$
$$\mu_2 = (z/y, x/y) = (y^1, y^2),$$
$$\mu_3 = (x/z, y/z) = (z^1, z^2).$$

The corresponding coordinate domains are those $\{p, -p\}$ for which $xp \neq 0$, $yp \neq 0$, and $zp \neq 0$, respectively.

Projective spaces of higher dimension can be defined analogously as opposite pairs on higher-dimensional spheres.

Problem 1.2.12. Just as the circle may be thought of as a half-closed interval $[0, 2\pi)$ with the end 0 bent around to fill the hole at 2π, the torus may be considered to be the "half-closed" square $[0, 2\pi) \times [0, 2\pi)$ with the closed sides folded over to fill the opposite open side in the *same* direction (see

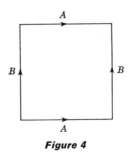

Figure 4

Figure 4). ("Direction" refers to direction in the plane R^2, not cyclic direction around the square.)

Problem 1.2.13. Show that the projective plane may be formed by folding the square so that the closed sides fill the open sides in the *opposite* direction (see

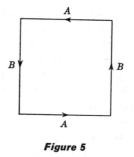

Figure 5

Figure 5). Another corner must be provided. To make the correspondence, stretch the square over a hemisphere with the edges laid along the bounding circle so that the corners divide the circle into four equal arcs. Since that stretching cannot be done so that the map at the corners is C^∞, this identification is only intended to be topological.

Problem 1.2.14.   By identifying one pair of opposite sides of the square in the same direction and the other in the opposite direction we get a two-dimensional manifold known as a *Klein bottle* (see Figure 6). The identification can be done

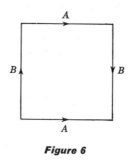

Figure 6

differentiably since the four corners of the square fit together nicely. Give an analytic definition of the Klein bottle in the form of **(1)** and **(2)** above, which has four charts pictured as having centers at the center of the original square, the corner of the original square, and the centers of the two sides.

The Klein bottle can be realized as a parametric surface in R^4 in much the same way as the torus in R^3. At each point of the circle of radius a in the xy plane there is now available a three-dimensional hyperplane in R^4 perpendicular to the circle. A smaller circle of radius $b < a$ can be rotated about a diameter at half the rate of revolution about the circle of radius a, giving a Klein bottle. The parametrization is given analytically as follows:

$$x = (a + b \sin v) \cos u,$$
$$y = (a + b \sin v) \sin u,$$
$$z = b \cos v \cos u/2,$$
$$w = b \cos v \sin u/2.$$

Points in the uv plane which are identified as indicated in Figure 6 are mapped into the same points in R^4 by these equations.

Remarks. The projective plane and the Klein bottle cannot be faithfully represented as surfaces in R^3 without "self-intersections." To describe what self-intersections are, we give the example of the disconnected manifold consisting of two copies of R^2 pictured in R^3 as two *intersecting* planes. The points along the line of intersection have a dual role, each being considered as two points, one in each copy of R^2. (This is the reason the planes form a disconnected manifold.) When such duplications are allowed, we say that the manifold is *immersed* rather than *imbedded* in R^3. In this sense the projective plane (*Boy's surface*) and the Klein bottle can be immersed as surfaces in R^3.

The three-dimensional projective space, RP^3, is the same, insofar as its manifold structure is concerned, as the set of all orthogonal matrices of order 3 having determinant $+1$. Since an orthogonal matrix of order 3 having determinant $+1$ is equivalent to a rotation of R^3 about the origin, projective 3-space is in turn the same manifold as the configuration space of an object in R^3 which has one fixed point but is otherwise free to rotate about any axis through the fixed point.

If an object is free to move in any way in space, we may determine its position by choosing a point in the object and specifying both where that point is placed in R^3 and how the object is rotated about that point relative to some initial position. Since these specifications are independent, the manifold of positions of a rigid object in space is $R^3 \times RP^3$.

1.3.   Differentiable Maps

If $F: M \to N$, where M and N are C^∞ manifolds, then we call F a C^∞ map if the coordinate expression for F consists of C^∞ maps on cartesian spaces. We now elaborate this statement into a complete definition, in particular making clear what is meant by "coordinate expressions."

Let $\mu_1: U \to R^d$ and $\mu_2: V \to R^e$ be C^∞ charts on M and N, so that U and V are open subsets of M and N, respectively. Assume that $F: M \to N$ is a continuous map, so that $W = F^{-1}V$ is an open subset of M (see Figure 7). Let $W_1 = \mu_1 W$, so that W_1 is an open set in R^d. The μ_1-μ_2 *coordinate expression*

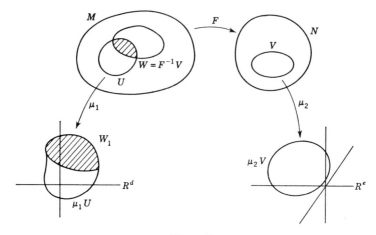

Figure 7

for F is the map $\mu_2 \circ F \circ \mu_1^{-1}: W_1 \to R^e$. The map *F* is C^∞ if all such coordinate expressions, for all admissible charts μ_1, μ_2, are C^∞ cartesian maps.

Proposition 1.3.1. *A map* $F: M \to N$ *is* C^∞ *if the* μ_α-μ_β *coordinate expressions for F are* C^∞ *for those* μ_α *in some atlas of M and those* μ_β *in some atlas of N.*

Proof. Let $\{\mu_\alpha: U_\alpha \to R^d \mid \alpha \in I\}$ and $\{\mu_\beta: V_\beta \to R^e \mid \beta \in J\}$ be atlases of *M* and *N*, respectively, such that for every $\alpha \in I$, $\beta \in J$, $\mu_\beta \circ F \circ \mu_\alpha^{-1}$ is a C^∞ map. Suppose that μ_1, μ_2 are any other charts as in the definition, so $\mu_2 \circ F \circ \mu_1^{-1}: W_1 \to R^e$. We must show that this is C^∞, but since being C^∞ is a local property it suffices to show it in a neighborhood of each point of W_1. If $m_1 \in W_1$, then there is an $\alpha \in I$ and $\beta \in J$ such that $\mu_1^{-1} m_1 = m \in U_\alpha$ and $n = Fm \in V_\beta$. By hypothesis, $\mu_\beta \circ F \circ \mu_\alpha^{-1}$ is a C^∞ cartesian map. But μ_1 and μ_2 are C^∞-related to μ_α and μ_β, respectively, so $\mu_\alpha \circ \mu_1^{-1}$ is defined and C^∞ in some neighborhood of m_1 and $\mu_2 \circ \mu_\beta^{-1}$ is defined and C^∞ in some neighborhood of $n_\beta = \mu_\beta n$.

The composition of C^∞ cartesian maps is C^∞, so that $\mu_2 \circ \mu_\beta^{-1} \circ \mu_\beta \circ F \circ \mu_\alpha^{-1} \circ \mu_\alpha \circ \mu_1^{-1}$ is a C^∞ map. However, it is defined on some neighborhood of m_1 and coincides with the restriction of $\mu_2 \circ F \circ \mu_1^{-1}$ on that neighborhood, so that $\mu_2 \circ F \circ \mu_1^{-1}$ is C^∞ in a neighborhood of m_1. ∎

In practice, verification that maps are C^∞ must be done by showing that the individual components of the coordinate expressions have continuous partial derivatives of all orders. These components are the functions $u^i \circ \mu_2 \circ F \circ \mu_1^{-1} = f^i$, $i = 1, \ldots, e$, which are real-valued functions of *d* real variables defined on an open subset W_1 or R^d.

If we let $y^i = u^i \circ \mu_2$ be the coordinate functions of μ_2 and $x^j = u^j \circ \mu_1$ be the coordinate functions of μ_1, then we have $y^i \circ F \circ \mu_1^{-1} = f^i$, or $y^i \circ F = f^i \circ \mu_1$. Applying this to $m \in W$,

$$y^i Fm = f^i \mu_1 m$$
$$= f^i(x^1 m, \ldots, x^d m).$$

It is customary to write this as an equation between functions in the form

$$y^i = f^i(x^1, \ldots, x^d), \tag{1.3.1}$$

but since this does not indicate the role of the map F itself, we prefer the more accurate version

$$y^i \circ F = f^i(x^1, \ldots, x^d). \tag{1.3.2}$$

These equations are also called the *coordinate expression for F*.

In particular, we may consider the case $N = R$ of real-valued functions on M. It is interesting that C^∞ functions need be defined directly only in this case, and the general definition of a C^∞ map then follows by means of the following proposition.

Proposition 1.3.2. *If* $F: M \to N$, *then* F *is* C^∞ *if for every* C^∞ *real-valued function* $y: V \to R$, *where* V *is an open submanifold of* N, $y \circ F$ *is a* C^∞ *real-valued function on the open submanifold* $F^{-1}V$ *of* M.

Proof. This follows trivially by taking as y, in turn, the coordinate functions y^i on $V \subseteq N$. ∎

A *diffeomorphism* from M onto N is a 1–1 onto C^∞ map $F: M \to N$ such that the inverse map $F^{-1}: N \to M$ is also C^∞. Two manifolds are *diffeomorphic* if there is a diffeomorphism from one to the other. This is the natural notion of isomorphism, or sameness, for manifolds. It is an equivalence relation. Two diffeomorphic manifolds are the same in all properties which concern only their structure as manifolds. In particular, they are topologically the same, that is, homeomorphic.

Examples.   (a) Let

$$F = \frac{u}{1 - u^2} : (-1, 1) \to R.$$

Then solving $x = Fu$ for u, bearing in mind that we must take the root of the quadratic equation which is between -1 and 1, we get

$$u = \frac{2x}{1 + \sqrt{(1 + 4x^2)}} = F^{-1}x.$$

Thus F has an inverse defined for all $x \in R$, so F is onto and 1–1. Moreover, both F and F^{-1} are quotients of C^∞ functions with nonzero denominator, and so are C^∞. Hence F is a diffeomorphism and R is diffeomorphic to the open interval $(-1, 1)$.

If (a, b) is any other open interval, then

$$\tfrac{1}{2}[(1 + u)b + (1 - u)a]: (-1, 1) \to (a, b)$$

is a diffeomorphism, so every connected open bounded submanifold of R is diffeomorphic to R. It is not difficult to see that the other connected open submanifolds of R, the open half-lines $(-\infty, b)$ and (a, ∞), are also diffeomorphic to R.

(b) If F is the map in Example **(a)**, then

$$F \times F: (-1, 1) \times (-1, 1) \to R^2$$

shows that an open square is diffeomorphic to the plane.

(c) Let x, y be the cartesian coordinates on R^2, and let u, v be the restrictions of x, y to the unit disk $D^2 = \{(x, y) \mid x^2 + y^2 < 1\}$, viewed as an open submanifold of R^2. Define G so that it is the same on radial lines as F above; that is, $G: D^2 \to R^2$ has coordinate expression

$$x \circ G = \frac{u}{1 - u^2 - v^2},$$
$$y \circ G = \frac{v}{1 - u^2 - v^2}.$$

The coordinate expression of the inverse map $G^{-1}: R^2 \to D^2$ is then

$$u \circ G^{-1} = \frac{2x}{1 + \sqrt{(1 + 4x^2 + 4y^2)}},$$
$$v \circ G^{-1} = \frac{2y}{1 + \sqrt{(1 + 4x^2 + 4y^2)}}.$$

These are C^∞, onto, and 1–1 for the same reasons that F and F^{-1} were, so G is a diffeomorphism and a circular disk is diffeomorphic to the plane and hence also to a square.

A topological manifold may have two different C^∞ atlases which are not C^∞-related, but the two C^∞ manifolds determined by these atlases can still be diffeomorphic. The catch is that the identity map is not a diffeomorphism. In fact, two C^∞ manifold structures on a manifold of dimension ≤ 4 are invariably diffeomorphic. On the other hand, any compact manifold of dimension ≥ 7 admits several nondiffeomorphic C^∞ manifold structures; that is, there can be a homeomorphism between two manifolds but no diffeomorphism.

For a simple example of different C^∞ structures which are still diffeomorphic, consider R with the standard structure, $\{u: R \to R\}$ as atlas (with one chart), and $M = R$ with the structure having $\{u^3: R \to R\}$ as atlas (again, one chart). Since an admissible chart is always a diffeomorphism on its open submanifold domain, the diffeomorphism from M onto R is the coordinate map of $M, u^3: M \to R$. The diffeomorphism going the other way is $u^{1/3}: R \to M$. The identity map $u: R \to M$ is C^∞, since the coordinate expression is $u^3 \circ u \circ u = u^3: R \to R$. The identity map $u: M \to R$ is not C^∞, since the coordinate expression is $u \circ u \circ u^{1/3} = u^{1/3}: R \to R$, which is not C^∞. Thus the identity map is not a diffeomorphism.

There are examples of nondiffeomorphic C^∞ structures on manifolds of dimension ≥ 7. These are not easy to describe, however.

Problem 1.3.1. Let $\mu: U \to R^d$ be an admissible chart of M. Then U is an open subset of M and $V = \mu U$ is an open subset of R^d, so U and V may be viewed as manifolds; specifically, they are open submanifolds of M and R^d, respectively. Show that $\mu: U \to V$ is a diffeomorphism.

Problem 1.3.2. Show that the composition of C^∞ maps is a C^∞ map.

Example. The parametrization of a sphere $F: R^2 \to R^3$, $F(u, v) = (\cos u \sin v, \sin u \sin v, \cos v)$, is a C^∞ map. The coordinate expression for F in terms of the standard coordinates are what is seen in the definition, and may be written in the alternative form

$$
\begin{aligned}
x \circ F &= \cos u \sin v, \\
y \circ F &= \sin u \sin v, \\
z \circ F &= \cos v.
\end{aligned}
\tag{1.3.3}
$$

If we view the same formula as defining a map $F: R^2 \to S^2$, it is still a C^∞ map. It we take as atlas on R^2 the identity chart (u, v) and as atlas on S^2 the six charts described in Section 1.2(**d**), then the six corresponding coordinate expressions for F are the formulas (1.3.3) taken two at a time and restricted to appropriate open subsets of R^2. The fact that F has singularities as a parametrization of the sphere has no bearing on it being a C^∞ map or not.

Problem 1.3.3. Let RP^2 be the projective plane, as in Section 1.2(**f**)(**2**), viewed as nonordered pairs $\{\{p, -p\} \mid p \in S^2\}$. Show that the 2–1 map $F: S^2 \to RP^2$, where $Fp = \{p, -p\}$, is C^∞.

Problem 1.3.4. If $F: S^2 \to RP^2$ is the same as in Problem 1.3.3, show that $G: M \to S^2$ is C^∞ if G is continuous and $F \circ G: M \to RP^2$ is C^∞. Find an example such that G is not continuous but $F \circ G$ is C^∞.

Problem 1.3.5. If S is a surface without singularities in R^3, show that the inclusion map $i: S \to R^3$ is a C^∞ map. Do both cases, level surfaces and parametric surfaces.

Problem 1.3.6. If $M = C \times C$, where C is the complex number field which as a manifold is the same as R^2, let complex multiplication be $F: M \to C$, $F(z, w) = zw$. Show that F is C^∞.

Problem 1.3.7. Let S^1 be viewed as the unit circle with center 0 in C. Then $S^1 \times S^1 \subset C \times C = M$. Let $G: S^1 \times S^1 \to S^1$ be the restriction of F in Problem 1.3.6. Show that G is C^∞.

Problem 1.3.8. Let $M = C - \{0\}$, the nonzero complex numbers and an open submanifold of R^2, and define $H: M \to M$ as $Hz = 1/z$. Show that H is C^∞.

Problem 1.3.9. Show that the *projections* $p: M \times N \to M$, $q: M \times N \to N$, $p(m, n) = m$, $q(m, n) = n$, and the *injections* $i_n: M \to M \times N$, $_mi: N \to M \times N$, $i_n m = {_m i n} = (m, n)$, are C^∞.

Problem 1.3.10. If $F: P \to M \times N$, show that F is C^∞ if $p \circ F: P \to M$ and $q \circ F: P \to N$ (p, q as in Problem 1.3.9) are C^∞.

1.4.   Submanifolds

A manifold M is *imbedded* in a manifold N if there is a 1–1 C^∞ map $F: M \to N$ such that at every $m \in M$ there is a neighborhood U of m and a chart of N at Fm, $\mu: V \to R^e$, $\mu = (y^1, \dots, y^e)$, such that $x^i = y^i \circ F|_U, i = 1, \dots, d$, are coordinates on U for M. The map F is then called an *imbedding* of M in N.

If the requirement that F be 1–1 is omitted but the requirement on obtaining coordinates for M from those of N by composition with F still holds, then M is said to be *immersed* in N and F is said to be an *immersion*. Another way of

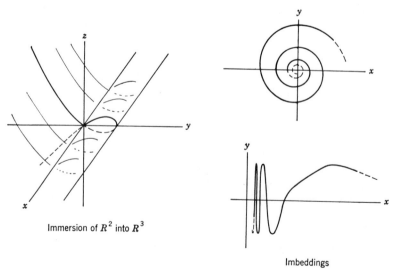

Immersion of R^2 into R^3

Imbeddings

Figure 8

stating this is to require that each point m of M be contained in an open submanifold U of M which is imbedded in N by F. Thus an immersion is a local imbedding (see Figure 8).

A *submanifold* of N is a subset FM, where $F: M \to N$ is an imbedding, provided with the manifold structure for which $F: M \to FM$ is a diffeomorphism.

The dimension of a submanifold is obviously not greater than the containing manifold's dimension. If it is equal to the dimension of the containing manifold, then the submanifold is nothing more than an open submanifold, which we have defined previously.

The topology of a submanifold need not be the induced topology from the larger manifold. Of course, the inclusion map is C^∞, in particular continuous, so that the open sets of the induced topology are open sets in the submanifold topology, but the submanifold topology can have many more open sets.

Examples. (a) Imbed an open segment in R^2 by bending it together in the shape of a figure 8, with the ends of the segment approaching the center of the segment at opposite sides of the cross (see Figure 9). In the induced topology

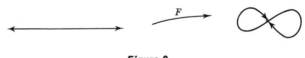

Figure 9

the neighborhoods of the center point always include a part of the two ends, but not in the submanifold topology.

(b) Let $Ft = (e^{it}, e^{i\alpha t}) \in S^1 \times S^1 \subset C \times C$, where α is an irrational real number. Then $F: R \to S^1 \times S^1$ is an imbedding. The line R is wound around the torus without coming back on itself, but filling the torus densely (see Figure 10). It crosses any open set in the torus infinitely many times, so the

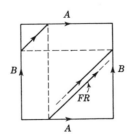

Figure 10

open sets in the induced topology have infinitely many pieces and are always
unbounded in R. This is quite unlike the standard topology on R.

A submanifold must be placed in the containing manifold in quite a special
way. For example, such things as cusps and corners are ruled out, even though
these may occur on the range of a C^∞, 1–1 map which is not an imbedding.
To describe carefully the special nature of a submanifold, we define a *coordinate slice* of dimension d in a manifold N of dimension e, to be a set of
points in a coordinate neighborhood U with coordinates $y^1, \ldots, y^e$ of the form
$\{m \mid m \in U, y^{d+1}m = c^{d+1}, \ldots, y^e m = c^e\}$, where the c^i are constants determining the slice. In other words, a coordinate slice is the image under the
inverse of a coordinate map of the part of a d-dimensional plane in R^e which
lies in the coordinate range.

Proposition 1.4.1.   *If M is a submanifold of N, then for every $m \in M$ there are
coordinates $y^1, \ldots, y^e$ for N in a neighborhood of m in N such that the coordinate slice corresponding to constants $c^{d+1} = y^{d+1}m, \ldots, c^e = y^e m$ is a
neighborhood of m in M and the restriction of $y^1, \ldots, y^d$ to that slice are
coordinates for M.*

Proof.   Let $F: P \to N$ be the imbedding such that $FP = M$. Choose
coordinates $z^1, \ldots, z^e$ for N in a neighborhood of m in N such that
$x^1 = z^1 \circ F|_U, \ldots, x^d = z^d \circ F|_U$, are coordinates at $p = F^{-1}m$ in coordinate
neighborhood $U \subset P$. Since F is C^∞ we may write

$$z^i \circ F = f^i(x^1, \ldots, x^d), \qquad i = 1, \ldots, e,$$

as in (1.3.2), where the f^i are C^∞ functions on an open set in R^d. It is clear that
$f^i(x^1, \ldots, x^d) = x^i$ for $i = 1, \ldots, d$, but the remaining f^i, $i > d$, need not be
so simple.
 Define

$$y^i = z^i, \qquad i \le d,$$
$$y^i = z^i - f^i(z^1, \ldots, z^d), \qquad i > d.$$

Then it is clear that

$$z^i = y^i, \qquad i \le d,$$
$$z^i = y^i + f^i(y^1, \ldots, y^d), \qquad i > d,$$

so the maps $\mu_1 = (z^1, \ldots, z^e)$ and $\mu_2 = (y^1, \ldots, y^e)$ are C^∞ related both ways.
The domain of the y^i's is included in that of the z^i's, so that we can claim that
μ_2 is an admissible chart without checking further relations with other coordinates. Moreover, FU is the coordinate slice $y^{d+1} = 0, \ldots, y^e = 0$, and the
restrictions of $y^1, \ldots, y^d$ to FU correspond to x^i under F, so are coordinates
for M on FU.   ∎

Remark. No claim is made that we can obtain *all* the points of M which lie in an N-neighborhood of m as members of a single coordinate slice. In fact, this is not possible in the case where m is the crossing point of the figure 8 in Example (**a**), or for any m in Example (**b**).

The converse of Proposition 1.4.1 is obvious from the definition; that is, if a subset has a manifold structure which is locally determined by coordinate slices with the nonconstant coordinates furnishing coordinates on the slice for the manifold structure of the subset, then the subset is a submanifold.

Whitney has proved that every manifold is diffeomorphic to a submanifold of R^e; if d is the dimension of the manifold, then we need take e no larger than $2d + 1$. Thus manifold theory can be considered to be the study of special subsets of cartesian spaces, if desired.

Example. (c) If $f: R^d \to R$ is a C^∞ function, then the nonsingular points of a level hypersurface

$$M = \{m \mid fm = c, df_m \neq 0\}$$

form a $(d - 1)$-dimensional submanifold on which the topology is the topology induced from R^d. Indeed, for each $m \in M$ one of the partial derivatives of f does not vanish at m, say, $\partial f/\partial u^d(m) \neq 0$. Then in some neighborhood of m we have that $(u^1, \ldots, u^{d-1}, f)$ is a coordinate system for R^d, because its jacobian determinant with respect to $(u^1, \ldots, u^d)$ is $\partial f/\partial u^d \neq 0$. In that neighborhood the points of M are those of the coordinate slice $f = c$. Hence M is a submanifold by the converse of Proposition 1.4.1. Note that the property disclaimed by the remark above, which is stronger than being a submanifold, is satisfied by these level hypersurfaces.

Problem 1.4.1. The map $F: R \to R^2$ given by $Ft = (t^2, t^3)$ is obviously C^∞. Why is it not an imbedding?

Problem 1.4.2. Show that the injections $i_n: M \to M \times N$ and $_m i: N \to M \times N$ (see Problem 1.3.9) are imbeddings and that the submanifolds $i_n M$, $_m iN$ of $M \times N$ have the induced topology.

Problem 1.4.3. Let $F: M \to N$ be any C^∞ map and define the graph of F to be $\{(m, Fm) \mid m \in M\} \subset M \times N$. Show that the graph of F is a submanifold of $M \times N$ with the induced topology and imbedding map $(i, F): M \to M \times N$ given by $(i, F)m = (m, Fm)$.

1.5. Differentiable Curves

In some contexts a curve is almost the same as a one-dimensional submanifold, but we prefer to deal only with curves which have a specific parametrization. Technically, then, changing the parametrization of a curve will give a different curve, but we shall often ignore the distinction and speak of a curve as if it were a set of points. Generally our curves will have a first and last point but we shall also consider curves with open ends.

A *differentiable curve* is a map of an interval of real numbers into a manifold such that there is an extension to an open interval which is a C^∞ map. The interval on which the curve is defined may be of any type, open, closed, half-open, bounded on both, one, or neither end. When the interval is open no proper extension is needed, but at a closed end we require that there be a C^∞ extension in a neighborhood of the end so that differentiability at that end will make sense.

If $\gamma: [a, b] \to M$ is a C^∞ curve, then, by definition, there must be a C^∞ extension $\bar{\gamma}: (a - c, b + c) \to M$, for some $c > 0$, such that $\gamma x = \bar{\gamma} x$ for every $x \in [a, b]$ (see Figure 11). We say that γa is the *initial point* of γ, γb is the

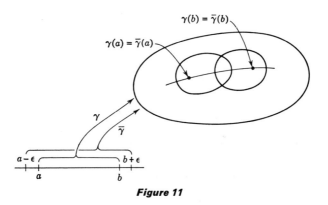

Figure 11

final point, and that γ is a C^∞ curve *from γa to γb*. A *closed curve* γ is one defined on a closed interval $[a, b]$ and for which $\gamma a = \gamma b$. A *simple closed curve* is a closed curve defined on $[a, b]$ which is 1–1 on $[a, b)$.

A C^∞ curve may double back on itself, have cusps, and come to a halt and start again, even turning a sharp corner in the process [see Example **(a)** below]. These features frequently prevent the curve from being an imbedding and prevent the range from being a one-dimensional submanifold. There may even be so many cusps that the curve cannot be chopped into finitely many pieces which are submanifolds.

In Example 1.4(**b**) there is a C^∞ curve $F: R \to S^1 \times S^1$ defined which comes arbitrarily close to every point in $S^1 \times S^1$. The range is a one-dimensional submanifold. In Problem 1.4.1 the C^∞ curve $F: R \to R^2$ has a cusp, since it comes into (0, 0), halts, and goes out on the same side of the x axis moving in the opposite direction.

Examples. (a) If f is the C^∞ function in Problem 1.1.1(a),
$$fx = \begin{cases} 0 & \text{if } x \le 0, \\ e^{-1/x} & \text{if } x > 0, \end{cases}$$
then $\gamma x = (fx, f(-x))$ defines a C^∞ curve $\gamma: R \to R^2$ which enters the origin $(0, 0)$ from the positive y axis, halts at $(0, 0)$, and exits via the positive x axis.

(b) Let $h: R \to R$ be a C^∞ probability distribution which vanishes outside the interval $(0, 1)$; for example,

$$hx = \begin{cases} 0 & \text{if } x \le 0 \text{ or } x \ge 1, \\ ce^{1/x(x-1)} & \text{if } 0 < x < 1, \end{cases}$$

where c is a positive normalizing constant chosen so that the area under the hump h is 1. Let g be the indefinite integral of h, $gx = \int_0^x h(t)dt$. We define a C^∞ function f which rises and falls periodically to level spots of length 1 at heights 0 and 1 by the specification

$$fx = \begin{cases} gx - g(x - 2) & \text{if } 0 \le x < 4, \\ f(x + 4) & \text{for all } x. \end{cases}$$

Then the curve $\gamma: R \to R^2$ given by $\gamma x = (fx, f(x + 1))$ is a C^∞ periodic parametrization of a square.

Remark. If a C^∞ curve is to turn a corner at a point m which is not simply a reversal of direction (as with a cusp), then the derivatives of all orders of the coordinate expressions must be 0 at m. For this to make sense, the tangent line (in some coordinate system) must have two different limits upon approaching m from opposite sides on the curve. Let μ be a coordinate map at m and $\mu \circ \gamma = (f^1, \ldots, f^d)$ the corresponding coordinate expression for the curve γ, and suppose $\gamma 0 = m$. If all the derivatives of all the f^i did not vanish at 0, then there is one of least order which is nonzero, say, the nth derivative of f^1. The limits of the slopes of the tangent lines to $\mu \circ \gamma$ relative to the first coordinate are

$$\lim_{t \to 0} \frac{\dfrac{df^i}{dt}}{\dfrac{df^1}{dt}} = \lim_{t \to 0} \frac{\dfrac{d^n f^i}{dt^n}}{\dfrac{d^n f^1}{dt^n}} = \frac{\dfrac{d^n f^i}{dt^n}(0)}{\dfrac{d^n f^1}{dt^n}(0)}.$$

Here the first equality follows from L'Hôpital's rule, the second from the continuity of the nth derivatives and the nonvanishing of the nth derivative of f^1. This shows that the limit is the same from both sides of 0, contrary to the condition that the curve turn a corner at m.

Problem 1.5.1. Specify a C^∞ parametrization of the polygonal curve in R^2 with vertices (a_i, b_i), $i = 0, 1, \ldots, n$.

A *continuous curve* from p to q in M is a continuous map $\gamma\colon [a, b] \to M$ such that $\gamma a = p, \gamma b = q$. There are many theorems of the sort "a continuous gadget may be approximated by a C^∞ gadget." We illustrate this in the case of curves.

Proposition 1.5.1. *If there is a continuous curve from p to q in M, then there is a C^∞ curve from p to q in M.*

Proof. Let $\gamma\colon [a, b] \to M$ be a continuous curve from p to q. At each γx there is a coordinate system which may be cut down so that the coordinate range in R^d is an open ball. Since the range of γ is compact, there are a finite number of these coordinate neighborhoods which cover the range of γ. Thus we may choose a partition of $[a, b]$, $x_0 = a < x_1 < \cdots < x_{n-1} < b = x_n$, such that for every i, γx_i and γx_{i+1} are in a common coordinate neighborhood. The corresponding straight line in R^d may be parametrized so that at each end all the derivatives of the coordinates with respect to the parameter are zero, similar to the way in which the sides of the square are parametrized in Example **(b)** above. By translating the parameters of these segments so that they match at each γx_i, we get a C^∞ parametrization for a curve from p to q which consists of a finite number of pieces which are, as point sets, straight line segments in terms of certain coordinates. The details are left as an exercise. Why did we specify balls for the coordinate ranges? ∎

Proposition 1.5.2. *If a C^∞ manifold M is connected, then every pair of points can be joined by a C^∞ curve. In particular, M is arc connected.*

Proof. Recall that M is connected means that the only subsets of M which are both open and closed are M itself and the empty set $\varnothing$. Thus it suffices to show that the points which can be joined to $p \in M$ by a C^∞ curve form a set S (obviously nonempty) which is both open and closed.

To show that S is closed we show that if $\{q_i\}$ is a sequence in S which converges in M, then $q = \lim q_i$ is in S. For any coordinate ball with center q there must be infinitely many q_i within that ball, so that by taking a curve from p to one of these q_i and chaining it to a segment (in the sense of the coordinates in the ball) from the q_i to q, and suitably altering the parametrization so it is C^∞ at the corner, we obtain a C^∞ curve from p to q. Thus S is closed.

On the other hand, if $q \in S$, then there is a coordinate ball around q, and any point in this coordinate ball may be joined to q by a segment, and then to p by a C^∞ curve. Thus S contains the coordinate ball. Since each point of S has a neighborhood contained in S, S is open. This shows that $S = M$, and so every point in M can be joined to p by a C^∞ curve. ∎

Thus for manifolds we need not distinguish between connectedness and arc connectedness. This is not true for topological subspaces of manifolds, however.

1.6.   Tangents

It is intuitively clear that a C^∞ curve should generally have a well-defined direction and speed. Of course, we have seen examples above in which a C^∞ curve turns a sharp corner and thus could have no single direction at the corner, but in those examples it will be found that the speed is zero at the corner. However, in any case the speed will not be defined absolutely, because there is no natural sense of distance on a manifold. The speed will be relative to the speed of other curves having the same direction. The notion of the tangent vector or velocity vector of a curve at a point is exactly this combination of direction and speed and no more. What is required is a definition of tangent vector which is operationally convenient and intuitively suggestive of the idea of direction and speed. We propose that an *operator* on C^∞ functions, the one which consists of taking the derivatives of all real-valued C^∞ functions along the curve with respect to the parameter, meets these requirements. This is similar to the operation of taking directional derivatives in R^3. In other words, we claim that if we are told how fast we are crossing the level surfaces of all functions, then we can determine the direction and speed of motion. Indeed, we actually need have such information for one set of coordinate functions only, but we avoid using this fact in our definition so as not to give the appearance of preferring one coordinate system over another. Such motivational arguments could be carried further and would lead us ultimately to the definition of tangents given below.

For every $m \in M$ we denote by $F^\infty(m)$ the collection of all C^∞ functions $f: U \to R$, where U is an open submanifold of M containing m. The set of functions $F^\infty(m)$ has considerable algebraic structure. If U, V are open sets containing m and $f: U \to R$, $g: V \to R$, then we define

$$f + g: U \cap V \to R \quad \text{and} \quad fg: U \cap V \to R$$

by

$$(f + g)n = fn + gn \quad \text{and} \quad (fg)n = (fn)(gn).$$

In particular, for $c \in R$ we have the constant function $c: M \to R$, $cm = c$ for every m. We have no notational distinction between c as a function and c as a real number. The following are then clear. For every $f \in F^\infty(m)$, $f + 0 = f$, $1f = f$. We also define $-f = (-1)f$. The commutative, associative, and distributive laws, which are the usual algebraic properties of addition, subtraction, and multiplication (but not division), are generally valid. However, since equality of functions requires equality of their domains, we have some exceptions to customary algebra: $f + (-f)$ and $0f$ are not 0 but rather $f + (-f) = 0f = 0|_U$, where $f: U \to R$. The function $0|_U$ differs from 0 in that it is defined only on U.

A *tangent at m ∈ M* is a function (operator)

$$t: F^\infty(m) \to R$$

such that for every $f, g \in F^\infty(m)$, $a, b \in R$,

(a) t is linear: $t(af + bg) = atf + btg$.

(b) t satisfies the product rule $t(fg) = (tf)gm + fm(tg)$.

[An operator on an algebraic system such as $F^\infty(m)$ which satisfies (a) and (b) is called a *derivation* of the system. Thus a tangent at $m \in M$ is a derivation of $F^\infty(m)$.]

Other synonyms for "tangent" are "tangent vector," "vector," "contravariant vector," the latter being the classical tensor terminology.

The set of all tangents at m will be denoted M_m, called the *tangent space at m*. We give M_m an algebraic structure, that of a vector space which is studied in detail in Chapter 2, by defining *addition, scalar multiplication,* and the *zero.* For $s, t \in M_m$, $a \in R$, we define

$$s + t \in M_m, \qquad as \in M_m, \qquad 0_m \subset M_m$$

by requiring for every $f \in F^\infty(m)$,

$$(s + t)f = sf + tf, \qquad (as)f = asf, \qquad 0_mf = 0.$$

It must be demonstrated that the things defined are in M_m, but these proofs are automatic.

Proposition 1.6.1. *For every $t \in M_m$ and constant function $c \in F^\infty(m)$, $tc = 0$.*

Proof. This follows from some simple computations using (a) and (b) with the constant functions c and 1:

$$ct1 = t(c1)$$
$$= (tc)1 + ct1$$
$$= tc + ct1.$$

Transposing $ct1$ gives $0 = tc$. ∎

Proposition 1.6.2. *If $f, g \in F^\infty(m)$ coincide on some neighborhood U of m, then for every $t \in M_m$, $tf = tg$.*

Proof. Let 1_U be the function which is constantly 1 on U and not defined elsewhere. Then the hypothesis on f and g can be written $1_Uf = 1_Ug$. By the product rule

$$t(1_Uf) = t1_U \cdot fm + 1tf$$
$$= t1_U \cdot fm + tf$$
$$= t(1_Ug)$$
$$= t1_U \cdot gm + tg$$
$$= t1_U \cdot fm + tg,$$

since $gm = fm$. Hence $tf = tg$. ∎

If γ is a C^∞ curve in M such that $\gamma c = m$, we define $\gamma_* c \in M_m$, the *tangent to γ at c*, by requiring for every $f \in F^\infty(m)$,

$$(\gamma_* c)f = \frac{df \circ \gamma}{du}(c).$$

We must show that $\gamma_* c$ actually is a tangent at m; that is, it satisfies **(a)** and **(b)**. For every $f, g \in F^\infty(m)$, $a \in R$, we have

$$(f + g) \circ \gamma u = f(\gamma u) + g(\gamma u)$$
$$= (f \circ \gamma + g \circ \gamma)u,$$

and similarly for fg and af, so we obtain

$$(f + g) \circ \gamma = f \circ \gamma + g \circ \gamma, \qquad (fg) \circ \gamma = (f \circ \gamma)(g \circ \gamma), \qquad (af) \circ \gamma = a(f \circ \gamma).$$

Rules **(a)** and **(b)** for $\gamma_* c$ now follow by applying the corresponding rules for $d/du(c)$ to the functions $f \circ \gamma$, $g \circ \gamma$, and to $a \in R$.

If c is an end of the interval on which γ is defined, we use the appropriate one-sided derivative, or we can replace γ in the expression on the right in the definition by any extension $\bar\gamma$ to an open interval.

Problem 1.6.1. **(a)** Suppose γ is a C^∞ curve such that $\gamma_* c = 0_m$. Let $\gamma_{**} c \colon F^\infty(m) \to R$ be defined by

$$2(\gamma_{**} c)f = \frac{d^2 f \circ \gamma}{du^2}(c).$$

Show that $\gamma_{**} c$ is a tangent at m.

(b) If $\gamma_* c \neq 0_m$, show that the formula for $\gamma_{**} c$ in **(a)** does not define a tangent at m.

If $\gamma_* c = 0_m$, then $\gamma_{**} c$ is called the *second-order tangent* to γ at c.

Problem 1.6.2. **(a)** Show that there is a C^∞ function $f \colon R \to R$ such that
$$fu = 1 \qquad \text{if } |u| < 1,$$
$$fu = 0 \qquad \text{if } |u| > 2.$$

(b) If x^i are coordinates at m such that $x^i m = 0$ and they are defined for $|x^i| < 3/a$, then $g \colon M \to R$ defined by

$$gn = \begin{cases} f(ax^1 n)f(ax^2 n) \cdots f(ax^d n) & \text{if } |x^i n| < 3/a, \\ 0 & \text{otherwise, including } n \text{ outside the } x^i \text{ domain,} \end{cases}$$

is C^∞ and the set on which it is nonzero can be made arbitrarily small by proper choices of a.

(c) If $h \in F^\infty(m)$, then there is $k \colon M \to R$, a C^∞ function, such that h and k coincide on some neighborhood of m.

(d) Let $F^\infty(M)$ be the real-valued C^∞ functions defined on all of M. Show that $F^\infty(m)$ may be replaced by $F^\infty(M)$ in the definition of a tangent at m without any essential change in the concept.

1.7. Coordinate Vector Fields

If $\mu = (x^1, \ldots, x^d)$ is a coordinate system at m, then for $f \in F^\infty(m)$ there is a coordinate expression for f, $f \circ \mu^{-1} = g: U \to R$, where U is an open set in R^d. As before [cf. (1.3.1) and (1.3.2)] we may also write $f = g \circ \mu = g(x^1, \ldots, x^d)$. The real-valued function g on U is C^∞, hence has partial derivatives with respect to the cartesian coordinates u^i on R^d. These partial derivatives will in turn be the coordinate expressions for some members of $F^\infty(m)$, which we define to be the partial derivatives of f with respect to the x^i. Specifically, the definition is

$$\partial_i f = \frac{\partial f}{\partial x^i} = \frac{\partial g}{\partial u^i} \circ \mu = \frac{\partial f \circ \mu^{-1}}{\partial u^i} \circ \mu.$$

Of the two notations, ∂_i and $\partial/\partial x^i$, we use the simpler, ∂_i, when only one coordinate system is involved. On R^2 and R^3 we shall use ∂_x, ∂_y, ∂_z rather than $\partial/\partial x$, $\partial/\partial y$, $\partial/\partial z$ or $\partial/\partial u^1$, $\partial/\partial u^2$, $\partial/\partial u^3$. The domain of these partial derivatives is the intersection of the domain of f with the coordinate neighborhood.

When viewed as a function on functions,

$$\partial_i: F^\infty(m) \to F^\infty(m)$$

satisfies properties much like a tangent at m, with appropriate modifications considering that the values are in $F^\infty(m)$, rather than in R. That is, for every $f, g \in F^\infty(m)$, $a, b \in R$,

(a) $\partial_i(af + bg) = a\,\partial_i f + b\,\partial_i g$,
(b) $\partial_i(fg) = f\partial_i g + g\partial_i f$.

These are easy to verify, since we know that $\partial/\partial u^i$ has the same properties. We call the operators ∂_i the *coordinate vector fields* of the coordinate system $(x^1, \ldots, x^d)$.

If application of ∂_i is followed by evaluation at m, the result is a tangent at m which we denote by $\partial_i(m) \in M_m$. That is, $\partial_i(m)f = \partial_i f(m)$ for every $f \in F^\infty(m)$. The tangent $\partial_i(m)$ is a tangent to the *i*th *coordinate curve* γ_i through m, which is defined by

$$\gamma_i u = \mu^{-1}(x^1 m, \ldots, x^{i-1}m, u, x^{i+1}m, \ldots, x^d m).$$

If we let $c = x^i m$, then for every $f \in F^\infty(m)$,

$$(\gamma_{i*}c)f = \frac{df \circ \gamma_i}{du}(c)$$

$$= \frac{d}{du}(c)f \circ \mu^{-1}(x^1 m, \ldots, u, \ldots, x^d m)$$

$$= \frac{\partial}{\partial u^i}(\mu m)f \circ \mu^{-1}$$

$$= \partial_i f(m).$$

Thus we have that $\gamma_{i*}c = \partial_i(m)$.

Besides thinking of ∂_i as a function $F^\infty(m) \to F^\infty(m)$ for every m in the coordinate neighborhood V, we may also consider ∂_i as a function $V \to \{M_m \mid m \in V\}$, assigning a tangent $\partial_i(m)$ at m to each $m \in V$.

Problem 1.7.1. For each sequence of numbers $a^i \in R$, $i = 1, \ldots, d$, there is a tangent at m,

$$t = \sum_{i=1}^d a^i \, \partial_i(m) \in M_m.$$

Show that there is a C^∞ curve γ with $\gamma 0 = m$ such that $\gamma_* 0 = t$.

Problem 1.7.2. Verify that $\partial_i x^j$ is the constant function

$$\delta_i^j = \begin{cases} 0 & \text{if } i \neq j, \\ 1 & \text{if } i = j. \end{cases}$$

The function δ_i^j of two integers i and j is called the *Kronecker delta*. Thus if $t = \sum_{i=1}^d a^i \, \partial_i(m)$, show that $tx^i = a^i$, so

$$t = \sum_{i=1}^d (tx^i) \, \partial_i(m).$$

Problem 1.7.3. If γ is a C^∞ curve and x^i coordinates at $m = \gamma c$, show that there are $a^i \in R$ such that

$$\gamma_* c = \sum_{i=1}^d a^i \, \partial_i(m).$$

We now show that the sort of tangent which Problems 1.7.1, 1.7.2, and 1.7.3 deal with is perfectly general, that is, the $\partial_i(m)$ form a *basis* for M_m. We shall discuss the concept of a basis of any vector space in Chapter 2. We also insert the result of Problem 1.7.2 in the following, which we call a theorem rather than a proposition because it is used so frequently later on.

Theorem 1.7.1. *For every tangent $t \in M_m$ there are unique constants a^i such that*

$$t = \sum_{i=1}^{d} a^i \, \partial_i(m),$$

namely,

$$a^i = tx^i.$$

Proof. It is convenient and no less general to assume $x^i m = 0$ for every i. Indeed, if $y^i = x^i + b^i$, b^i constants, then $\partial/\partial y^i = \partial/\partial x^i$. To obtain the desired result we need a first-order finite Taylor expansion for $f \in F^\infty(m)$, of a special form. Specifically, we claim there are $f^i \in F^\infty(m)$, $i = 1, \ldots, d$, such that on some neighborhood of m

$$f = fm + \sum_{i=1}^{d} x^i f_i.$$

Assuming for the moment that this expansion exists, the rest follows by easy computations. Applying $\partial_j(m)$ to both sides of the above equation we get

$$\partial_j f(m) = 0 + \sum_{i=1}^{d} [\partial_j x^i(m) f_i m + x^i m \, \partial_j f_i(m)]$$

$$= f_j m.$$

Having found what the $f_j m$ are, we can now evaluate tf:

$$tf = t(fm) + \sum_{i=1}^{d} t(x^i f_i)$$

$$= 0 + \sum_{i=1}^{d} [(tx^i) f_i m + x^i m \, tf_i]$$

$$= \sum_{i=1}^{d} (tx^i) \, \partial_i(m) f,$$

where we have used the fact that t has value 0 on constants.

Since t and $\sum_{i=1}^{d} (tx^i) \, \partial_i(m)$ give the same result when applied to every $f \in F^\infty(m)$, they are equal. If $t = \sum_{i=1}^{d} a^i \, \partial_i(m)$, then $tx^j = \sum_{i=1}^{d} a^i \, \partial_i x^j(m) = a^j$, showing the a^i are unique.

It remains to show the existence of the first-order Taylor expansion of the form stated. This need only be done for the coordinate expression of f, since such Taylor expansions can be transferred back and forth by composition with μ or μ^{-1}. Thus if

$$g = a + \sum_{i=1}^{d} u^i g_i,$$

where a is constant, and $g, g_j \in F^\infty(0)$, $0 = (0, \ldots, 0)$ the origin in R^d, then

$$f = g \circ \mu = a + \sum_{i=1}^{d} (u^i \circ \mu) g_i \circ \mu$$

$$= a + \sum_{i=1}^{d} x^i f_i,$$

defining $f_i = g_i \circ \mu$.

It is convenient to introduce notation defining sum and scalar multiples in R^d; that is, for $p = (p^1, \ldots, p^d)$, $q = (q^1, \ldots, q^d) \in R^d$, $b \in R$ define $bp = (bp^1, \ldots, bp^d)$ and $p + q = (p^1 + q^1, \ldots, p^d + q^d)$.

For $g \in F^\infty(0)$ there is a neighborhood U of 0 such that whenever $p \in U$, then g is defined on all bp for $0 \le b \le 1$; that is, g is defined on the segment from 0 to p. We deal only with such p. Then by the chain rule,

$$\frac{d}{ds} g(sp) = \sum_{i=1}^{d} \frac{\partial g}{\partial u^i} \frac{du^i(sp)}{ds}$$

$$= \sum_{i=1}^{d} \frac{\partial g}{\partial u^i} p^i,$$

since $u^i(sp) = p^i s$. By the fundamental theorem of calculus, $h1 = h0 + \int_0^1 \frac{dh}{ds} ds$ which, when applied to $hs = g(sp)$, yields

$$gp = g0 + \int_0^1 \sum_{i=1}^{d} p^i \frac{\partial g}{\partial u^i} (sp) \, ds$$

$$= g0 + \sum_{i=1}^{d} p^i \int_0^1 \frac{\partial g}{\partial u^i} (sp) \, ds.$$

We define $g_i p = \int_0^1 \frac{\partial g}{\partial u^i} (sp) \, ds$. Then the formula just obtained is

$$g = g0 + \sum_{i=1}^{d} u^i g_i,$$

valid on a neighborhood of 0.

To show that $g_i \in F^\infty(0)$, we invoke a theorem from advanced calculus: If $h(p, s)$ is a C^1 function of $(p, s) \in R^{d+1}$ and $kp = \int_0^1 h(p, s) \, ds$, then k is C^1 and $\frac{\partial k}{\partial u^i} (p) = \int_0^1 \frac{\partial h}{\partial u^i} (p, s) \, ds$. Repeated applications of this theorem and the chain rule to functions of the form $h(p, s) = \dfrac{\partial^n g}{\partial u^{i_n} \cdots \partial u^{i_1}} (sp)$ shows that g_i is C^∞. ∎

The a^i are called the *components* of $t = \sum_{i=1}^{d} a^i \partial_i(m)$ with respect to the coordinates x^i.

Remarks. The Taylor expansion given is slightly different in nature when C^k functions are expanded, in that we can only assert that f_i is C^{k-1}. This loss of one degree of differentiability prevents us from using the same definition of tangents for C^k manifolds, because axioms (**a**) and (**b**) will allow many operators which cannot be applied to C^{k-1} functions and in particular are not of the form $\sum_{i=1}^d a^i \partial_i(m)$. The resolution is to define tangents as being only those which have this form; that is, the tangent space is *spanned* by the $\partial_i(m)$, or to require that a tangent be the tangent to some C^k curve. Both processes lead to the result we have achieved—that each tangent space M_m is a vector space of dimension d, the same dimension as M, and has bases given by the coordinate vector fields.

For C^∞ manifolds we summarize the various equivalent ways of viewing what tangents are.

(**a**) For our definition we have taken a tangent to be a real-valued *derivation* of $F^\infty(m)$; that is, $t\colon F^\infty(m) \to R$, satisfying (**a**) and (**b**) of Section 1.6.

(**b**) For any coordinate system (x^i) at m the tangent space at m consists, by Theorem 1.7.1, of expressions of the form $\sum_{i=1}^d a^i \partial_i(m)$.

(**c**) The tangent space at m consists of tangents to C^∞ curves, $\gamma_* c$, where $\gamma c = m$, by (**b**) and Problem 1.7.3.

(**d**) The classical tensor definition of tangents (see Proposition 1.7.1) is essentially the same as (**b**) combined with the rule for relating the components with respect to different coordinate systems. The tangent, from this point of view, is not considered to be an operator on functions but is rather the sequence of components a^i assigned to the coordinate system (x^i). A formal definition on these lines is quite formidable, since one must consider a tangent as being a function which assigns to each coordinate system (x^i) at m the sequence a^i, and satisfying the transformation law. In applications this definition is quite convenient to use, and it also is the most obvious generalization of the definition of a tangent vector in R^d as being a directed line segment. For these reasons we do not anticipate that the classical definition will entirely disappear, at least not very soon.

An immediate application of the formula in Theorem 1.7.1 is a version of the chain rule for a manifold: For two coordinate systems (x^i) and (y^i),

$$\frac{\partial}{\partial y^i} = \sum_{j=1}^d \frac{\partial x^j}{\partial y^i} \frac{\partial}{\partial x^j}.$$

From this the *law of transformation of tangents* is obtained. If

$$t = \sum_{i=1}^d a^i \frac{\partial}{\partial x^i}(m) = \sum_{i=1}^d b^i \frac{\partial}{\partial y^i}(m),$$

then

$$a^i = \sum_{j=1}^{d} b^j \frac{\partial x^i}{\partial y^j}(m). \tag{1.7.1}$$

To establish the equivalence of the classical definition, which incorporates this law in its statement, and ours, the following proposition is offered.

Proposition 1.7.1. *Let t be a function on the charts at m which assigns to each such chart a sequence of d real numbers and such that if $t(x^i) = (a^i)$, $t(y^i) = (b^i)$, then (a^i) and (b^i) are related by formula (1.7.1), for all pairs of charts (x^i) and (y^i). Then for all such pairs*

$$\sum_{i=1}^{d} a^i \frac{\partial}{\partial x^i}(m) = \sum_{i=1}^{d} b^i \frac{\partial}{\partial y^i}(m).$$

Problem 1.7.4. Show that the notation of partial derivatives is misleading in the following way. If dimension $M > 1$, there are coordinates x^i and y^i such that $x^1 = y^1$ but for which $\partial/\partial x^1 \neq \partial/\partial y^1$. Thus the coordinate vector $\partial/\partial x^1$ does not depend merely on the function x^1 (as the notation might lead one to believe) but also on the remaining functions $x^2, \ldots, x^d$. If $x^2, \ldots, x^d$ are changed, then $\partial/\partial x^1$ may change even though x^1 remains the same.

Problem 1.7.5. Show that for $d > 1$ the Taylor expansion is not unique. Also show that there are higher-order expansions of the type used above.

1.8.   Differential of a Map

Let us denote the union of all the tangent spaces to a manifold M by TM; that is, TM is the collection of all tangents to m at all points $m \in M$. In Chapter 3 we shall spell out a natural manifold structure for TM, which makes TM into a manifold of dimension $2d$ on which the coordinates are, roughly, the d coordinates of a system of M joined with the d components of a tangent with respect to the coordinate vector basis. Then TM is called the *tangent bundle* of M. However, for our purposes here it will be sufficient to regard TM as a set.

Now suppose $\mu: M \to N$ is a C^∞ map. Then there is induced a map $\mu_*: TM \to TN$, called the *differential of μ*. Alternative names for μ_* are the *prolongation of μ to TM* and the *tangent map of μ*. We shall give two definitions and prove that they are the same.

(a) If $t \in TM$, then $t \in M_m$ for some $m \in M$ and there is a C^∞ curve γ such that $\gamma_* 0 = t$. Since $\gamma: R \to M$ and $\mu: M \to N$, the composition $\mu \circ \gamma: R \to N$ is a C^∞ curve in N. We define

$$\mu_* t = (\mu \circ \gamma)_* 0. \tag{1.8.1}$$

This definition could conceivably depend on the choice of γ, not just on t, but we shall not show independence of choice directly, since it follows from equivalence with the second definition.

   (b) If $t \in TM$, it is sufficient to say how $\mu_* t \in TN$ operates on $F^\infty(n)$, where $n = \mu m$ and $t \in M_m$. If $f \in F^\infty(n)$, we then have $f \circ \mu \in F^\infty(m)$, so the following definition makes sense:

$$(\mu_* t)f = t(f \circ \mu). \qquad (1.8.2)$$

With this definition it must be demonstrated that $\mu_* t$ is actually a tangent at n; that is, it is a derivation of $F^\infty(n)$. Again, this will follow from the proof of equivalence, since we know that $(\mu \circ \gamma)_* 0$ is a tangent at $\mu \gamma 0 = \mu m = n$.

   Proof of Equivalence. In the notation of **(a)** and **(b)** we have

$$\begin{aligned}
[(\mu \circ \gamma)_* 0]f &= \frac{d}{du}(0)[f \circ (\mu \circ \gamma)] \\
&\quad - \frac{d}{du}(0)[(f \cap \mu) \cap \gamma] \\
&= (\gamma_* 0)(f \circ \mu) \\
&= t(f \circ \mu).
\end{aligned}$$

Thus the right side of (1.8.1) applied to f is the same as the right side of (1.8.2). Hence the two definitions agree.

   Coordinate Expressions. In terms of components with respect to coordinate vector fields, μ_* is expressed by means of the jacobian matrix of μ. Suppose that x^i, $i = 1, \ldots, d$, are coordinates at m and y^α, $\alpha = 1, \ldots, e$, are coordinates at $n = \mu m$. Then in a neighborhood of m, μ has the coordinate expression

$$y^\alpha \circ \mu = f^\alpha(x^1, \ldots, x^d).$$

If $t \in M_m$, then we may write $t = \sum_{i=1}^d a^i \, \partial_i(m)$. Let $b^\alpha = (\mu_* t)y^\alpha$. Then from Theorem 1.7.1 we have

$$\mu_* t = \sum_{\alpha=1}^e b^\alpha \frac{\partial}{\partial y^\alpha}(n).$$

We evaluate b^α by means of definition (1.8.2), since $y^\alpha \in F^\infty(n)$:

$$\begin{aligned}
b^\alpha &= (\mu_* t)y^\alpha \\
&= t(y^\alpha \circ \mu) \\
&= \sum_{i=1}^d a^i \, \partial_i(y^\alpha \circ \mu)(m). \qquad (1.8.3)
\end{aligned}$$

   The coefficients of the a^i in this expression are arranged into a rectangular $e \times d$ array, with α constant on rows and i constant on columns.

$$J = \begin{pmatrix} \partial_1 y^1 \circ \mu & \partial_2 y^1 \circ \mu & \cdots & \partial_d y^1 \circ \mu \\ \vdots & & & \vdots \\ \partial_1 y^e \circ \mu & \cdots & \cdots & \partial_d y^e \circ \mu \end{pmatrix}.$$

This rectangular array is called the *jacobian matrix of* μ with respect to the coordinates x^i and y^α. The formula for μ_* in terms of coordinate components, (1.8.3), is the matrix theory definition of the product of $e \times d$ matrix J by $d \times 1$ column matrix (a^i), producing the $e \times 1$ column matrix (b^α).

Problem 1.8.1. Show that μ_* is *linear* on M_m; that is, for $s, t \in M_m$ and $a \in R$:
 (a) $\mu_*(s + t) = (\mu_* s) + (\mu_* t)$.
 (b) $\mu_*(at) = a\mu_* t$.

Problem 1.8.2. If $\mu: M \to N$ and $\tau: N \to P$ are C^∞ maps, prove the *chain rule*:

$$(\tau \circ \mu)_* = \tau_* \circ \mu_*. \tag{1.8.4}$$

By expressing (1.8.4) in terms of coordinates, justify the name "chain rule."

Special Cases. The cases for which M or N is R deserve additional mention, since they concern the important notions of C^∞ curves and real-valued functions, respectively. This special treatment is based on the fact that R has a natural coordinate, the identity coordinate u, and hence a distinguished basis for tangents at c, $d/du(c)$, for every $c \in R$.

In the case $M = R$ we have a curve $\gamma: R \to N$. To say what γ_* is it is sufficient to say what it does to $d/du(c)$, since the effect on other tangents $a[d/du(c)]$ is then known by linearity [(b) in Problem 1.8.1]. A curve in R for which the tangent is d/du is the identity curve $u: R \to R$, so by definition (1.8.1),

$$\gamma_* \frac{d}{du}(c) = (\gamma \circ u)_* c$$
$$= \gamma_* c.$$

The latter expression is the previous definition, from Section 1.7, of the tangent vector to the curve γ, so our notation is in reasonably close agreement.

In the case $N = R$ we have a real-valued C^∞ function $f: M \to R$. If $t \in M_m$ and $c = fm$, to say what $f_* t$ is we must find its component with respect to basis $d/du(c)$ of R_c. By Theorem 1.7.1,

$$f_* t = a \frac{d}{du}(c),$$

where

$$a = (f_* t)u$$
$$= t(u \circ f) \qquad \text{by (1.8.2)}$$
$$= tf.$$

Thus

$$f_* t = (tf) \frac{d}{du}(c).$$

We redefine the *differential* of $f: M \to R$ to be the component tf of $f_* t$ and change the notation to

$$(df)t = tf. \tag{1.8.5}$$

Thus $df: TM \to R$ replaces $f_*: TM \to TR$ in our subsequent usage.

On each tangent space M_m, $df: M_m \to R$ is a linear, real-valued function. In Chapter 2 we define the dual space V^* of a vector space V to be the collection of all linear, real-valued functions on the vector space. In this terminology, the differential of a real-valued function gives a member of the dual space M_m^* of the tangent space M_m for each m. For manifolds, the dual space M_m^* of M_m is called the *cotangent space at m*, or the *space of differentials at m*, or, in the classical terminology, the space of *covariant vectors at m*.

Problem 1.8.3. If x^i are coordinates on $M, f: M \to R$, show that the classical formula

$$df = \sum_{i=1}^{d} \partial_i f \, dx^i$$

is a consequence of (1.8.5).

Problem 1.8.4. Show that dx^i, $i = 1, \ldots, d$ is the *dual basis* to ∂_i, $i = 1, \ldots, d$; that is (see Section 2.7),

$$(dx^i) \, \partial_j = \delta_j^i.$$

Problem 1.8.5. Let $\mu: R^2 \to R^2$ be defined by $\mu = (x^2 + 2y^2, 3xy)$. Find the matrix of μ_* at $(1, 1)$ with respect to coordinates x, y in each place. Use this to evaluate

$$\mu_* \left(\frac{\partial}{\partial x} (1,1) + 3 \frac{\partial}{\partial y} (1,1) \right)$$

by matrix multiplication.

CHAPTER **2**

Tensor Algebra

2.1. Vector Spaces

In Chapter 1 we saw that the set of tangent vectors at a point m of a manifold M has a certain algebraic structure. In this chapter we present and study this structure abstractly, but it should be borne in mind that the tangent spaces of manifolds are the principal examples.

A *vector space* or *linear space* V (over R) is a set with two operations, *addition*, denoted by $+$, which assigns to each pair $v, w \in V$, a third element, $v + w \in V$, and *scalar multiplication*, which assigns to each $v \in V$ and $a \in R$ an element $av \in V$, and having a distinguished element $0 \in V$ such that the following axioms are satisfied. These axioms hold for all $v, w, x \in V$ and all $a, b \in R$.

(1) The *commutative law* for $+$: $v + w = w + v$.
(2) The *associative law* for $+$: $(v + w) + x = v + (w + x)$.
(3) Existence of *identity* for $+$: $v + 0 = v$.
(4) Existence of *negatives*: There is $-v$ such that $v + (-v) = 0$.
(5) $a(v + w) = av + aw$.
(6) $(a + b)v = av + bv$.
(7) $(ab)v = a(bv)$.
(8) $1v = v$.

The elements of V are called *vectors*. Not all the properties of the real numbers are needed for the theory of vector spaces (only those called the field axioms), so to allow easy generalization to other fields, the real numbers are called *scalars* in this context. In particular, certain topics in the study of real vector spaces are facilitated by an extension to the complex numbers as scalars.

Axioms (2) and (7) justify the elimination of parentheses in the expressions; that is, we define $v + w + x = (v + w) + x$ and $abv = (ab)v$. Strictly speaking, the right sides of **(5)** and **(6)** also need parentheses, but there is only one reasonable interpretation. We define $v - w = v + (-w)$.

Remark. Formally, addition and scalar multiplication are functions $+ : V \times V \to V$ and $\cdot : R \times V \to V$.

We shall use freely the following propositions, the proofs of which are automatic.

(a) If $0^- \in V$ is such that $v + 0^- = v$ for some v, then $0^- = 0$; that is, 0 is uniquely determined by its property **(3)**.

(b) For every $v \in V$, $0v = 0$. In this equation the 0 on the left is the scalar 0, the 0 on the right is the vector 0.

(c) If $v + w = 0$, then $w = -v$; that is, inverses are unique.

(d) For all $v, w \in V$, there is a unique $x \in V$ such that $v + x = w$, namely, $x = w - v$.

(e) For every $a \in R$, $a0 = 0$. In this equation both 0's are the vector 0.

(f) If $a \in R$, $v \in V$, and $av = 0$, then either $a = 0 \in R$ or $v = 0 \in V$.

(g) For every $v \in V$, $(-1)v = -v$.

Problem 2.1.1.   Let $V = R \times R$ and define

$$(a, b) + (c, d) = (a + c, b + d),$$
$$c(a, b) = (ca, b).$$

Show that all the axioms except **(6)** hold for V. What does this tell you about the proof of **(b)**?

Example. Let $V = R^d$ and define

$$(a^1, \dots, a^d) + (b^1, \dots, b^d) = (a^1 + b^1, \dots, a^d + b^d),$$
$$c(a^1, \dots, a^d) = (ca^1, \dots, ca^d).$$

Then R^d is a vector space. In particular, we have that R is a vector space under the usual operations of addition and multiplication. The complex numbers C may be viewed as R^2 and the rules for addition and multiplication of complex numbers by real numbers agree with the operations just given on R^d in the case $d = 2$. Thus C is a vector space over R. If we allow multiplication by complex scalars, then C is a different vector space, this time over C instead of R.

Problem 2.1.2.   Show that the set of C^∞ functions $F^\infty(M)$ on a C^∞ manifold M form a vector space over R.

Problem 2.1.3.   Let V be the first quadrant of R^2, that is, $V = \{(x, y) \mid x \geq 0$ and $y \geq 0\}$. With addition and scalar multiplication defined as in the example above, how does V fail to be a vector space?

Problem 2.1.4. Let R^+ denote the set of positive real numbers. Define the "sum" of two elements of R^+ to be their product in the usual sense, and scalar multiplication by elements of R to be $\cdot : R \times R^+ \to R^+$ given by $\cdot(r, p) = p^r$. With these operations show that R^+ is a vector space over R.

Direct Sums. If V and W are vector spaces, then we construct a new vector space from $V \times W$ by defining

$$(v, w) + (v', w') = (v + v', w + w'),$$
$$c(v, w) = (cv, cw).$$

We denote this new vector space by $V + W$ and call it the *direct sum* of V and W. The operation of forming direct sums can be defined in an obvious way for more than two summands. The summands need not be different.

Problem 2.1.5. Show that the example of R^d above is the d-fold direct sum of R with itself.

2.2. Linear Independence

Let V be a vector space. A finite set of vectors, say $v_1, \ldots, v_r$, are *linearly dependent* if there are scalars $a^1, \ldots, a^r$, not all zero, such that $\sum_{i=1}^r a^i v_i = 0$. An infinite set is *linearly dependent* if some finite subset is linearly dependent. A set of vectors is *linearly independent* if it is not linearly dependent.

A sum of the form $\sum_{i=1}^r a^i v_i$, where $v_i \in V$ and a^i are scalars, is called a *linear combination* of $v_1, \ldots, v_r$. If at least one a^i is not zero, the linear combination is called *nontrivial*; the linear combination with all $a^i = 0$ is called *trivial*. Thus a set of vectors is linearly dependent iff there is a nontrivial linear combination of the vectors which equals the zero vector. Other forms of the definition of linear (in)dependence which are used are as follows.

Proposition 2.2.1. *The following statements are equivalent to the set S being linearly independent.*

  (a) *The only 0 linear combination of vectors in S is trivial.*
  (b) *If $v_i \in S$, then $\sum_{i=1}^r a^i v_i = 0$ implies $a^i = 0$, $i = 1, \ldots, r$.*
  (c) *If $v_i \in S$ and a^i are scalars, not all 0, then $\sum_{i=1}^r a^i v_i \neq 0$.*
  (d) *If $v_i \in S$, a^i are scalars, not all 0, then $\sum_{i=1}^r a^i v_i = 0$ leads to a contradiction.*

Proposition 2.2.2. *A set S is linearly dependent iff there are distinct $v_0, v_1, \ldots, v_r \in S$ such that v_0 is a linear combination of $v_1, \ldots, v_r$.*

Proof. If S is linearly dependent, then there are $v_0, \ldots, v_r \in S$ and scalars $a^0, \ldots, a^r$, not all zero, such that $\sum_{i=0}^{r} a^i v_i = 0$. Renumbering if necessary, we may assume $a^0 \neq 0$. Then $v_0 = \sum_{i=1}^{r} (-a^i/a^0) v_i$.

Conversely, if $v_0, \ldots, v_r \in S$ and $v_0 = \sum_{i=1}^{r} b^i v_i$, then $\sum_{i=0}^{r} a^i v_i = 0$, where $a^0 = 1, a^i = -b^i, i = 1, \ldots, r$ are not all zero, so S is linearly dependent. ∎

As simple consequences we note that two vectors are linearly dependent iff one is a multiple of the other; we cannot say each is a multiple of the other, since one of them may be 0. If a set S includes 0, then it is linearly dependent regardless of the remaining members. Geometrically, for vectors in R^3, two vectors are linearly dependent iff they are parallel. Three vectors are linearly dependent iff they are parallel to a plane. Four vectors in R^3 are always linearly dependent.

The maximum number of linearly independent vectors in a vector space V is called the *dimension* of V and is denoted $\dim_R V$. Of course, there may be no finite maximum, in which case we write $\dim_R V = \infty$; this means that for every positive integer n there is a linearly independent subset of V having n elements. (We shall not concern ourselves with refinements which deal with orders of infinity.) If a vector space admits two distinct fields of scalars (for example, a complex vector space may be considered to be a real vector space also), then the dimension depends on the field in question. We indicate which field is used by a subscript on "dim."

In particular, $\dim_R V = 2 \dim_C V$, provided addition and scalar multiplication in V are the same and compatible with the inclusion $R \subset C$. However, this situation is exceptional for us, and when there is no danger of confusion we shall write "dim V" for "$\dim_R V$."

Problem 2.2.1. If V is a vector space over both C and R, and S is a subset of V linearly independent over C, show that the set $S \cup iS$, consisting of all $v \in S$ and iv, where $v \in S$, and thus having twice the number of elements as S, is linearly independent over R.

Problem 2.2.2. Show that the dimension of R^d is at least d.

Problem 2.2.3. If S is a linearly independent subset of V, T a linear independent subset of W, then the subset

$$S \times \{0\} \cup \{0\} \times T = \{(v, 0) \mid v \in S\} \cup \{(0, w) \mid w \in T\}$$

of the direct sum $V + W$ is linearly independent. Thus

$$\dim V + W \geq \dim V + \dim W.$$

Problem 2.2.4. Let F^k be the vector space of all C^k functions defined on R. Show that the subset of all exponential functions $\{e^{\alpha x} \mid \alpha \in R\}$ is linearly independent. *Hint:* Proceed by induction on the number of terms in a null linear combination. Eliminate one term between such a sum and its derivative.

Closely related to the dimension, or maximum number of linearly independent vectors, is the notion of a maximal linearly independent subset. A set S is a *maximal linearly independent subset* or *basis* of a vector space V if S is linearly independent and if for every $v \notin S$, $S \cup \{v\}$ is linearly dependent. By Proposition 2.2.2 this means that v is a linear combination of some $v_1, \ldots, v_k \in S$. Thus we have

Proposition 2.2.3. *A subset S of V is a basis iff*:
(**a**) *S is linearly independent.*
(**b**) *Every element of V is a linear combination of elements of S.*

Remark. We mention without proof that a basis always exists. This is obvious if dim V is finite but otherwise requires transfinite induction.

Proposition 2.2.4. *If S is a basis, then the linear combination expressing $v \in V$ in terms of elements of S is unique, except for the order of terms.*

Proof. Suppose that $v \in V$ can be expressed in two ways as a linear combination of elements of S. These two linear combinations will involve only a finite number k of the members of S, say $v_1, \ldots, v_k$. Then the combinations are

$$v = \sum_{i=1}^{k} a^i v_i, \qquad v = \sum_{i=1}^{k} b^i v_i.$$

Thus

$$v - v = 0 = \sum_{i=1}^{k} (a^i - b^i) v_i.$$

Since S is linearly independent, Proposition 2.2.1(**b**) yields $a^i - b^i = 0$, $i = 1, \ldots, k$, that is, $a^i = b^i$, as desired. ∎

If S is a basis of V, then for each $v \in V$ the unique scalars occurring as coefficients in the linear combination of elements of S expressing v are called the *components of v with respect to the basis S*. We take the viewpoint that a component of v is assigned to each element of S; however, only finitely many components are nonzero.

Remark. In vector spaces only linear combinations with a finite number of terms are defined, since no meaning has been given to limits and convergence.

Vector spaces in which a notion of limit is defined and satisfies certain additional relations (pun intended) is called a *topological vector space*. When this further structure is derived from a positive definite inner product, the space is called a *Hilbert space*. We shall not consider vector spaces from a topological viewpoint, even though in finite-dimensional real vector spaces the topology is unique.

Problem 2.2.5.   Prove: A subset S of V is a basis iff every element of V can be expressed uniquely as a linear combination of elements of S.

Proposition 2.2.5.   *If S is a linearly independent subset and T is a basis of V, then there is a subset U of T such that $S \cup U$ is a basis.*

Proof. We prove this only in the case where dim V is finite.

Some member of T is not a linear combination of members of S, for otherwise every $v \in V$ would be a linear combination of elements of T and hence, by substitution, of elements of S, and S would already be a basis. Thus we may adjoin an element v_1 of T to S, obtaining a larger linearly independent set $S_1 = S \cup \{v_1\}$. Continuing in this way k times we reach a point where all members of T are linear combinations of elements of $S_k = S \cup \{v_1, \ldots, v_k\}$, which is then a basis by our first argument. ∎

Note that U is not unique.

Proposition 2.2.6.   *All bases have the same number of elements, the dimension of V.*

Proof. Again, and for similar reasons, we assume dim V is finite.

Suppose S and T are bases having k and d elements, respectively, and that $k < d = \dim V$. Let $T = \{t_1, \ldots, t_d\}$. Then $T_1 = \{t_2, \ldots, t_d\}$ is not a basis so there is $s_1 \in S$ such that $\{s_1, t_2, \ldots, t_d\}$ is a basis. Similarly, $\{s_1, t_3, \ldots, t_d\}$ is not a basis, so there is $s_2 \in S$ such that $\{s_1, s_2, t_3, \ldots, t_d\}$ is a basis. Continuing in this way we must exhaust S before we run out of members of T, obtaining that $S \cup \{t_{k+1}, \ldots, t_d\}$ is a basis. This contradicts the fact that S is a basis, since no set containing S properly can be linearly independent. ∎

Problem 2.2.6.   In the above proof why is only one member of S needed to fill out T_1 to give a basis? Why must a different member of S be taken at each step?

Problem 2.2.7.   Show that dim $V + W = \dim V + \dim W$. (Direct sum.)

Problem 2.2.8.   Show that dim $R^d = d$.

Example.   Let M be a d-dimensional C^∞ manifold, $m \in M$, and $x^1, \ldots, x^d$ coordinates at m. Then Theorem 1.7.1 says that tangents at m can be expressed uniquely as linear combinations of the $\partial_i(m)$. Thus the $\partial_i(m)$ are a basis of the tangent space M_m. In particular, the dimensions of the tangent spaces to M are all equal to d, the manifold dimension.

2.3. Summation Convention

At this point it is convenient to introduce the *(Einstein) summation convention*. This makes it possible to indicate sums without dots $(\cdots)$ or a summation symbol $\sum$. Thus $a^i e_i$ will be our new notation for $\sum_{i=1}^{d} a^i e_i$. The summation symbols also will be omitted in double, triple, etc., sums when the sum index occurs twice, usually once up and once down.

To use the summation convention it must be agreed upon beforehand what letters of the alphabet are to be used as *sum indices* and through what range they are to vary. We shall frequently use $h, \ldots, n, p, \ldots, v$ as sum indices and the range will usually be the dimension of the basic vector space or manifold.

One effect of the sum convention is to make the chain rule for partial derivatives have the appearance of a cancellation, as in the single-variable case. Thus in the formula

$$\frac{\partial}{\partial x^i} = \frac{\partial y^j}{\partial x^i} \frac{\partial}{\partial y^j}$$

it appears that "∂y^j" is being canceled. Another effect is to make it more difficult to express some simple things, for example, one arbitrary term of a sum $a^i e_i$. This difficulty usually occurs only in more mathematical (rather than routine) arguments, and is handled either by using a previously agreed upon nonsum index, say A, and merely writing $a^A e_A$, or by indicating the suppression of the sum convention directly, for example,

$$a^i e_i \qquad (i \text{ not summed}).$$

In normal usage of the sum convention a sum index will not occur more than twice in a term. When it does it usually means some error has been made. A common error of this type occurs when indices are being substituted without sufficient attention to detail, and usually produces an even number of occurrences of an index. In some cases it takes application of the distributive law to put a formula in proper sum convention form. For example, $a^i(e_i + f_i)$ has three occurrences of i, but it is natural to write it as $a^i e_i + a^i f_i$, which makes sense. Such undefined uses will be allowed as long as they are not so complicated that they confuse.

To illustrate the use of the sum convention we discuss the relation between two bases of a d-dimensional vector space V. Let $\{e_i\}$ and $\{f_i\}$ be two bases of V. Then each e_i has an expression in terms of the f_j and vice versa,

$$e_i = a_i^j f_j,$$
$$f_i = b_i^j e_j.$$

The d^2 numbers a_i^j are customarily arranged in a square array, called the $d \times d$ matrix of change from basis $\{f_j\}$ to basis $\{e_i\}$, so that j is constant on rows, i is constant on columns.† This arrangement is indicated by placing parentheses on a_i^j:

$$(a_i^j) = \begin{pmatrix} a_1^1 & a_2^1 & \cdots & a_d^1 \\ \vdots & & & \\ a_1^d & a_2^d & \cdots & a_d^d \end{pmatrix}.$$

Substituting $f_j = b_j^k e_k$ in $e_i = a_i^j f_j$ we obtain

$$e_i = a_i^j b_j^k e_k.$$

Comparing with the obvious formula $e_i = \delta_i^k e_k$ and applying the uniqueness of components with respect to the basis $\{e_i\}$ we obtain

$$a_i^j b_j^k = \delta_i^k = \begin{cases} 0 & \text{if } i \neq k, \\ 1 & \text{if } i = k. \end{cases} \tag{2.3.1}$$

Similarly, by reversing e_i and f_j,

$$b_i^j a_j^k = \delta_i^k. \tag{2.3.2}$$

When two matrices (a_i^j) and (b_i^j) are related by formulas (2.3.1) and (2.3.2), they are called *inverses* of each other. Thus we have proved

Proposition 2.3.1.   *The two matrices of change from one basis of a vector space to another and back are inverses of one another.*

Now suppose that we have two $d \times d$ matrices (a_i^j) and (b_i^j) which satisfy *one* of the two relations above, say, (2.3.1). Let $\{e_i\}$ be any basis of V, a d-dimensional vector space, and define d vectors f_i by $f_i = b_i^j e_j$. Then by (2.3.1),

$$a_k^i f_i = a_k^i b_i^j e_j = \delta_k^j e_j = e_k.$$

That is, the e_i can be expressed in terms of the f_i. Since any $v \in V$ can be expressed in terms of the e_i, the same is true for the f_i. All the f_i, hence all $v \in V$, can be expressed in terms of a maximum number of linearly independent f_i. In other words, the f_i contain a basis. But they are d in number, so the f_i are a basis, and (a_i^j) and (b_i^j) are the change of basis matrices between $\{e_i\}$ and $\{f_i\}$. Now the other relation (2.3.2) follows as before. We have proved a theorem which in the following form is entirely about square matrices.

---

† The conventions of matrix algebra would then seem to call for viewing the e_i and f_i as forming $1 \times d$ rows and writing $e_i = f_j a_i^j$, but scalars customarily precede vectors.

Proposition 2.3.2. *Let* (a_i^j) *be a* $d \times d$ *matrix such that there is a* $d \times d$ *matrix* (b_i^j) *satisfying* $a_i^j b_j^k = \delta_i^k$. *Then the matrices* (a_i^j) *and* (b_i^j) *are inverses of each other; that is,* $b_i^j a_j^k = \delta_i^k$.

Problem 2.3.1. Evaluate $\delta_i^k \delta_k^l$.

Problem 2.3.2. Show that the relation between components of a vector with respect to two different bases is the reverse of the relation between the bases themselves, both in the index of the matrix which is summed and in which matrix is used in each direction.

Problem 2.3.3. If V is a finite-dimensional vector space of dimension d, show that a subset S of V is a basis iff (**a**) every $v \in V$ is a linear combination of elements of S and (**b**) there are d elements in S.

2.4. Subspaces

A nonempty subset W of a vector space V is called a *subspace* of V if W is closed under addition and scalar multiplication, that is, if $w + x \in W$ and $aw \in W$ for every $w, x \in W$ and $a \in R$.

Problem 2.4.1. A subspace W of a vector space V is a vector space with operations obtained by simply restricting the operations of V to W.

To make it clear that operations which make a subset a vector space need not make it a subspace, Problem 2.1.4 gives an example of a subset R^+ of R which is not a subspace, but which has operations defined making it a vector space. In fact, the reader should be able to show easily that the only subspaces of $R = R^1$ are the singleton subset $\{0\}$ and all of R itself.

The proofs of the following are automatic.

Proposition 2.4.1. *The intersection of any collection of subspaces is a subspace.*

Proposition 2.4.2. *If W is a subspace of V and E is a subset of W, then E is linearly independent as a subset of the vector space W iff E is linearly independent as a subset of the vector space V.*

Proposition 2.4.3. *If W is a subspace of V, then there exist bases of V of the form $E \cup F$, where E is a basis of W.*

(Choose a basis E of W and apply Proposition 2.2.5.)

Proposition 2.4.4. *If W is a subspace of V, then dim $W \leq$ dim V.*

Proposition 2.4.5. *If S is any subset of a vector space V, then there is a unique subspace W of V containing S and which is contained in any subspace containing S, namely, W is the intersection of all subspaces containing S.*

The minimal subspace containing a subset S, which is referred to in Proposition 2.4.5, is called the subspace *spanned* by S. We also say S *spans* W. In particular, a basis of V spans V. Many of the propositions of Section 2.2 can be abbreviated by proper use of this terminology.

Problem 2.4.2. If W is a subspace of V, then there is a subspace X of V such that V is essentially the direct sum $W + X$. More precisely, every element of v of V can be written uniquely as $v = w + x$, where $w \in W$ and $x \in X$. The complementary space X is not unique except in the cases where W is all of V or W is 0 alone.

Geometrically, the subspaces of R^3 are 0, the lines through 0, the planes through 0, and R^3 itself, of dimensions 0, 1, 2, and 3, respectively.

If W and X are subspaces of V, then the subspace spanned by $W \cup X$ is called the *sum* of W and X and is denoted $W + X$. Although the notation is the same, "sum" is a broader notion than "direct sum." The sum $W + X$ is *direct* iff $W \cap X = 0$. This differs slightly from our previous definition of direct sum in that here W, X, and $W + X$ are all parts of the given space V, whereas before only W and X were given and their direct sum had to be constructed by specifying a vector-space structure on $W \times X$. If the sum is direct in the new sense, then $W + X$ may be naturally identified with the old version of direct sum $W \times X$ by the correspondence $(w, x) \leftrightarrow w + x$.

We leave the development of the elementary properties of the sum of subspaces as problems.

Problem 2.4.3. The sum $W + X$ consists of all sums of the form $w + x$, where $w \in W$ and $x \in X$. The decomposition of $z \in W + X$ as $z = w + x$ is unique iff the sum is direct.

Problem 2.4.4. A basis E of V can be chosen so that it is a disjoint union $E = E_0 \cup E_1 \cup E_2 \cup E_3$, where

E_0 is a basis of $W \cap X$,
$E_0 \cup E_1$ is a basis of W,
$E_0 \cup E_2$ is a basis of X,
$E_0 \cup E_1 \cup E_2$ is a basis of $W + X$.

Problem 2.4.5. If $\dim(W + X)$ is finite, then

$$\dim(W + X) + \dim(W \cap X) = \dim W + \dim X.$$

2.5. Linear Functions

Let V and W be vector spaces and $f: V \to W$. We call f a *linear function* or *linear transformation* of V into W if for all $v_1, v_2 \in V$ and $a \in R$:

(**a**) $f(v_1 + v_2) = fv_1 + fv_2$.

(**b**) $f(av_1) = afv_1$.

A linear function $f: V \to W$ is said to be an *isomorphism* of V onto W if f is 1–1 onto. The term isomorphism means that in terms of their properties as vector spaces, V and W are not distinguishable even though vectors in V are realized differently from those in W. In this case V and W are said to be *isomorphic* and we write $V \simeq W$.

Problem 2.5.1. The zero of V is mapped into the zero of W by a linear function $f: V \to W$.

Problem 2.5.2. If $f: V \to W$ is an isomorphism, then dim $V = $ dim W.

If $f: V \to W$ is a linear function, then we call $fV \subset W$ the *image space* of f and $f^{-1}\{0\} \subset V$ the *null space* of f.

Problem 2.5.3. The image space and null space of $f: V \to W$ are subspaces of W and V, respectively.

Problem 2.5.4. The linear function $f: V \to W$ is 1–1 iff $f^{-1}\{0\} = \{0\}$.

Proposition 2.5.1. *If $f: V \to W$ is a linear function, then*

$$dim\ V = dim\ fV + dim\ f^{-1}\{0\}.$$

Proof. Choose a basis E of $f^{-1}\{0\}$ and extend E to a basis $E \cup E_1 = E_2$ of V. We claim that f is 1–1 on E_1 and fE_1 is a basis of fV. For the first fact, if $e_1, e_2 \in E_1$ and $fe_1 = fe_2$, then

$$
\begin{aligned}
f(e_1 - e_2) &= f(e_1 + [-e_2]) \\
&= fe_1 + f(-e_2) \\
&= fe_1 + f[(-1)e_2] \\
&= fe_1 + (-1)fe_2 \\
&= fe_1 - fe_2 \\
&= 0.
\end{aligned}
$$

Thus $e_1 - e_2 \in f^{-1}\{0\}$. Hence $e_1 - e_2$ is a linear combination of elements of E, but this contradicts the linear independence of E_2.

If $w \in fV$, then there is $v \in V$ such that $w = fv$. The expression for v in terms of the basis E_2 is

$$v = \sum_i a_i e_i + \sum_j b_j \bar{e}_j,$$

where $e_i \in E$ and $\bar{e}_j \in E_1$. Then by the linearity of f,

$$fv = \sum_i a_i f e_i + \sum_j b_j f \bar{e}_j$$
$$= 0 + \sum_j b_j f \bar{e}_j;$$

that is, fv is a linear combination of members of fE_1, so fE_1 spans fV.

Finally, fE_1 is linearly independent. For if $\sum_j b_j f \bar{e}_j = 0$, then $f(\sum_j b_j \bar{e}_j) = 0$, and $\sum_j b_j \bar{e}_j \in f^{-1}\{0\}$. Hence $\sum_j b_j \bar{e}_j$ is a linear combination of elements of E,

$$\sum_j b_j \bar{e}_j = \sum_i a_i e_i,$$

or

$$\sum_i (-a_i) e_i + \sum_j b_j \bar{e}_j = 0.$$

Since E_2 is linearly independent, all the coefficients are 0, so in particular $b_j = 0$ for all j.

Now we have

$$\dim V = \text{number of elements of } E_2 = N(E_2)$$
$$= N(E) + N(E_1)$$
$$= \dim f^{-1}\{0\} + \dim fV.$$

Note that the statement and proof are valid if $\dim V = \infty$, with the proper interpretation. ∎

Corollary.   *If dim V = dim W and this dimension is finite, then the following are equivalent.*

(a) $f: V \to W$ *is an isomorphism.*
(b) *f is onto.*
(c) *f is 1-1.*

Proposition 2.5.2.   *A linear function is uniquely determined by its values on a basis. Given a set of values in 1-1 correspondence with the elements of a basis of V, there is a unique linear function having these values as its values on the basis.*

Proof. Let $f: V \to W$ be a linear function and $\{e_i\}$ a basis of V. We are to show that f is determined by the values $fe_i \in W$. For any $v \in V$ we have the unique coordinate expression $v = a^i e_i$; since f is linear,

$$fv = f(a^i e_i)$$
$$= a^i f e_i.$$

Thus fv depends on the values fe_i as well as the a^i, which depend on v and the e_i. An alternative way of stating this result is that if $g: V \to W$ is a linear function such that $ge_i = fe_i$ for all i, then $gv = fv$ for all $v \in V$.

On the other hand, given a set of vectors $w_i \in W$ (the w_i need not be linearly independent, or nonzero; indeed, they may be all equal to each other), the formula

$$fv = f(a^i e_i)$$
$$= a^i w_i$$

defines a linear function. Indeed,

$$f(v + \bar{v}) = f(a^i e_i + \bar{a}^i e_i)$$
$$= (a^i + \bar{a}^i)w_i$$
$$= fv + f\bar{v},$$
$$f(av) = f(aa^i e_i)$$
$$= aa^i w_i$$
$$= afv. \quad \blacksquare$$

Problem 2.5.5.   If dim V = dim W, then V and W are isomorphic.

Remark. The isomorphism desired in Problem 2.5.5 is far from being unique, depending on choices of bases of V and W. Occasionally, further structure will give more conditions on an isomorphism, which will determine it uniquely, as, for example, in the case of a finite-dimensional space and its second dual (see Section 2.9). In such a situation we say that the isomorphism is *natural*, as opposed to arbitrary.

2.6.   Spaces of Linear Functions

The set of linear functions $f, g, \ldots$ of V into W forms a vector space which we denote $L(V, W)$. We define the sum of linear functions f and g by

$$(f + g)v = fv + gv$$

and the scalar product of $a \in R$ and f by

$$(af)v = a(fv)$$

for all $v \in V$. It is trivial to verify that $f + g$ and af are again linear functions and that $L(V, W)$ is a vector space under these operations.

We now examine what form linear functions and their operations take in terms of components with respect to bases. Suppose that the dimensions are finite, say dim $W = d_1$ and dim $V = d_2$, and that $\{e_i\}$ is a basis of V, $\{\bar{e}_\alpha\}$ is a basis of W. The index α will be used as a sum index running from 1 to d_1, and

i will run from 1 to d_2. For a linear function $f: V \to W$ we may write the coordinate expressions for fe_i as

$$fe_i = a_i^\alpha \bar{e}_\alpha. \tag{2.6.1}$$

By Proposition 2.5.2, f determines and is uniquely determined by its basis values $a_i^\alpha \bar{e}_\alpha$, and hence by the matrix (a_i^α) and the bases $\{e_i\}$ and $\{\bar{e}_\alpha\}$. The scalars a_i^α are made into a $d_1 \times d_2$ *matrix* by arranging them in a rectangular array with α constant on rows (α is the *row index*) and i constant on columns (i is the *column index*). We say that (a_i^α) is the *matrix of f with respect to* $\{e_i\}$ and $\{\bar{e}_\alpha\}$.

Just as we may think of the components of a vector with respect to a basis as the coordinates of the vector, we may think of the entries of the matrix as coordinates of the linear transformation. Thus a choice of bases gives *coordinatizations* of V, W, and $L(V, W)$, that is, 1–1 mappings onto R^{d_2}, R^{d_1}, and the set of $d_1 \times d_2$ matrices, respectively. The first two coordinatizations are vector space isomorphisms, and it is natural to define a vector space structure on the set of $d_1 \times d_2$ matrices so that the third coordinatization also be a vector space isomorphism. It is easy to see that the definitions must be

$$(a_i^\alpha) + (b_i^\alpha) = (a_i^\alpha + b_i^\alpha), \tag{2.6.2}$$
$$a(a_i^\alpha) = (aa_i^\alpha). \tag{2.6.3}$$

Remark. We have previously encountered square matrices in expressing the change of basis in a vector space (Section 2.3). Even in the case $V = W$ these are different uses of square matrices. It is common to confuse a matrix with the object it coordinatizes, thus thinking of a matrix as being a basis change in the one case and a linear function in the other case. A similar confusion is frequently allowed between a vector and its coordinates. In either situation, coordinate change or linear function action, we have two sets of scalars for each vector: in the first case its coordinates with respect to each of two bases, in the second case the coordinates with respect to a single basis of v and fv. The two uses of coordinates are described as the *alias* and *alibi* viewpoints, respectively.

A linear function can be described in terms of components and matrices. That is, given the components of $v \in V$, say $v = v^i e_i$, we write down the formulas for the components of $w = fv = w^\alpha \bar{e}_\alpha$. These formulas follow directly from (2.6.1):

$$f(v^i e_i) = v^i fe_i$$
$$= v^i a_i^\alpha \bar{e}_\alpha$$
$$= w^\alpha \bar{e}_\alpha.$$

So by uniqueness of components we have

$$w^\alpha = a_i^\alpha v^i. \tag{2.6.4}$$

These formulas are taken as the definition of multiplication of a $d_1 \times d_2$ matrix $A = (a_i^\alpha)$ and a $d_2 \times 1$ (column) matrix $\bar{v} = (v^i)$ to yield a $d_1 \times 1$ (column) matrix $\bar{w} = (w^\alpha)$, indicated by

$$\bar{w} = A\bar{v}. \tag{2.6.5}$$

Let us denote the isomorphism $v \to \bar{v}$ by $E \colon V \to R^{d_2}$, and, similarly, $w \to \bar{w}$ by $\bar{E} \colon W \to R^{d_1}$. Then the relation between f and A is conveniently expressed by means of the *commutative* diagram,

$$
\begin{array}{ccc}
V & \xrightarrow{f} & W \\
{\scriptstyle E}\downarrow & & \downarrow{\scriptstyle E} \\
R^{d_2} & \xrightarrow{A} & R^{d_1}
\end{array}
$$

where by "commutative" we mean that the same result is obtained from following either path indicated by arrows. In a formula this means

$$\bar{E} \circ f = A \circ E. \tag{2.6.6}$$

Since E and $\bar{E}$ are isomorphisms, they have inverses, so (2.6.6) may be solved for f or A, giving formulas expressing Proposition 2.5.2 again.

$$f = \bar{E}^{-1} \circ A \circ E, \tag{2.6.7}$$
$$A = \bar{E} \circ f \circ E^{-1}. \tag{2.6.8}$$

Since the matrix of a linear function $f \colon V \to W$ consists of $d_1 d_2$ scalars which entirely determine f, and since any matrix determines a linear function, we should expect that the dimension of the space $L(V, W)$ of linear functions of V into W is $d_1 d_2$. That this is the case is given by the following.

Proposition 2.6.1.   (a) *If dim $V = d_2$ and dim $W = d_1$, then dim $L(V, W) = d_1 d_2$.*

(b) *If $\{e_i\}$ is a basis for V and $\{\bar{e}_\alpha\}$ a basis for W, then a basis for $L(V, W)$ is $\{E_\beta^j\}$, where E_β^j is the linear function defined* (see Proposition 2.5.2) *by giving its values on a basis as*

$$E_\beta^j e_i = \delta_i^j \bar{e}_\beta.$$

(c) *If (f_i^α) is the matrix of $f \in L(V, W)$, then the expression for f in terms of the basis $\{E_\beta^j\}$ is*

$$f = f_i^\alpha E_\alpha^i.$$

Proof. The E_β^j are linearly independent. For if $a_j^\beta E_\beta^j = 0$, then

$$a_j^\beta E_\beta^j e_i = a_j^\beta \delta_i^j \bar{e}_\beta = a_i^\beta \bar{e}_\beta = 0,$$

and so by the linear independence of the $\bar{e}_\alpha$, $a_i^\alpha = 0$ for $\alpha = 1, \ldots, d_1$ and $i = 1, \ldots, d_2$.

Since, by the definition of (f_i^α),

$$fe_i = f_i^\alpha \bar{e}_\alpha$$

and also

$$f_j^\alpha E_\alpha^j e_i = f_j^\alpha \delta_i^j \bar{e}_\alpha$$
$$= f_i^\alpha \bar{e}_\alpha,$$

it follows that $f = f_i^\alpha E_\alpha^i$. Thus the E_β^j also span $L(V, W)$, so they are a basis. ∎

Problem 2.6.1. Show that the matrix of E_β^j is $(\delta_i^j \delta_\beta^\alpha)$.

If V, W, and X are vector spaces with bases $\{e_i\}$, $\{f_\alpha\}$, and $\{g_A\}$, respectively, and $F: V \to W$ and $G: W \to X$ are linear functions, then $Fe_i = F_i^\alpha f_\alpha$, $Gf_\alpha = G_\alpha^A g_A$, and

$$(G \circ F)e_i = G(F_i^\alpha f_\alpha)$$
$$= F_i^\alpha G(f_\alpha)$$
$$= G_\alpha^A F_i^\alpha g_A.$$

Thus the matrix of the composition $G \circ F$ is the *product* of the matrices:

$$(G_\alpha^A)(F_i^\beta) = (G_\alpha^A F_i^\alpha).$$

The following propositions can be proved by manipulating the matrices as well as the linear functions.

Proposition 2.6.2. *Matrix multiplication is associative; that is, for matrices A, B, C such that $A(BC)$ exists, $A(BC) = (AB)C$.*

Proof. A, B, C correspond to linear functions F, G, H and their products correspond to the various compositions of F, G, H. Since $(F \circ G) \circ H = F \circ (G \circ H)$ for any functions, the corresponding formula for matrices is also valid. ∎

Proposition 2.6.3. (a) *If $F, G: V \to W$ and $H: W \to X$ are linear functions, and A, B, C are the corresponding matrices with respect to some bases, then*

$$H \circ (F + G) = H \circ F + H \circ G,$$
$$C(A + B) = CA + CB.$$

(b) *Similarly, for linear functions $F: V \to W$, $G, H: W \to X$ and their matrices A, B, C,*

$$(G + H) \circ F = G \circ F + H \circ F,$$
$$(C + B)A = CA + BA.$$

Proof. (a) For every $v \in V$ we have

$$H \circ (F + G)v = H(Fv + Gv)$$
$$= HFv + HGv$$
$$= (H \circ F + H \circ G)v.$$

The proof for (**b**) is not much different. The formulas for the matrices follow from the correspondence between matrices and linear functions.   ∎

2.7.   Dual Space

The vector space $L(V, R)$ is called the *dual space* of V. It is denoted V^*.

Proposition 2.7.1.   *If dim V is finite, then dim $V^* = $ dim V.*

This follows immediately from Proposition 2.6.1. However, if dim V is infinite, then dim $V^* > $ dim V, provided the usual interpretation is given to the meaning of inequalities between various orders of infinity. (Precisely, if dim V is infinite, then there is a 1–1 correspondence between a basis of V and a *subset* of a basis of V^*, but it is impossible to have a 1–1 correspondence between a basis of V^* and all or any part of a basis of V.)

Henceforth, unless specifically denied, we shall assume that the vector spaces we deal with have finite dimension.

There is a natural basis for R—the number 1. Thus, according to Proposition 2.6.1, for each basis $\{e_i\}$ of V there is a unique basis $\{\varepsilon^i\}$ of V^* such that

$$\varepsilon^i e_j = \delta_j^i. \tag{2.7.1}$$

The linear functions $\varepsilon^i \colon V \to R$ defined by (2.7.1) are called the *dual basis* to the basis $\{e_i\}$.

Now suppose that $\{f_i\}$ is another basis of V and that $\{\varphi^i\}$ is the dual basis to the basis $\{f_i\}$. Then by the definition of the dual basis we have

$$\varphi^i f_j = \delta_j^i. \tag{2.7.2}$$

The f_i are given in terms of the e_i by a matrix (a_i^j), and vice versa by the inverse matrix (b_i^j):

$$f_i = a_i^j e_j, \tag{2.7.3}$$
$$e_i = b_i^j f_j. \tag{2.7.4}$$

Then

$$\varepsilon^i f_j = \varepsilon^i a_j^k e_k$$
$$= a_j^k \delta_k^i$$
$$= a_j^i$$
$$(a_k^i \varphi^k) f_j = a_k^i \delta_j^k$$
$$= a_j^i.$$

Since ε^i and $a_k^i \varphi^k$ have the same values on the basis f_j,

$$\varepsilon^i = a_k^i \varphi^k. \tag{2.7.5}$$

Hence also

$$\varphi^i = b^i_k \varepsilon^k. \tag{2.7.6}$$

The content of (2.7.3) to (2.7.6) may be expressed verbally as

Proposition 2.7.2. *The matrix of change of dual bases is the inverse of the matrix of change of bases. However, the sum takes place on rows in one case, columns in the other.*

2.8.   Multilinear Functions

Let V_1, V_2, and W be vector spaces. A map $f\colon V_1 \times V_2 \to W$ is called *bilinear* if it is linear in each variable; that is,

$$f(av_1 + \bar{a}\bar{v}_1, v_2) = af(v_1, v_2) + \bar{a}f(\bar{v}_1, v_2),$$
$$f(v_1, av_2 + \bar{a}\bar{v}_2) = af(v_1, v_2) + \bar{a}f(v_1, \bar{v}_2).$$

The extension of this definition to functions of more than two variables is simple, and such functions are called *multilinear functions*. In the case of r variables we sometimes use the more specific term *r-linear*, and the defining relation is

$$f(v_1, \ldots, av_i + \bar{a}\bar{v}_i, \ldots, v_r) = af(v_1, \ldots, v_i, \ldots, v_r) + \bar{a}f(v_1, \ldots, \bar{v}_i, \ldots, v_r).$$

Suppose that $\tau \in V^*$ and $\theta \in W^*$; that is, τ and θ are linear real-valued functions on V and W, respectively. Then we obtain a bilinear real-valued function $\tau \otimes \theta\colon V \times W \to R$ by the formula

$$\tau \otimes \theta(v, w) = (\tau v)(\theta w).$$

This bilinear function is called the *tensor product* of τ and θ, and we read it "τ tensor θ."

Multilinear functions may be multiplied by scalars and two multilinear functions of the same kind (having the same domain and range space) may be added, in each case resulting in a multilinear function of the same kind. Thus the r-linear functions mapping $V_1 \times V_2 \times \cdots \times V_r$ into W form a vector space, which we denote $L(V_1, \ldots, V_r; W)$.

Problem 2.8.1.   Prove that tensor products $\tau \otimes \theta \in L(V, W; R)$, where $\tau \in V$ and $\theta \in W$, span $L(V, W; R)$. However, show that except in very special cases $L(V, W; R)$ does not consist entirely of tensor products $\tau \otimes \theta$; that is, usually there are members of $L(V, W; R)$ which can only be expressed as sums of two or more such $\tau \otimes \theta$. Determine the special cases.

2.9.   Natural Pairing

If V is a vector space and $\tau \in V^*$, then by definition τ is a function on V; that is, τv is a function of the V-valued variable v. We can twist our viewpoint around and consider v as a function of the V^*-valued variable τ, with value τv again. When we take this latter viewpoint, v is a linear function on V^* and hence a member of V^{**}. More precisely, v and the function τv of τ are not really the same, but we merely have a way of proceeding from v to an element of V^{**}. However, we choose to ignore the difference and regard V as being included in V^{**} by this change-of-viewpoint procedure. This identification of V with part (or all) of V^{**} is called the *natural imbedding* of V into V^{**}; it is natural because it only depends on the vector-space structure itself, not on any choice of basis or other machinery.

Theorem 2.9.1.   *The natural imbedding of V into V^{**} is an isomorphism of V with V^{**}.*

Proof. For this proof we must distinguish between $v \in V$ and its natural image in V^{**}, which we shall denote $\bar{v} \in V^{**}$. That is, $\bar{v}\tau = \tau v$ defines $\bar{v} \colon V^* \to R$ for each $v \in V$. The map $v \to \bar{v}$ is clearly linear. To show that it is 1–1 we only need show that if $\bar{v} = 0$, then $v = 0$. Suppose $v \neq 0$. Then v may be included in a basis $\{e_i\}$, with $v = e_1$. Let $\{\varepsilon^i\}$ be the dual basis. Then $\bar{v}\varepsilon^1 = \varepsilon^1 v = \varepsilon^1 e_1 = 1 \neq 0$, so $\bar{v} \neq 0 \in V^{**}$. It follows from the Corollary to Proposition 2.5.1 that V and V^{**} are isomorphic under this mapping, since their dimensions are the same by Proposition 2.7.1.   ∎

Remark. If dim $V = \infty$, the natural imbedding is still 1–1 by the same proof, but it is never onto V^{**}, and so it is not an isomorphism.

Problem 2.9.1.   Show that the dual basis to the dual basis to a basis $\{e_i\}$ is simply the natural imbedding $\{\bar{e}_i\}$ of $\{e_i\}$ into V^{**}.

The two viewpoints contrasted above, considering τv first as a function τ of v, then as a function v of τ are both asymmetric, giving preference to one or the other of τ and v. A third viewpoint now eliminates this asymmetry. That is, we consider τv as being a function of two variables v and τ, which we shall denote

$$\langle \ , \ \rangle \colon V \times V^* \to R,$$

defined by

$$\langle v, \tau \rangle = \tau v.$$

The function $\langle \ , \ \rangle$ is called the *natural pairing* of V and V^* into R. It is an easy verification to show that $\langle \ , \ \rangle$ is bilinear.

If $\{e_i\}$ is a basis of V, $\{\varepsilon^i\}$ the dual basis, $v = a^i e_i$, $\tau = b_i \varepsilon^i$, then

$$\begin{aligned}
\langle v, \tau \rangle &= b_i \varepsilon^i (a^j e_j) \\
&= b_i a^j \delta^i_j \\
&= b_i a^i.
\end{aligned}$$

Thus in terms of a basis and its dual basis, evaluating the natural pairing consists in taking the sum of products of corresponding components. The natural pairing is sometimes called the *scalar product* of vectors and dual vectors.

2.10.   Tensor Spaces

Let V be a vector space. The scalar-valued multilinear functions with variables all in either V or V^* are called *tensors over V* and the vector spaces they form are called the *tensor spaces over V*. The numbers of variables from V^* and V are called the *type numbers* or *degrees* of the tensor, with the number of variables from V^* called the *contravariant degree*, the number of V the *covariant degree*. Thus for a multilinear function on $V^* \times V \times V$ the type is $(1, 2)$.

We shall not need to consider distinctions between tensors of the same type based on different orderings of the V^* and V variables. In fact, we shall generally agree to place all the V^* variables before the V variables, so that tensors which are functions on $V \times V^* \times V$ will be replaced by those defined on $V^* \times V \times V$. Sometimes it will be necessary to permute variables to achieve the preferred order, in which case the order of the V^* variables and the order of the V variables must be retained. If there is some relation between a tensor and the tensor with its V^* variables (or V variables) permuted in a certain fashion, then the tensor is said to have a *symmetry property*. Special cases are discussed in Sections 2.15 to 2.19. Besides the main topics of these sections see also Problems 2.16.6 and 2.17.4. A general study of symmetry classes of tensors requires more group theory than we can give here.

The space of multilinear functions on $V^* \times V \times V$ is denoted

$$V \otimes V^* \otimes V^* = T^1_2(V).$$

The reversal of factors with $*$'s and without is intentional, and is explained by the fact that it generalizes the case of tensors of degree 1. In fact, by definition V^* consists of linear functions on V; by Theorem 2.9.1, V may be considered to be the same as V^{**}, the linear functions on V^*. In general tensors of type (r, s) form a vector space denoted by $T^r_s = V \otimes \cdots \otimes V \otimes V^* \otimes \cdots \otimes V^*$ (V: r times, V^*: s times) and consist of multilinear functions on

$$V^* \times \cdots \times V^* \times V \times \cdots \times V$$

(V^*: r times, V: s times).

A tensor of type $(0,0)$ is defined to be a scalar, so $T_0^0 = R$. A tensor of type $(1, 0)$ is sometimes called a *contravariant vector* and one of type $(0, 1)$ a *covariant vector*. A tensor of type $(r, 0)$ is sometimes called a *contravariant tensor* and one of type $(0, s)$ is sometimes called a *covariant tensor*.

The notation introduced in Section 2.8 is consistent with what we have just done. In fact, if $v \in V$ and $\tau \in V^*$, then $v \otimes \tau \in V \otimes V^*$, for $v \otimes \tau$ was defined to be a bilinear function on $V^* \times V$. However, as noted in Problem 2.8.1, the space $V \otimes V^*$ does not consist merely of such tensor products $v \otimes \tau$ (except in a special case) but rather of sums of such terms.

2.11.   Algebra of Tensors

As part of the vector space structure, we have that tensors of the same type can be added and multiplied by scalars. Now we shall define the *tensor product* of tensors of possibly different types. The *tensor product* of tensor A of type (r, s) and tensor B of type (t, u) is a tensor $A \otimes B$ of type $(r + t, s + u)$ defined, as a function on $(V^*)^{r+t} \times V^{s+u}$, by

$$A \otimes B(\tau^1, \ldots, \tau^{r+t}, v_1, \ldots, v_{s+u})$$
$$= A(\tau^1, \ldots, \tau^r, v_1, \ldots, v_s)B(\tau^{r+1}, \ldots, \tau^{r+t}, v_{s+1}, \ldots, v_{s+u}).$$

This generalizes the definition of $v \otimes \tau$ given in Section 2.8.

The associative law and the distributive laws for tensor product are true and easily verified. That is,

$$(A \otimes B) \otimes C = A \otimes (B \otimes C),$$
$$A \otimes (B + C) = A \otimes B + A \otimes C,$$
$$(A + B) \otimes C = A \otimes C + B \otimes C,$$

whenever the types of A, B, C are such that these formulas make sense.

Problem 2.11.1. If $v, w \in V$ are linearly independent, show that $v \otimes w \neq w \otimes v$. Hence the tensor product is not generally commutative.

2.12.   Reinterpretations

Tensors generally admit several interpretations in addition to the definitive one of being a multilinear function with values in R. For tensors arising in applications or from mathematical structures it is rarely the case that the multilinear function interpretation of a tensor is the most meaningful in a physical or geometric sense. Thus it is important to be able to pass from one interpretation to another. The number of interpretations increases rapidly as a function of the degrees.

Let us first examine how such other interpretations are obtained for a tensor A of type $(1, 1)$. For a fixed $\tau \in V^*$, $A(\tau, v)$ is a linear function of $v \in V$. Let us denote it by $A_1\tau \in V^*$ so that

$$\langle v, A_1\tau \rangle = A(\tau, v). \tag{2.12.1}$$

Since A is bilinear, the function $A_1\tau$ is linear as a function of τ, so we have a linear function

$$A_1: V^* \to V^*, \tau \to A_1\tau.$$

Thus for each tensor of type $(1, 1)$ we have a corresponding linear function of V^* into itself.

Conversely, if $B: V^* \to V^*$ is a linear function, then we can define a tensor A of type $(1, 1)$ by

$$A(\tau, v) = \langle v, B\tau \rangle.$$

This A is a tensor since B is linear and $\langle \ , \ \rangle$ is bilinear. It is easily seen that $B = A_1$, so that the procedure goes both ways.

Similarly, for a fixed $v \in V$, $A(\tau, v)$ is a linear function of $\tau \in V^*$ which we denote $A_2v \in V^{**} = V$. Again we have

$$\langle A_2v, \tau \rangle = A(\tau, v),$$

and $A_2: V \to V$ is linear. Moreover, the converse is essentially the same; that is, for each linear $B: V \to V$ there is a unique $A \in T_1^1$ such that $B = A_2$.

These reinterpretations are *natural* since no choices were made to define them. Thus we have

Theorem 2.12.1. *The vector spaces T_1^1, $L(V, V)$, and $L(V^*, V^*)$ are naturally isomorphic.*

This natural isomorphism is also quite obvious in terms of components with respect to bases. If $\{e_i\}$ is a basis of V, $\{\varepsilon^i\}$ the dual basis, then the d^2 elements $e_i \otimes \varepsilon^j$ ($d = \dim V$) form a basis for T_1^1. Indeed, if $A_1^1 \in T$, let $A_j^i = A(\varepsilon^i, e_j)$. Then $A = A_j^i e_i \otimes \varepsilon^j$ by the following theorem, which generalizes Proposition 2.5.2.

Theorem 2.12.2. *A tensor is determined by its values on a basis and its dual basis. These values are the components of the tensor with respect to the tensor products of basis and dual basis elements, which form bases of the tensor spaces.*

Proof. Let $A \in T_s^r$ and $A_{j_1 \ldots j_s}^{i_1 \ldots i_r} = A(\varepsilon^{i_1}, \ldots, \varepsilon^{i_r}, e_{j_1}, \ldots, e_{j_s})$. Then for any $\tau^1, \ldots, \tau^r \in V^*$ and $v_1, \ldots, v_s \in V$, we have

$$\tau^p = a_i^p \varepsilon^i, \qquad v_q = b_q^i e_i, \qquad p = 1, \ldots, r, \quad q = 1, \ldots, s$$

$$
\begin{aligned}
A(\tau^1, \ldots, \tau^r, v_1, \ldots, v_s) &= a_{i_1}^1 \ldots a_{i_r}^r b_1^{j_1} \ldots b_s^{j_s} A(\varepsilon^{i_1}, \ldots, \varepsilon^{i_r}, e_{j_1}, \ldots, e_{j_s}) \\
&= a_{i_1}^1 \ldots a_{i_r}^r b_1^{j_1} \ldots b_s^{j_s} A_{j_1 \ldots j_s}^{i_1 \ldots i_r} \\
&= A_{j_1 \ldots j_s}^{i_1 \ldots i_r} \langle e_{i_1}, \tau^1 \rangle \ldots \langle e_{i_r}, \tau^r \rangle \langle v_1, \varepsilon^{j_1} \rangle \ldots \langle v_s, \varepsilon^{j_s} \rangle \\
&= A_{j_1 \ldots j_s}^{i_1 \ldots i_r} e_{i_1} \otimes \ldots \otimes e_{i_r} \otimes \varepsilon^{j_1} \otimes \ldots \\
&\qquad\qquad\qquad\qquad \otimes \varepsilon^{j_s}(\tau^1, \ldots, \tau^r, v_1, \ldots, v_s).
\end{aligned}
$$

Thus

$$A = A_{j_1 \ldots j_s}^{i_1 \ldots i_r} e_{i_1} \otimes \ldots \otimes e_{i_r} \otimes \varepsilon^{j_1} \otimes \ldots \otimes \varepsilon^{j_s},$$

since they have the same values as functions on $V^{*r} \times V^s$.

The proof that the $e_{i_1} \otimes \ldots \otimes e_{i_r} \otimes \varepsilon^{j_1} \otimes \ldots \otimes \varepsilon^{j_s}$ are linearly independent is left as an exercise. ∎

Corollary.   *The dimension of T_s^r is d^{r+s}.*

Now returning to tensors of type $(1, 1)$, the coordinate form of the interpretations is given in the following.

Theorem 2.12.3.   *If $A = A_j^i e_i \otimes \varepsilon^j$, a member of T_1^1, then the A_j^i are:*

(a) *The components of A as a member of T_1^1 with respect to the basis $\{e_i \otimes \varepsilon^j\}$.*

(b) *The matrix entries of the matrix of A_1 with respect to the basis $\{\varepsilon^i\}$ of V^*, with i the column index, j the row index in (A_j^i).*

(c) *The matrix entries of the matrix of A_2 with respect to the basis $\{e_i\}$ of V, with i the row index, j the column index in (A_j^i).*

Proof. Part (a) follows directly from the definition of components. For (b) we have, by (2.12.1),

$$
\begin{aligned}
\langle e_j, A_1 \varepsilon^i \rangle &= A(\varepsilon^i, e_j) \\
&= A_j^i,
\end{aligned}
$$

by Theorem 2.12.2. But if (B_j^i) is the matrix of A_1, then $A_1 \varepsilon^i = B_k^i \varepsilon^k$, so

$$
\begin{aligned}
\langle e_j, A_1 \varepsilon^i \rangle &= \langle e_j, B_k^i \varepsilon^k \rangle \\
&= B_k^i \delta_j^k \\
&= B_j^i.
\end{aligned}
$$

In this, i is the column index of (B_j^i), hence also of (A_j^i).

For (c) we have

$$\langle A_2 e_j, \varepsilon^i \rangle = A_j^i,$$

and if (C_j^i) is the matrix of A_2, $A_2 e_j = C_j^i e_i$, where j is the column index of $(C_j^i) = (A_j^i)$. ∎

In terms of a basis the action of A on V or V^* may be viewed as a "partial evaluation": For $v = a^i e_i \in V$,

$$
\begin{aligned}
A_2 v &= (A^i_j e_i \otimes \varepsilon^j)_2 v \\
&= A^i_j a^k (e_i \otimes \varepsilon^j)_2 e_k \\
&= A^i_j a^k e_i \langle e_k, \varepsilon^j \rangle \\
&= A^i_j a^j e_i.
\end{aligned}
$$

It is as though we evaluated v on the ε^j part and left the rest of A unaltered. Similarly, for $\tau \in V^*$, $\tau = b_i \varepsilon^i$,

$$
A_1 \tau = A^i_j b_i \varepsilon^j.
$$

The way the indices are arranged makes this procedure practically automatic.

Problem 2.12.1.   Since $\langle \ , \ \rangle$ is a bilinear function on $V \times V^*$ it is a tensor of type $(1, 1)$. What are the corresponding functions $\langle \ , \ \rangle_1 : V^* \to V^*$ and $\langle \ , \ \rangle_2 : V \to V$? What are the components of $\langle \ , \ \rangle$?

Higher degree tensors have other interpretations in an analogous way. These other interpretations take the form of multilinear functions of V^* and V into tensor spaces. For example, a tensor A of type $(1, 2)$ may be considered as a map $A_{2,3} : V \times V \to V$. The subscripts 2 and 3 indicate that the variables (v, w) of $V \times V$ become the 2nd and 3rd variables of A, leaving the 1st variable, in V^*, of A open. Thus each $(v, w) \in V \times V$ yields a linear function on V^*, that is, a member of $V^{**} = V$. In terms of coordinates it is again a partial evaluation: If

$$
A = A^i_{jk} e_i \otimes \varepsilon^j \otimes \varepsilon^k,
$$

then

$$
A_{2,3}(v, w) = A^i_{jk} \langle v, \varepsilon^j \rangle \langle w, \varepsilon^k \rangle e_i \in V.
$$

The other interpretations of $A \in T^1_2$ are

$$
\begin{aligned}
A_{1,2} &: V^* \times V \to V^*, \\
A_{1,3} &: V^* \times V \to V^*, \\
A_1 &: V^* \to V^* \otimes V^*, \\
A_2 &: V \to V \otimes V^*, \\
A_3 &: V \to V \otimes V^*.
\end{aligned}
$$

The range of, say, A_2 may be viewed, as in Theorem 2.12.1, as $L(V, V)$. Thus A may be interpreted as an object which assigns linearly to each $v \in V$ a linear transformation $(A_2 v)$ of V into V. The matrix of $(A_2 v)_2$ is $(A^i_{kj} \langle v, \varepsilon^k \rangle)$, where i is the row index, j the column index.

Problem 2.12.2.   The components of a tensor product are the products of the components of the factors. That is,

$$
(A \otimes B)^{i_1 \ldots i_{r+t}}_{j_1 \ldots j_{s+u}} = A^{i_1 \ldots i_r}_{j_1 \ldots j_s} B^{i_{r+1} \ldots i_{r+t}}_{j_{s+1} \ldots j_{s+u}}.
$$

Problem 2.12.3.   What type of tensor can be interpreted as a multilinear map $V^s \to V$?

2.13.   Transformation Laws

The components of a tensor A are functions of the basis as well as the super-script and subscript entries, the indices. The way which the components depend on the basis is determined by the matrix of change of basis, its inverse, and certain rules for using these matrices, which depend on the type of the tensor and are called the *transformation law* for the tensor of that type.

We shall indicate the functional dependence of a tensor on the basis, when more than one basis is being considered, by a superscript generically related to the basis. Thus for a tensor of type $(1, 2)$, the components with respect to basis $\{e_i\}$ and its dual $\{\varepsilon^i\}$ will be denoted by $A^{e;i}_{jk}$, and are given, according to Theorem 2.12.2, by

$$A^{e;i}_{jk} = A(\varepsilon^i, e_j, e_k).$$

Note that we are using A as a function in two ways, once as a multilinear function on $V^* \times V \times V$, the other as a function of four variables: the basis e and the three integer variables i, j, and k.

Now let $\{f_i\}$ be another basis, $\{\varphi^i\}$ its dual basis, which are related to the first bases by

$$f_i = a^j_i e_j,$$
$$\varphi^i = b^i_j \varepsilon^j.$$

The components of A with respect to the new basis are

$$\begin{aligned}
A^{f;i}_{jk} &= A(\varphi^i, f_j, f_k) \\
&= A(b^i_m \varepsilon^m, a^n_j e_n, a^p_k e_p) \\
&= b^i_m a^n_j a^p_k A(\varepsilon^m, e_n, e_p) \qquad\qquad (2.13.1) \\
&= b^i_m a^n_j a^p_k A^{e;m}_{np}.
\end{aligned}$$

This equation is the classical *law of transformation of the components of the tensor of A of type* $(1, 2)$.

The alterations necessary for obtaining the laws for other types should be obvious and will not be written out here.

If V is the tangent space at a point m of a manifold, $V = M_m$, and the bases are obtained as coordinate vector fields with respect to two systems of coordinates (x^i) and (y^i) at m, then

$$e_i = \frac{\partial}{\partial x^i}, \qquad \varepsilon^i = dx^i,$$

$$f_i = \frac{\partial}{\partial y^i}, \qquad \varphi^i = dy^i,$$

$$a_i^j = \frac{\partial x^j}{\partial y^i}, \qquad b_j^i = \frac{\partial y^i}{\partial x^j},$$

all evaluated at m. Then the transformation law (2.13.1) above has a form to be found in most standard works:

$$A_{jk}^{y,i} = A_{np}^{x,m} \frac{\partial y^i}{\partial x^m} \frac{\partial x^n}{\partial y^j} \frac{\partial x^p}{\partial y^k}. \tag{2.13.2}$$

Problem 2.13.1. Let dim $V = 3$, $A = A^{e,i}_{\ j} e_i \otimes \varepsilon^j$, where

$$(A^{e,i}_{\ j}) = \begin{pmatrix} 2 & 0 & 1 \\ 0 & 3 & -1 \\ 1 & -1 & 0 \end{pmatrix};$$

let $\{f_1 = e_1 + e_2, f_2 = 2e_2, f_3 = -e_2 + e_3\}$ be a new basis and $\{\varphi^i\}$ the new dual basis. Let $v = -e_1 + 2e_3$ and $\tau = 5\varepsilon^1 - 2\varepsilon^2 + \varepsilon^3$.

(a) Evaluate $A(\tau, v)$.

(b) Evaluate $A_2 v$.

(c) Evaluate $A_1 \tau$.

(d) Find the expression for the φ^i in terms of the ε^i.

(e) Find the expressions for v and τ in the new basis.

(f) Find the new components $A^{f,i}_{\ j}$.

(g) Verify that det $(A^{e,i}_{\ j}) = $ det $(A^{f,i}_{\ j})$ and that tr $(A^{e,i}_{\ j}) = $ tr $(A^{f,i}_{\ j})$. ["det" abbreviates "determinant"; the *trace* of a matrix (A^i_j) is the sum of the main diagonal terms, tr $(A^i_j) = A^i_i$; see Section 2.14.] Note that $(A^{e,i}_{\ j})$ is symmetric, $(A^{f,i}_{\ j})$ is not.

(h) Do parts (a), (b), and (c) over in terms of the new basis, showing that the result is the same.

With respect to a given basis of V, we may simply speak of a tensor by giving its components. In fact, this is the classical treatment of tensors. The classical definition of a tensor is that it is a function of $1 + r + s$ variables, that is the basis (or coordinate system) as one variable, r contravariant (upper) indices, and s covariant (lower) indices, which satisfies the transformation law of a tensor of type (r, s) for each pair of bases [that is, equation (2.13.2) when $r = 1$ and $s = 2$]. Then one would speak of "the tensor A^i_{jk}." Having the variables i, j, and k as part of the symbol denoting the tensor A is comparable to having the variable x as part of the symbol for a function f, as in $f(x)$, which we have all seen. Practically, no harm is done, only the logic is slightly strained.

Problem 2.13.2. If A is a tensor of type $(1, 1)$ and A has the same components with respect to every basis, show that A is a multiple of $\langle\ ,\ \rangle$; that is, $A^i_j = \alpha \delta^i_j$ for some $\alpha \in R$.

Problem 2.13.3. If A is a tensor of type (r, s) such that the components of A are the same with respect to every basis, show that either $A = 0$ or $r = s$.

2.14. Invariants

Scalar-valued functions of tensors frequently are described in terms of the components of the tensors with respect to a certain basis. If these values do not depend on the basis employed, the functions are called *invariants*, or, more precisely, *scalar invariants*. One may also speak of *tensor invariants* when the values are tensors themselves rather than scalars.

As an illustration of these concepts we define an invariant of tensors of type $(1, 1)$, the *trace*, which is a well-known invariant of matrices. We have already seen how these tensors may be considered as matrices. If $A = A^i_j e_i \otimes \varepsilon^j$ we define

$$trace\ of\ A = \operatorname{tr} A = A^i_i,$$

that is, the sum of the main diagonal elements of the matrix (A^i_j). It is not a priori evident that we have defined something which depends on A only, since the A^i_j depend not only on A but also on the basis $\{e_i\}$. To show that tr A is a number determined entirely by A itself and not by the e_i as well, we must show invariance; that is, if A is expressed in terms of another basis $\{f_i\}$, then the corresponding formula in the new components gives the same number as before. Thus we write $A = A'^i_j f_i \otimes \varphi^j = A^{e\cdot i}_j e_i \otimes \varepsilon^j$ and show that $A'^i_i = A^{e\cdot i}_i$. Using the same notation for change of basis as before, (2.7.3) to (2.7.6), we have the transformation law

$$A'^n_m = A^{e\cdot i}_j a^j_m b^n_i,$$

from which it follows that

$$A'^i_i = A^{e\cdot p}_j a^j_i b^i_p = A^{e\cdot p}_j \delta^j_p = A^{e\cdot i}_i.$$

We have proved

Proposition 2.14.1. *The trace of a tensor of type* $(1, 1)$ *is an invariant.*

To show that not every expression in terms of the components of a tensor need be an invariant, consider the following example. Suppose $d = 2$ and consider

$$A = e_1 \otimes e_1 + e_1 \otimes e_2,$$

a tensor of type $(2, 0)$. The expression A_{ii}^e in this case is $A_{11}^e + A_{22}^e = 1 + 0 = 1$. Now consider the new basis given by $e_1 = f_1 + f_2$ and $e_2 = f_2$. Then

$$A = (f_1 + f_2) \otimes (f_1 + f_2) + (f_1 + f_2) \otimes f_2$$
$$= f_1 \otimes f_1 + 2f_1 \otimes f_2 + f_2 \otimes f_1 + 2f_2 \otimes f_2,$$

from which we get $A_{ii}^f = A_{11}^f + A_{22}^f = 1 + 2 = 3$.

When a quantity is defined without reference to a basis, there is no question that it is an invariant. Sometimes this is difficult to do, and so one must establish invariance. An invariant (basis-free) definition of the determinant of a linear transformation A, $\det A$, will be given below after we study exterior algebra (2.19). The more common definitions of determinant are given in terms of components, either by means of sums of products with signs attached or inductively on the dimension by means of the rule for row or column expansion. For these definitions invariance under change of basis is another step beyond the definition. Indeed, one of the best procedures in demonstrating invariance of the componentwise definitions is to establish equivalence with the invariant definition from exterior algebra.

Besides invariants of one variable, we may also consider invariants of several variables. An invariant is called *linear* or *multilinear* if it is linear in its variable or each of its variables, as the case may be. Thus the dual space V^* of a vector space V may be described as the vector space of linear invariants on V. Moreover, the tensors over V of type (r, s) are the $(r + s)$-linear invariants on $V^{*r} \times V^s$. An invariant I is of *degree p* if it is a linear invariant of the p-fold tensor product of the variable with itself, that is,

$$IA = J(\underbrace{A \otimes \cdots \otimes A}_{p \text{ times}}),$$

where J is a linear invariant. The determinant is an invariant of degree d on tensors of type $(1, 1)$.

An important class of linear invariants are the contractions. These are not real-valued invariants except in the case of the trace, which they generalize. A contraction assigns to a tensor of type (r, s) another tensor of type $(r - 1, s - 1)$. They are essentially traces with respect to two of the indices, one contravariant, one covariant, while the others are held fixed. The formal definition follows.

The *contraction of a tensor A of type (r, s) with respect to contravariant index p $(\leq r)$ and covariant index q $(\leq s)$* is the tensor of type $(r - 1, s - 1)$ having components

$$B_{j_1 \ldots j_{s-1}}^{i_1 \ldots i_{r-1}} = A_{j_1 \ldots j_{q-1} k j_q \ldots j_{s-1}}^{i_1 \ldots i_{p-1} k i_p \ldots i_{r-1}}.$$

Problem 2.14.1. Contractions are invariants.

Problem 2.14.2. How many different contractions does a tensor of type (3, 2) have?

Problem 2.14.3. Show that $(\text{tr } A)^p$, where A is a tensor of type (1, 1), is an invariant of A of degree p. [The fact that it is an invariant follows from the fact that tr A is an invariant. The question is whether $(\text{tr } A)^2$ is a linear function of the coefficients of $A \otimes A$, etc.]

Problem 2.14.4. Show that the product of two $d \times d$ matrices is a bilinear invariant of the two matrices, all viewed as tensors of type (1, 1). (One way of doing this is to show that the matrix product is a contraction of the tensor product. Tensor product of two tensors was given an invariant definition, and its bilinearity is expressed in part by the distributive laws.)

2.15. Symmetric Tensors

A tensor A is *symmetric in the pth and qth contravariant indices* if the components with respect to *every* basis are unchanged when these indices are interchanged. A tensor A is *symmetric in the pth and qth variables* if its values as a multilinear function are unchanged when these variables are interchanged. (Of course, the two variables interchanged must be of the same type.)

Theorem 2.15.1. *The following three conditions on a tensor A are equivalent.*

(a) *A is symmetric in the pth and qth contravariant indices.*

(b) *A is symmetric in the pth and qth variables.*

(c) *The components of A with respect to some single basis are unchanged when the pth and qth contravariant indices are interchanged.*

Proof. First of all, we note that (c) is obviously a special case of (a) and, moreover, since components are obtained by substituting basis elements for the variables in A as a multilinear function, (a) is a special case of (b). Thus it suffices to show that (c) has (b) as a consequence. For simplicity we let $p = 1$, $q = 2$, and A be of type (3, 1). Then by (c) we have that there is a basis $\{e_i\}$ such that for every i, j, k, and m,

$$A_m^{ijk} = A_m^{jik}.$$

Then for any $\tau_1, \tau_2, \tau_3 \in V^*$ and $v \in V$, with components τ_{1i}, τ_{2i}, τ_{3i}, and v^i, respectively, we have

$$\begin{aligned}
A(\tau_1, \tau_2, \tau_3, v) &= A(\tau_{1i}\varepsilon^i, \tau_{2j}\varepsilon^j, \tau_{3k}\varepsilon^k, v^m e_m) \\
&= \tau_{1i}\tau_{2j}\tau_{3k}v^m A(\varepsilon^i, \varepsilon^j, \varepsilon^k, e_m) \\
&= \tau_{1i}\tau_{2j}\tau_{3k}v^m A_m^{ijk} \\
&= \tau_{2j}\tau_{1i}\tau_{3k}v^m A_m^{jik} \\
&= A(\tau_2, \tau_1, \tau_3, v),
\end{aligned}$$

by following the previous steps backward with pairs i, j and $1, 2$ transposed. ∎

Since this theorem shows there is no difference, we shall abandon the distinction between indicial and variable symmetry, and refer to the property by the first name only. In our terminology we have relied upon our agreement to place variables from V^* before those from V. Thus in defining the corresponding concept for covariant indices, symmetry in the pth and qth covariant indices will be equivalent to symmetry in the $(r + p)$th and $(r + q)$th variables. Obviously the analogous theorem holds for covariant indices.

A tensor is *symmetric in its contravariant indices* or *contravariant symmetric* if it is symmetric in every pair of contravariant indices, and similarly for *covariant symmetric*. A tensor is *symmetric* if it is both contravariant symmetric and covariant symmetric, although this concept is usually limited to purely contravariant [type $(r, 0)$ for some r] or purely covariant [type $(0, s)$ for some s] tensors. By convention (or a strict logical interpretation of the definition) we agree that tensors of degree 0 or 1 are symmetric.

It is not possible to have an invariant definition of symmetry in one contravariant and one covariant index. The example of Problem 2.13.1 shows that symmetry in mixed indices is not invariant under change of basis. The following problem shows how restrictive such symmetry is.

Problem 2.15.1. If a tensor A of type $(1, 1)$ is symmetric in its indices with respect to every basis, that is, $A_j^i = A_i^j$, then A is a multiple of the identity tensor, $A_j^i = \alpha \delta_j^i$.

2.16.   Symmetric Algebra

The symmetric tensors of type $(r, 0)$ form a subspace S^r of T_0^r; those of type $(0, s)$ form a subspace S_s of T_s^0. In general, a symmetric tensor is given by the components $A^{i_1 \cdots i_r}$, where $i_1 \leq \cdots \leq i_r$; the other components are given by symmetry, and symmetry gives no relations among the components with nondecreasing indices. Thus one basis of S^r is obtained by letting basis elements be those for which all these special components are 0 except one, which we let be 1.

The product of two symmetric tensors is not usually symmetric. For example, if $A = A^{ij} e_i \otimes e_j$ and $B = B^{ij} e_i \otimes e_j$ are symmetric tensors of type $(2, 0)$, $A \otimes B$ is not generally a symmetric tensor of type $(4, 0)$. Indeed, $A^{ij} B^{kl}$ need not equal $A^{ik} B^{jl}$. To define a multiplication of symmetric tensors which results in a symmetric tensor, we first define a *symmetrization operation* $A \to A_s$ given by the formula

$$A_s(\tau^1, \ldots, \tau^r) = \frac{1}{r!} \sum_{(i_1, \ldots, i_r)} A(\tau^{i_1}, \ldots, \tau^{i_r}) \qquad (2.16.1)$$

where the sum is taken over the $r!$ permutations of the integers $1, \ldots, r$. Here A is a tensor of type $(r, 0)$ and $\tau^1, \ldots, \tau^r$ are any elements in V^*. The τ^i need not be all different, so in case two or more are identical some of the permutations will not change the sequence.

It is easily checked that A_s is a symmetric tensor of type $(r, 0)$. For example, when $r = 3$,

$$A_s(\alpha, \beta, \gamma) = \tfrac{1}{6}[A(\alpha, \beta, \gamma) + A(\beta, \gamma, \alpha) + A(\gamma, \alpha, \beta) \\ + A(\beta, \alpha, \gamma) + A(\alpha, \gamma, \beta) + A(\gamma, \beta, \alpha)]. \tag{2.16.2}$$

Problem 2.16.1. Write out the formula for A_s analogous to (2.16.2) in the cases $r = 2$ and $r = 4$.

Problem 2.16.2. Show that the components of A_s are given in terms of the components of A by a formula similar to (2.16.1).

Problem 2.16.3. Let $s(d, r)$ be the dimension of S^r, the space of symmetric tensors over a vector space of dimension d. From the above remarks $s(d, r)$ is the number of different choices of r integers $i_1, \ldots, i_r$ such that

$$1 \leq i_a \leq i_{a+1} \leq d$$

for each a. Show that

(a) $s(1, r) = 1$, $s(d, 1) = d$.
(b) $s(d + 1, r) = s(d, r) + s(d + 1, r - 1)$.
(c) $s(d, r) = \begin{pmatrix} d + r - 1 \\ r \end{pmatrix}$
$$= (d + r - 1)!/[r!(d - 1)!],$$

the binomial coefficient.

Problem 2.16.4. If A is symmetric then $A_s = A$.

The *symmetric product* of symmetric tensors $A \in S^p$ and $B \in S^q$ is the symmetric tensor $(A \otimes B)_s \in S^{p+q}$. We denote this product by AB. For example,

$$e_1 e_1 = (e_1 \otimes e_1)_s = \tfrac{1}{2}(e_1 \otimes e_1 + e_1 \otimes e_1) = e_1 \otimes e_1,$$
$$e_1 e_2 = (e_1 \otimes e_2)_s = \tfrac{1}{2}(e_1 \otimes e_2 + e_2 \otimes e_1) = e_2 e_1,$$
$$(e_1 e_2)e_3 = \tfrac{1}{2}(e_1 \otimes e_2 + e_2 \otimes e_1)e_3$$
$$= \tfrac{1}{2}(e_1 \otimes e_2 \otimes e_3)_s + \tfrac{1}{2}(e_2 \otimes e_1 \otimes e_3)_s$$
$$= \tfrac{1}{6}(e_1 \otimes e_2 \otimes e_3 + e_1 \otimes e_3 \otimes e_2 + e_2 \otimes e_1 \otimes e_3$$
$$\qquad + e_2 \otimes e_3 \otimes e_1 + e_3 \otimes e_1 \otimes e_2 + e_3 \otimes e_2 \otimes e_1)$$
$$= e_1(e_2 e_3) = e_1(e_3 e_2) = \cdots.$$

In general, symmetric multiplication is

(a) Commutative: $AB = BA$.
(b) Associative: $(AB)C = A(BC)$.
(c) Distributive: $(A + B)C = AC + BC$.

By means of the commutative, associative, and distributive laws of symmetric multiplication, any symmetric tensor may be expressed as a sum of terms of the form $c(e_1)^{n_1}(e_2)^{n_2}\cdots(e_d)^{n_d}$, where $\{e_i\}$ is a basis of V and $c \in R$. In other words, a symmetric tensor may be expressed as a polynomial in d indeterminates, and symmetric multiplication is the same as multiplication of polynomials. In working with symmetric tensors it is much more convenient to use symmetric product notation and its properties than the $\otimes$ notation.

The following theorem is stated without proof, except for the case $r = 2$, which is important for the relation between bilinear and quadratic forms discussed in Section 2.21.

Theorem 2.16.1. *For every $A \in T_0^r$, A_s is the unique symmetric tensor such that for every $\tau \in V^*$,*

$$A_s(\tau, \ldots, \tau) = A(\tau, \ldots, \tau). \qquad (2.16.3)$$

Remarks. It is clear that (2.16.3) is true, since the sum (2.16.1) has $r!$ identical terms when $\tau^i = \tau$ for each i, but what is not evident is that (2.16.3) determines A_s completely. For the case $r = 2$ the determination of A_s by (2.16.3) is given by (2.21.1) and (2.21.2), letting $b = A_s$. If (2.16.3) had been used as the definition of A_s, then besides verifying that such an A_s exists we would have to check that it was unique.

The polynomial obtained from a symmetric tensor A of type $(r, 0)$ is *homogeneous of degree r*; that is, the sum of the exponents of the e_i is r for every term. The scalar-valued functions P on V^* given in the form

$$P\tau = A(\tau, \ldots, \tau),$$

where A is a tensor of type $(r, 0)$, are called *homogeneous polynomial functions of degree r on V^**. (They are identical with the scalar invariants on V^* of degree r, as defined in Section 2.14.) A polynomial function on V^* is a sum of such P with different degrees. The polynomial functions are therefore in 1–1 correspondence with sums of symmetric tensors of different contravariant degrees.

Problem 2.16.5. Applying Theorem 2.16.1, prove the commutative, associative, and distributive laws for symmetric multiplication.

Problem 2.16.6. Let A be a tensor of type $(0, 3)$ having the "symmetries" $A_{ijk} + A_{jki} + A_{kij} = 0$ and $A_{ijk} = -A_{ikj}$. If $d = 3$, find how many components of A are independent, choose an independent set, and express the others in terms of the chosen ones.

2.17. Skew-Symmetric Tensors

The definitions of skew-symmetry in tensors follow those for symmetry except that interchange of a pair of indices or variables changes the sign of the tensor instead of leaving it unchanged. The following theorem is the analogue of Theorem 2.15.1 and the proof is practically the same.

Theorem 2.17.1.   *The following three conditions on a tensor A are equivalent*:

(a) *A is skew-symmetric in the pth and qth contravariant indices.*

(b) *A is skew-symmetric in the pth and qth variables.*

(c) *The components of A with respect to some single basis are changed in sign only when the pth and qth contravariant indices are interchanged.*

However, for skew-symmetric tensors a further characterization is possible, as follows.

Theorem 2.17.2.   *The tensor A is skew-symmetric in contravariant indices p and q iff for all $\tau \in V^*$, insertion of τ for both the pth and qth variables of A gives the value 0 irrespective of the remaining variable values*:

$$A(\alpha^1, \ldots, \alpha^{p-1}, \tau, \alpha^p, \ldots, \alpha^{q-2}, \tau, \alpha^{q-1}, \ldots, v_1, \ldots, v_s) = 0,$$

for all $\alpha^i, \tau \in V^*, v_i \in V$.

Proof. We give the proof for A of type $(3, 1)$ with $p = 1$ and $q = 2$. The proof in the other cases is not essentially different.

If A is skew-symmetric in contravariant indices 1 and 2 then $A(\tau, \tau, \alpha, v) = -A(\tau, \tau, \alpha, v)$ by interchanging variables 1 and 2, so by transposing and dividing by 2, we get $A(\tau, \tau, \alpha, v) = 0$.

On the other hand, if A gives 0 whenever variables 1 and 2 are equal, then

$$\begin{aligned}
0 &= A(\alpha + \beta, \alpha + \beta, \gamma, v) \\
&= A(\alpha, \alpha, \gamma, v) + A(\alpha, \beta, \gamma, v) + A(\beta, \alpha, \gamma, v) + A(\beta, \beta, \gamma, v) \\
&= 0 + A(\alpha, \beta, \gamma, v) + A(\beta, \alpha, \gamma, v),
\end{aligned}$$

so

$$A(\alpha, \beta, \gamma, v) = -A(\beta, \alpha, \gamma, v). \quad \blacksquare$$

Problem 2.17.1. If a tensor A of type $(3, 0)$ is symmetric in variables 1 and 2 and skew-symmetric in variables 1 and 3, then $A = 0$.

Problem 2.17.2 (cf. Problem 2.15.1). If a tensor A of type $(1, 1)$ is skew-symmetric in its two indices for every choice of basis, then $A = 0$. Thus skew-symmetry in mixed indices is no more sensible than symmetry.

Problem 2.17.3. Let A be a tensor of type $(r, 0)$ which is skew-symmetric in all pairs of variables; that is, A is skew-symmetric. If $\tau^1, \ldots, \tau^r \in V^*$ are linearly dependent, show that $A(\tau^1, \ldots, \tau^r) = 0$.

Problem 2.17.4. Let A be a tensor of type $(0, 4)$ which satisfies the following symmetries (these are the symmetries of a riemannian curvature tensor; see Section 5.11):

(1) $A_{ijkl} = -A_{jikl}$.
(2) $A_{ijkl} = -A_{ijlk}$.
(3) $A_{ijkl} + A_{iklj} + A_{iljk} = 0$.

[Equation **(3)** is called the *cyclic sum identity* of a curvature tensor.]

(a) Show that A satisfies the following symmetry also:

(4) $A_{ijkl} = A_{klij}$.
(b) If $A(v, w, v, w) = 0$ for all v and $w \in V$, then $A = 0$.
(c) If B and C are tensors of type $(0, 4)$ satisfying the symmetries **(1)**, **(2)**, and **(3)** and if $B(v, w, v, w) = C(v, w, v, w)$ for all v and $w \in V$, then $B = C$. (*Hint:* Let $A = B - C$.)

Problem 2.17.5. Let B be a symmetric tensor of type $(0, 2)$. Define a tensor A of type $(0, 4)$ by
$$A_{ijkl} = B_{ik}B_{jl} - B_{il}B_{jk}.$$

(a) Show that A satisfies the symmetries **(1)**, **(2)**, and **(3)** of the curvature tensor, given in Problem 2.17.4.
(b) If $B(v, v) > 0$ whenever $v \neq 0$, show that $A(v, w, v, w) > 0$ whenever v and w are linearly independent. Note that
$$A(v, w, v, w) = B(v, v)B(w, w) - B(v, w)^2.$$

Problem 2.17.6. If A is skew-symmetric in some pair of variables, show that $A_s = 0$.

2.18.   Exterior Algebra

The analogue of symmetric multiplication of symmetric tensors for skew-symmetric tensors is called the *exterior* (or: *alternating, Grassmann, wedge*) *product*, and the resulting algebra is called *exterior (Grassmann) algebra*. The symbol for this product is a wedge, $\wedge$, and we employ this symbol to denote the space of skew-symmetric tensors of type $(r, 0)$, $\bigwedge^r V$. The skew-symmetric tensor space of type $(0, s)$ is denoted by $\bigwedge^s V^*$.

In general, $A \in \bigwedge^r V$ is given by its components $A^{i_1 \cdots i_r}$, where

$$i_1 < i_2 < \cdots < i_r.$$

Such increasing sequences are in 1–1 correspondence with the partitions of the first d integers into two parts with r and $d - r$ members. The number of such partitions is the binomial coefficient $\binom{d}{r}$. From this, or directly (cf. Problem 2.17.3), it is evident that dim $\bigwedge^r V = 0$ if $r > d$; that is, only the 0 tensor is skew-symmetric for degrees greater than d. Moreover, we have the following.

Theorem 2.18.1.   *The dimension of $\bigwedge^r V$ is $\binom{d}{r}$, where $d = \dim V$.*

If $j_1, \ldots, j_r$ is a permutation of $i_1, \ldots, i_r$ the component $A^{j_1 \cdots j_r}$ of a skew-symmetric tensor is either the same or the negative of $A^{i_1 \cdots i_r}$. A permutation of symbols may be obtained from a sequence of *transpositions* (interchanges of pairs). This can be done in many ways. For a given permutation, the number of transpositions is either even or odd, in which case we say that the *sign of the permutation* is 1 or -1, respectively, and denote this by *sgn* π, the sign of the permutation π. For example, if the symbols are 1, 2, 3 and it is required to put them in the order 3, 1, 2, then we use the abbreviation $(3, 1, 2)$ for the permutation and write sgn$(3, 1, 2) = 1$, since this permutation requires 2 or 4, etc., transpositions: $(1, 2, 3) \rightarrow (1, 3, 2) \rightarrow (3, 1, 2)$ or $(1, 2, 3) \rightarrow (2, 1, 3) \rightarrow (2, 3, 1) \rightarrow (3, 2, 1) \rightarrow (3, 1, 2)$. Skew-symmetry may then be expressed by the requirement that permutation of the variables (indices) has the effect of multiplying the tensor values (components) by the sign of the permutation.

Problem 2.18.1.   If $\pi = (i_1, \ldots, i_r)$ is a permutation of $(1, \ldots, r)$, define the *number of inversions* of π to be $s\pi = s_1 + s_2 + \cdots + s_r$, where $s_\alpha =$ the number of i_β such that $\beta < \alpha$ and $i_\beta > i_\alpha$. Thus $s_1 = 0$, since there are no $\beta < 1$, and in general, $s_\alpha < \alpha$. Show that

(a) If π differs from μ by the transposition of two adjacent indices, then $s\pi$ differs from $s\mu$ by 1 or -1.

(b) If i_α and i_β have k indices between them, the transposition of i_α and i_β can be accomplished by a sequence of $2k + 1$ transpositions of adjacent indices.

(c) If π differs from μ by a transposition of i_α and i_β which have k indices between them, then $s\pi - s\mu$ is an odd integer of magnitude $\leq 2k + 1$.

(d) If π and $(1, \ldots, r)$ are placed one above the other and equal symbols are joined by line segments, then the number of intersecting pairs of segments is $s\pi$. For example $\begin{array}{c}(4, 1, 3, 2)\\ \diagdown\kern-0.4em\diagup\kern-0.9em\diagdown\kern-0.4em\diagup \\ (1, 2, 3, 4)\end{array}$ has four intersections and

$$s(4, 1, 3, 2) = 0 + 1 + 1 + 2 = 4.$$

The *alternating operator* $A \to A_a$ is a linear function $T_s^0 \to \bigwedge^s V^*$, for each s, which assigns to each tensor its *skew-symmetric part*, using a formula similar to the symmetrizing operator except for signs. For $v_1, \ldots, v_s \in V$ we define

$$A_a(v_1, \ldots, v_s) = \frac{1}{s!} \sum_{(i_1, \ldots, i_s)} \mathrm{sgn}(i_1, \ldots, i_s) A(v_{i_1}, \ldots, v_{i_s}), \qquad (2.18.1)$$

where the sum runs over all $s!$ permutations of $(1, \ldots, s)$. It is easily checked that A_a is skew-symmetric. There is an obvious version for contravariant tensors as well. If A is already skew-symmetric, then $A - A_a$.

The *exterior product* is now defined by the formula

$$A \wedge B = (A \otimes B)_a,$$

where A and B are skew-symmetric covariant (or contravariant) tensors. It has the following properties.

(a) *Associativity.* $(A \wedge B) \wedge C = A \wedge (B \wedge C)$.

(b) *Anticommutativity.* If A is of degree p and B is of degree q, then

$$A \wedge B = (-1)^{pq} B \wedge A.$$

In particular, for all $\alpha, \beta \in V^*$, $\alpha \wedge \beta = -\beta \wedge \alpha$.

(c) *Distributivity.* $(A + B) \wedge C = A \wedge C + B \wedge C$.

The reader is asked to check these properties, at least in special cases.

In working with skew-symmetric tensors it is much more convenient to use the exterior product notation and its properties rather than regressing to a use of the tensor product symbol $\otimes$.

If $\{\varepsilon^i\}$ is a basis of V^*, then a basis of $\bigwedge^s V^*$ is given by $\{\varepsilon^{i_1} \wedge \cdots \wedge \varepsilon^{i_s}\}$, where $i_1, \ldots, i_s$ are arbitrary increasing sequences; that is,

$$1 \leq i_1 < \cdots < i_s \leq d.$$

For $d = 3$, the dimensions of $\bigwedge^0 V^*$, $\bigwedge^1 V^*$, $\bigwedge^2 V^*$, and $\bigwedge^3 V^*$ are 1, 3, 3, and 1, respectively. Since $\bigwedge^1 V^*$ and $\bigwedge^2 V^*$ have the same dimension, it is possible to choose an isomorphism between them; we shall see later how such an isomorphism arises naturally as a consequence of an inner product structure (Section 2.22). In fact, if ε^1, ε^2, ε^3 is a basis of V^*, then $\varepsilon^2 \wedge \varepsilon^3$, $\varepsilon^3 \wedge \varepsilon^1$, $\varepsilon^1 \wedge \varepsilon^2$ is a basis of $\bigwedge^2 V^*$, and we let them correspond in this order. That is, we make the correspondence $\varepsilon^i \leftrightarrow \varepsilon^j \wedge \varepsilon^k$ if (i, j, k) is an even permutation of $(1, 2, 3)$. When the wedge product is compounded with this isomorphism we get an operation just like the vector product in euclidean space. Indeed,

$$a_i\varepsilon^i \wedge b_j\varepsilon^j = (a_2b_3 - a_3b_2)\varepsilon^2 \wedge \varepsilon^3 + (a_3b_1 - a_1b_3)\varepsilon^3 \wedge \varepsilon^1 + (a_1b_2 - a_2b_1)\varepsilon^1 \wedge \varepsilon^2$$

$$\leftrightarrow \det \begin{pmatrix} \varepsilon^1 & \varepsilon^2 & \varepsilon^3 \\ a_1 & a_2 & a_3 \\ b_1 & b_2 & b_3 \end{pmatrix}.$$

Note that $a_ib_j - a_jb_i$ are the components of the vector product of the vectors $a_i\varepsilon^i$ and $b_i\varepsilon^i$. Recall, however, that the vector product is defined in a euclidean vector space, that is, when the concept of a length is given in addition to the vector space structure. In particular, the ε^i must be orthogonal unit vectors with the correct orientation.

It is only when d is 3 that the wedge product corresponds to the vector product, that is, to the product of vectors which yields a vector of the same type. For when $d \neq 3$, $\dim \bigwedge^2 V = \binom{d}{2} = d(d-1)/2 \neq d$. In spite of this the wedge product, insofar as integration theory (see Chapter 4) is concerned, is the proper generalization of the vector product.

Problem 2.18.2. For $\tau \in V^*$, $\theta \in \bigwedge^2 V^*$, $v, w, x \in V$ show that

$$\tau \wedge \theta(v, w, x) = [\tau(v)\theta(w, x) + \tau(w)\theta(x, v) + \tau(x)\theta(v, w)]/3.$$

Problem 2.18.3. Find the symmetric and skew-symmetric parts of

$$A = e_1 \otimes e_1 \otimes e_2 + e_3 \otimes e_1 \otimes e_1.$$

Must you know that e_1, e_2, and e_3 are linearly independent? Is A the sum of its symmetric and skew-symmetric parts?

Problem 2.18.4. A set $v_1, \ldots, v_p$ in V is linearly independent iff
$$v_1 \wedge \cdots \wedge v_p \neq 0.$$

Problem 2.18.5. If $\{e_i\}$ is a basis, $d \geq 4$, then the vectors $3e_1 + e_2 + 2e_3 + 2e_4$, $4e_1 + 5e_2 + 7e_3 + e_4$, and $-2e_1 + 3e_2 + 3e_3 - 3e_4$ are linearly dependent.

Problem 2.18.6. If $v \in V$, $v \neq 0$, and $f \in \bigwedge^p V$, then $v \wedge f = 0$ iff there is $g \in \bigwedge^{p-1} V$ such that $f = v \wedge g$. (*Hint:* Use a basis such that $v = e_1$.)

Problem 2.18.7. (Cartan's Lemma). Let $\{e_i\}$, $i = 1, \ldots, d$, be a basis of V, and let $v_i \in V$, $i = 1, \ldots, p$ such that $\sum_{i=1}^{p} e_i \wedge v_i = 0$. Then there are scalars A_{ij} such that

$$v_i = \sum_{j=1}^{p} A_{ij} e_j \quad \text{and} \quad A_{ij} = A_{ji}.$$

A tensor $A \in \bigwedge^p V$ is called *decomposable* if there are $v_1, \ldots, v_p \in V$ such that $A = v_1 \wedge \cdots \wedge v_p$. Otherwise A is called *indecomposable*.

Problem 2.18.8. If $\dim V \leq 3$, then every $A \in \bigwedge^p V$ is decomposable. If $\dim V > 3$ and $\{e_i\}$ is a basis, then $e_1 \wedge e_2 + e_3 \wedge e_4$ is indecomposable.

Problem 2.18.9. If $A \in \bigwedge^2 V$, then A is decomposable iff $A \wedge A = 0$, or equivalently, iff for all i, j_1, j_2, and j_3,

$$A^{ij_1} A^{j_2 j_3} - A^{ij_2} A^{j_1 j_3} + A^{ij_3} A^{j_1 j_2} = 0.$$

Problem 2.18.10. $A \in \bigwedge^3 V$ is decomposable iff

$$A^{i_1 i_2 j_1} A^{j_2 j_3 j_4} - A^{i_1 i_2 j_2} A^{j_1 j_3 j_4} + A^{i_1 i_2 j_3} A^{j_1 j_2 j_4} - A^{i_1 i_2 j_4} A^{j_1 j_2 j_3} = 0.$$

Problem 2.18.11. Generalize Problem 2.18.10 to the case $A \in \bigwedge^p V$.

Problem 2.18.12. All $A \in \bigwedge^{d-1} V$ are decomposable.

Problem 2.18.13. This collection of facts concerns the relation between subspaces of V and exterior algebra. Grassmann originally founded the subject because of these facts and a desire to study the structure of subspaces.

(a) If W is a p-dimensional subspace of V, then $\bigwedge^p W$ is a one-dimensional subspace of decomposable elements of $\bigwedge^p V$.

(b) If Y is a one-dimensional subspace of $\bigwedge^p V$ consisting only of decomposable elements, then $Y = \bigwedge^p W$ for some p-dimensional subspace W of V.

Let W and X be subspaces of V of dimensions p and q, respectively, $w \in \bigwedge^p W$, $x \in \bigwedge^q X$, $w \neq 0$, $x \neq 0$.

(c) $X \subset W$ iff there is a decomposable y such that $w = x \wedge y$. What freedom of choice is there for y?

(d) $X \cap W = 0$ iff $w \wedge x \neq 0$.

(e) If $X \cap W = 0$, then $w \wedge x$ is a basis of $\bigwedge^{p+q}(W + X)$.

(f) $W = \{v \mid v \in V \text{ and } v \wedge w = 0\}$.

Problem 2.18.14. Let B be a tensor of type $(0, 4)$ such that for every $v, w \in V$,

$$B(v, w, v, w) = -B(w, v, v, w) = -B(v, w, w, v).$$

(a) If $v \wedge w = x \wedge y$, then $B(v, w, v, w) = B(x, y, x, y)$.

(b) Is B necessarily skew-symmetric in the first two variables?

Problem 2.18.15. When acting on T_0^2 the symmetric and alternating operators are linear transformations $\mathscr{S}, \mathscr{A}: T_0^2 \to T_0^2$, and thus may be regarded as tensors, $\mathscr{S}, \mathscr{A} \in T_2^2$.

(a) Find the components of $\mathscr{S}, \mathscr{A} \in T_2^2$ with respect to a basis and show that they are the same with respect to every basis.

(b) If $C \in T_2^2$ is a tensor such that the components of C are the same with respect to every basis, show that there are scalars α, β such that $C = \alpha\mathscr{S} + \beta\mathscr{A}$.

2.19. Determinants

The reason for the use of exterior algebra in integration theory is the built-in determinant-producing feature which makes the appearance of the jacobian of a transformation (the jacobian determinant) automatic. We state this in the form of a theorem. But first we need a preliminary remark.

If W is a one-dimensional vector space, then a linear transformation of W into W is equivalent to multiplication by a scalar. Indeed, the matrix is a 1×1 matrix, obviously the same as a scalar. We wish to apply this to the one-dimensional space $\bigwedge^d V$, where $\dim V = d$.

If $A: V \to V$ is a linear function, then a *homomorphic extension* of A to skew-symmetric tensor spaces is a linear function $A: \bigwedge^p V \to \bigwedge^p V$, for each p, such that

$$A(v_1 \wedge \cdots \wedge v_p) = Av_1 \wedge \cdots \wedge Av_p \qquad (2.19.1)$$

for all $v_1, \ldots, v_p \in V$. Let $A\alpha = \alpha$ for $\alpha \in \bigwedge^0 V = R$. (Note that we have not distinguished between A and its extension notationally.)

Theorem 2.19.1. *For each linear function $A: V \to V$ there is a unique homomorphic extension.*

Proof. Let $\{e_i\}$ be a basis of V. For $i_1 < \cdots < i_p$, we define

$$A(e_{i_1} \wedge \cdots \wedge e_{i_p}) = Ae_{i_1} \wedge \cdots \wedge Ae_{i_p} \qquad (2.19.2)$$

and extend A uniquely to $\bigwedge^p V$ by linearity, à la Proposition 2.5.2. It is clear that (2.19.2) follows from (2.19.1), so that the homomorphic extension must certainly be unique and equal to the one we have defined. However, we have not shown existence, since that requires the satisfaction of (2.19.1) for all vectors v_i, not just some special e_i. For such arbitrary v_i we let $v_i = a_i^j e_j$.

Then by the properties of the wedge product (the distributive and associative laws),

$$v_1 \wedge \cdots \wedge v_p = a_1^{j_1} \cdots a_p^{j_p} e_{j_1} \wedge \cdots \wedge e_{j_p}.$$

Note that in this sum we do not have $j_1 < \cdots < j_p$, or even $j_\alpha \neq j_\beta$ if $\alpha \neq \beta$. If we group the terms and use anticommutativity we would get the components of $v_1 \wedge \cdots \wedge v_p$ with respect to the basis $\{e_{i_1} \wedge \cdots \wedge e_{i_p} \mid i_1 < \cdots < i_p\}$ of $\bigwedge^p V$. However, this is unnecessary, since we can use (2.19.2) and the linearity of A to show directly how to evaluate A on $e_{j_1} \wedge \cdots \wedge e_{j_p}$. Indeed, by linearity, since $A0 = 0$, we have the cases where some $j_\alpha = j_\beta$, $\alpha \neq \beta$, since then $e_{j_1} \wedge \cdots \wedge e_{j_p} = 0$ and $Ae_{j_1} \wedge \cdots \wedge Ae_{j_p} = 0$. For the case where the j_α are distinct, the same permutation π is required on the indices of $e_{j_1} \wedge \cdots \wedge e_{j_p}$ as on $Ae_{j_1} \wedge \cdots \wedge Ae_{j_p}$ to produce the increasing order, say $i_1, \ldots, i_p$, in which case

$$\begin{aligned} A(e_{j_1} \wedge \cdots \wedge e_{j_p}) &= A(\pm e_{i_1} \wedge \cdots \wedge e_{i_p}) \\ &= \pm A(e_{i_1} \wedge \cdots \wedge e_{i_p}) \\ &= \pm Ae_{i_1} \wedge \cdots \wedge Ae_{i_p} \\ &= Ae_{j_1} \wedge \cdots \wedge Ae_{j_p}. \end{aligned}$$

Here the $+$ sign is used if π is even, the $-$ sign if π is odd. This extension of (2.19.2) to the case of arbitrary $i_1, \ldots i_r$ now combines with linearity to give (2.19.1):

$$\begin{aligned} A(v_1 \wedge \cdots \wedge v_p) &= A\left(a_1^{j_1} \cdots a_p^{j_p} e_{j_1} \wedge \cdots \wedge e_{j_p}\right) \\ &= a_1^{j_1} \cdots a_p^{j_p} A(e_{j_1} \wedge \cdots \wedge e_{j_p}) \\ &= a_1^{j_1} \cdots a_p^{j_p} Ae_{j_1} \wedge \cdots \wedge Ae_{j_p} \\ &= (Aa_1^{j_1} e_{j_1}) \wedge \cdots \wedge (Aa_p^{j_p} e_{j_p}) \\ &= Av_1 \wedge \cdots \wedge Av_p. \quad\blacksquare \end{aligned}$$

Remark. The above definition and theorem can be easily modified for the more general linear function $A: V \to W$.

Theorem 2.19.2. *Let $A: V \to V$ be linear. Then the restriction of the homomorphic extension of A to $\bigwedge^d V$ consists of multiplying by the determinant $\cdot$ the matrix of A with respect to any basis. In particular, the determinant of the matrix of a tensor A of type $(1, 1)$ is an invariant.*

Proof. Let $\{e_i\}$ be a basis of V. Then $e_1 \wedge \cdots \wedge e_d$ is a basis of $\bigwedge^d V$, so that what we claim is

$$A(e_1 \wedge \cdots \wedge e_d) = \det (A_j^i) e_1 \wedge \cdots \wedge e_d,$$

where $Ae_j = A_j^i e_i$. Thus since A is homomorphic,

$$\begin{aligned} A(e_1 \wedge \cdots \wedge e_d) &= Ae_1 \wedge \cdots \wedge Ae_d \\ &= A_1^{i_1} e_{i_1} \wedge \cdots \wedge A_d^{i_d} e_{i_d} \\ &= A_1^{i_1} \cdots A_d^{i_d} e_{i_1} \wedge \cdots \wedge e_{i_d}. \end{aligned}$$

This sum has d^d terms. However, each term having two of the i_j's the same may be dropped, since then $e_{i_1} \wedge \cdots \wedge e_{i_d} = 0$. In the other terms, one for each permutation $(i_1, \ldots, i_d)$ of $(1, \ldots, d)$, the factors of $e_{i_1} \wedge \cdots \wedge e_{i_d}$ may be transposed, getting a $-$ sign for each transposition, until they appear in the order $e_1, \ldots, e_d$. The total sign produced is sgn $(i_1, \ldots, i_d)$. Hence the factor multiplying $e_1 \wedge \cdots \wedge e_d$ is

$$\sum_{(i_1, \ldots, i_d)} A_1^{i_1} \cdots A_d^{i_d} \text{ sgn } (i_1, \ldots, i_d), \qquad (2.19.3)$$

that is, the determinant of (A_j^i). ∎

Problem 2.19.1.   (a) Show that the coefficient of $e_2 \wedge \cdots \wedge e_d$ in $A(e_2 \wedge \cdots \wedge e_d)$ is the minor of (A_j^i) obtained by deleting the first row and column and taking the determinant.

(b) Obtain the column expansion of det (A_j^i) on the first column by considering the formula

$$A(e_1 \wedge \cdots \wedge e_d) = Ae_1 \wedge A(e_2 \wedge \cdots \wedge e_d).$$

The *determinant of a linear function* $A: V \to V$ is the determinant of any of its matrices: det $A = $ det (A_j^i).

Corollary 1.   (a) *Let A and B be linear functions $V \to V$. Then det $(A \circ B) = $ det A det B, where $A \circ B$ is the linear function given by $(A \circ B)v = A(Bv)$.*

(b) *The determinant of the product of two square matrices is the product of their determinants.*

Proof. (a) Apply B to $e \in \bigwedge^d V$ and then apply A to the result. Thus

$$\begin{aligned} A \circ Be &= A \text{ (det } B)e \\ &= (\text{det } B)Ae \\ &= (\text{det } B)(\text{det } A)e. \end{aligned}$$

But by the theorem $A \circ Be = (\text{det } A \circ B)e$, so det $A \circ B = $ det A det B.

The second part, (b), is left as an exercise. ∎

Problem 2.19.2.   In the proof of Corollary 1 it was assumed that the homomorphic extension of a composition is the composition of the homomorphic extensions. Find which step of the proof uses this and prove it.

Corollary 2.   *If we regard the determinant of a matrix to be a function of the columns $C_1 = (A_1^i), \ldots, C_d = (A_d^i)$, that is, det $(A_j^i) = f(C_1, \ldots, C_d)$, then f is determined by the properties*

(a) f is d-linear.

(b) *If the same column occurs twice the f-value is 0, or, equivalently, if two columns are interchanged the sign of the f-value is changed.*

(c) *The f-value of the identity matrix (δ^i_j) is 1.*

Proof. Consider the columns as members of the vector space R^d. Then (a) and (b) say that f is a skew-symmetric tensor of type $(0, d)$ over R^d, $f \in \bigwedge^d R^{d*}$. But $\bigwedge^d R^{d*}$ is one-dimensional, so f is determined by its component with respect to any basis, a single scalar. Hence, by (c), f is uniquely determined. ∎

Problem 2.19.3. If $A: V \to V$, then A may also be interpreted as a linear function $A^*: V^* \to V^*$. Indeed, A^* may be defined by $\langle Av, \tau \rangle = \langle v, A^*\tau \rangle$ for all $v \in V$ and $\tau \in V^*$. Using (2.19.3) or otherwise, show that $\det A^* = \det A$.

A^* is called the *dual of A*. Other names in common use are the *adjoint* and the *transpose* of A.

2.20.   Bilinear Forms

A *bilinear form* on V is a tensor of type $(0, 2)$, that is, a bilinear function $b: V \times V \to R$. According to Section 2.12, such a form may be interpreted in two ways as a linear function, $b_1: V \to V^*$ or $b_2: V \to V^*$. Specifically, if $\{e_i\}$ is a basis of V, $\{\varepsilon^i\}$ the dual basis, $b = b_{ij}\varepsilon^i \otimes \varepsilon^j$, and $v = v^i e_i \in V$, then

$$\begin{aligned} b_1 v &= b_{ij}\langle v, \varepsilon^i \rangle \varepsilon^j \\ &= (b_{ij}v^i)\varepsilon^j, \end{aligned}$$

and

$$\begin{aligned} b_2 v &= b_{ij}\varepsilon^i \langle v, \varepsilon^j \rangle \\ &= (b_{ij}v^j)\varepsilon^i. \end{aligned}$$

In classical language, the operation of passing from $v \in V$, with components v^i, to $b_1 v \in V^*$, with components $v_j = b_{ij}v^i$, is called *lowering the index of v by means of the bilinear form b*. This operation does not make much sense unless the indices can be raised again, that is, unless the function b_1 has an inverse.

If b_1 has an inverse, then b is called *nondegenerate*.

Other means of describing this important property are given in the following proposition.

Proposition 2.20.1. *A bilinear form b is nondegenerate iff*

(a) *For every $v \in V$, $v \neq 0$, there is some $w \in V$ such that $b(v, w) \neq 0$, or*

(b) *the matrix of components (b_{ij}) is nonsingular, that is, has an inverse matrix and/or determinant $(b_{ij}) \neq 0$, or*

(c) *b_2 has an inverse.*

Proof. The matrix of $b_1: V \to V^*$ with respect to bases $\{e_i\}$ and $\{\varepsilon^i\}$ is (b_{ij}); the matrix of b_2 is the transpose of that of b_1. From these facts we get the equivalence of nondegeneracy, **(b)**, and **(c)**.

If b_1 is nondegenerate, then for every $v \in V$, $v \neq 0$, we have $b_1 v \neq 0$. Hence there is $w \in V$ such that $\langle w, b_1 v \rangle \neq 0$; that is, $b(v, w) \neq 0$. Thus **(a)** holds.

Conversely, if **(a)** holds, then for every $v \in V$, $v \neq 0$, there is some $w \in V$ such that $\langle w, b_1 v \rangle = b(v, w) \neq 0$, so we must have $b_1 v \neq 0$. Thus b_1 maps nonzero elements into nonzero elements, and hence is an isomorphism because dim $V = $ dim V^*. ∎

We shall not be concerned much with general bilinear forms, only with symmetric and skew-symmetric ones. Let us note, however, that every bilinear form b may be written uniquely as a sum of a symmetric and a skew-symmetric one. So we have $b = b_s + b_a$, where

$$b_s(v, w) = [b(v, w) + b(w, v)]/2,$$
$$b_a(v, w) = [b(v, w) - b(w, v)]/2.$$

Problem 2.20.1.   Show that the determinant of (b_{ij}) is not an invariant of b, although the property of that determinant being nonzero, and indeed, either positive or negative, is invariant.

Problem 2.20.2.   Show by example that b_s and b_a can both be degenerate even though b is nondegenerate.

Problem 2.20.3.   If (b^{ij}) is the matrix inverse to (b_{ij}), show that the inverse to the operation of lowering indices by means of b is given by $v_j \to b^{ij} v_i$. The indices of any tensor may be raised or lowered by means of b. Show that if the indices of b^{ij} are lowered the result is b_{ij}.

2.21.   Quadratic Forms

A *quadratic form* on V is a quadratic invariant, that is, an invariant of degree 2, with variable in V, or, what is the same, a quadratic polynomial function on V.

To every quadratic form q there is an *associated symmetric bilinear form b*, defined by

$$b(v, w) = [q(v + w) - qv - qw]/2. \qquad (2.21.1)$$

Conversely, to every symmetric bilinear form b there is an *associated quadratic form q*, defined by

$$qv = b(v, v). \qquad (2.21.2)$$

Problem 2.21.1.   Show that each of the formulas (2.21.1) and (2.21.2) may be derived from the other.

In terms of a basis $\{\varepsilon^i\}$ of V^*, q is given by $qv = a_{ij}\langle v, \varepsilon^i\rangle\langle v, \varepsilon^j\rangle$, where $a_{ij} = a_{ji} \in R$, or we may simply write

$$q = a_{ij}\varepsilon^i\varepsilon^j,$$

where $\varepsilon^i\varepsilon^j$ is to be considered as a product of real-valued functions on V. On the other hand, if we view $\varepsilon^i\varepsilon^j$ as the symmetric product of the covariant vectors ε^i and ε^j then a formula with the same appearance gives the associated bilinear form b:

$$\begin{aligned}
b &= a_{ij}\varepsilon^i\varepsilon^j \\
&= a_{ij}(\varepsilon^i \otimes \varepsilon^j)_s \\
&= \tfrac{1}{2}a_{ij}(\varepsilon^i \otimes \varepsilon^j + \varepsilon^j \otimes \varepsilon^i) \\
&= a_{ij}\varepsilon^i \otimes \varepsilon^j,
\end{aligned}$$

since $a_{ij} = a_{ji}$.

A quadratic form q is *positive definite* if $qv > 0$ for every $v \neq 0$. We then say that b is positive definite also. A familiar example is the dot product in three-dimensional euclidean vector space:

$$q(a\mathbf{i} + b\mathbf{j} + c\mathbf{k}) = a^2 + b^2 + c^2.$$

With respect to the standard unit orthogonal basis $\mathbf{i}$, $\mathbf{j}$, $\mathbf{k}$, the matrix (a_{ij}) is (δ_{ij}), the identity matrix.

A quadratic form q is:

(a) *Negative definite if $qv < 0$ for every $v \neq 0$.*
(b) *Definite if q is either positive or negative definite.*
(c) *Positive semidefinite if $qv \geq 0$ for every v.*
(d) *Negative semidefinite if $qv \leq 0$ for every v,*
(e) *Semidefinite if q is either positive or negative semidefinite.*

The same terms are used for symmetric bilinear forms and for their component matrices, the definitions being given in terms of the associated quadratic form.

A nondegenerate symmetric bilinear form is called an *inner product*; sometimes this term is also taken to mean positive definite as well.

Proposition 2.21.1.   *A definite bilinear form is nondegenerate.*

Proof. For every $v \neq 0$, $b(v, v) \neq 0$, so there is w, namely, $w = v$ such that $b(v, w) \neq 0$. Thus by Proposition 2.20.1(a), b is nondegenerate. ∎

Examples.   On R^2 we have the following quadratic forms:

(a) Positive definite: $q(x, y) = x^2 + y^2$.
(b) Negative definite: $q(x, y) = -x^2 + xy - y^2$.

(c) Nondegenerate, indefinite: $q(x, y) = xy = \frac{1}{4}[(x + y)^2 - (x - y)^2]$, or $q(x, y) = x^2 - y^2$.

(d) Positive semidefinite, degenerate: $q(x, y) = x^2$.

Problem 2.21.2. Show that a nondegenerate semidefinite form is definite.

We say that $v, w \in V$ are *orthogonal* (*perpendicular*) *with respect to b* if $b(v, w) = 0$. If v is orthogonal to itself, that is, if $b(v, v) = 0$, then v is called a *null vector* of b. If b is definite, then the only null vector is 0. The converse is true by the following.

Proposition 2.21.2.   *If b is not definite, then there is a nonzero null vector.*

Proof. Since b is not positive definite, there is some $v \neq 0$ such that $b(v, v) \leq 0$. Similarly, there is a vector $w \neq 0$ such that $b(w, w) \geq 0$, since b is not negative definite. Consider the vectors of the form

$$z = \alpha v + (1 - \alpha)w,$$

where $0 \leq \alpha \leq 1$. These vectors are all nonzero unless v and w are linearly dependent, in which case $v = \beta w$, so then $b(v, v) = \beta^2 b(w, w) \geq 0$ and hence $b(v, v) = 0$. Otherwise

$$b(z, z) = \alpha^2 b(v, v) + 2\alpha(1 - \alpha)b(v, w) + (1 - \alpha)^2 b(w, w)$$

is a continuous function of α having values $b(w, w) \geq 0$ when $\alpha = 0$ and $b(v, v) \leq 0$ when $\alpha = 1$, so there is some α for which $b(z, z) = 0$. ∎

In a three-dimensional euclidean vector space, **i, j, k** form an *orthonormal basis*, that is, a set of mutually orthogonal vectors, each of unit length. The existence and use of such a basis leads to many computational simplifications. We ask, naturally, whether such a basis exists relative to a symmetric bilinear form b on an arbitrary vector space V of dimension d: that is, does there exist a basis $\{e_i\}$ of V such that $b(e_i, e_j) = \delta_{ij}$? Actually, this is a little too much to ask, and, in fact, implies that b is positive definite. To cover all cases we must allow a more general normal form, and accordingly give the following definition.

A basis $\{e_i\}$ of V is *orthonormal with respect to b* if (**a**) for $i \neq j$, $b(e_i, e_j) = 0$ and (**b**) each $b(e_i, e_i)$ (not summed on i) is one of the three values 1, -1, and 0.

The values $b(e_i, e_i)$ are called the *diagonal terms* of b and when the other components of b are all 0, b is said to be *diagonal* with respect to the basis $\{e_i\}$. A process for finding such bases is called *diagonalization*. In terms of such an orthonormal basis the associated quadratic form q assigns to v a sum and difference of squares of the components of v:

$$qv = \sum_i b(e_i, e_i)(v^i)^2.$$

Thus the procedures for finding orthonormal bases are also called *reducing a quadratic form to a sum and difference of squares.*

In terms of an orthonormal basis the interpretation of b as a function $b_1 = b_2: V \to V^*$ assumes as simple a component form as possible, since its matrix is $(b(e_i, e_j))$. What this means is that if $\{\varepsilon^i\}$ is the dual basis to an orthonormal basis $\{e_i\}$, then $b_1 e_i =$ either ε^i, $-\varepsilon^i$, or 0, depending on the value of $b(e_i, e_i)$. The relation $b_1 = b_2$ follows from the symmetry of b. The converse is also true: If $b_1 e_i =$ either ε^i, $-\varepsilon^i$, or 0, then $\{e_i\}$ is an orthonormal basis.

The main theorem of this section is the existence of an orthonormal basis. In the case of a positive (or negative) definite form an alternative proof in the form of an explicit diagonalization procedure, the *Gram-Schmidt process*, is also given.

Theorem 2.21.1.   *For every bilinear form b on V there is an orthogonal basis. The numbers of positive, negative, and zero diagonal components with respect to any orthonormal basis are the same, and hence are invariants of b.*

Proof. This will proceed by induction on d, the dimension of V. If $d = 1$, then either $b = 0$ and we take any basis for the orthonormal basis, or $b_{11} = b(f_1, f_1) \neq 0$ for some $f_1 \in V$. We let $e_1 = (1/\sqrt{|b_{11}|})f_1$ and an easy computation shows that $b(e_1, e_1) = \pm 1$.

Now suppose that every symmetric bilinear form on $d - 1$ or less dimensional vector spaces has an orthonormal basis, and that we are given b on a d-dimensional space V. If $b = 0$, then any basis of V is orthonormal. If $b \neq 0$, then we claim there is a vector $v \in V$ such that $b(v, v) \neq 0$. For indeed, there are vectors $v, w \in V$ such that $b(v, w) \neq 0$, and if both $b(v, v) = 0$ and $b(w, w) = 0$, then

$$b(v + w, v + w) = b(v, v) + 2b(v, w) + b(w, w)$$
$$= 2b(v, w) \neq 0.$$

Accordingly, let $v \in V$ be such that $\alpha = b(v, v) \neq 0$ and define $e_d = (1\sqrt{|\alpha|})v$, so that $b(e_d, e_d) = \pm 1$.

Now let $W = v^\perp = \{w \in V \mid b(v, w) = 0\}$, that is, the set of all vectors orthogonal to v, called "v perp." Then W is a subspace, for if $a \in R$, $w_1, w_2 \in W$, then

$$b(v, \alpha w_1) = \alpha b(v, w_1) = \alpha 0 = 0,$$
$$b(v, w_1 + w_2) = b(v, w_1) + b(v, w_2) = 0 + 0 = 0,$$

so $\alpha w_1 \in W$ and $w_1 + w_2 \in W$. Moreover, $W \neq V$, since $v \notin W$. Hence $\dim W < d$ and since the restriction of b to W is a symmetric bilinear form, our induction hypothesis gives a basis $e_1, \ldots, e_k$ of W such that $b(e_i, e_j) = a_i \delta_{ij}$, where each $a_i = 1, -1,$ or 0, $i, j = 1, \ldots, k$.

We claim that $k = d - 1$ and $e_1, \ldots, e_d$ is an orthonormal basis of V. The fact that $e_1, \ldots, e_k, e_d$ are orthonormal is clear from the construction, since $b(e_i, e_d) = 0$ for $i < d$ because $e_i \in W$. It remains to show that $\{e_i\}$ is a basis of V, for which it suffices to show they span V, since their number is $k + 1 \leq d$. Let $x \in V$, $\alpha = b(e_d, e_d)b(x, e_d)$, and $v = \beta e_d$. Then

$$b(x - \alpha e_d, v) = \beta[b(x, e_d) - \alpha b(e_d, e_d)]$$
$$= \beta b(x, e_d)(1 - b(e_d, e_d)^2)$$
$$= 0,$$

since $b(e_d, e_d)^2 = 1$. Thus $x - \alpha e_d \in W$, so that there are a^i such that $x = \sum_{i=1}^k a^i e_i + \alpha e_d$. This shows that $\{e_i\}$ spans V.

To show that the numbers of positive, negative, and zero diagonal components $b(e_i, e_i) = a_i$ are invariants of b, not depending on the choice of orthonormal basis, we give invariant characterizations of them.

(a) The number of $a_i = 0$ is the dimension of the subspace

$$N = \{w \mid b(v, w) = 0 \text{ for all } v \in V\},$$

the *null space* of b. Indeed the corresponding e_i are a basis of N: If $w \in N$, $w = w^i e_i$, then $0 = b(e_i, w) = a_i w^i$ (not summed on i), so $w^i = 0$ whenever $a_i \neq 0$; that is, w is a linear combination of those e_i for which $a_i = 0$. Conversely, if $a_i = 0$, then $e_i \in N$.

(b) The number of $a_i = 1$ is the dimension of a *maximal positive definite subspace* for b. Such subspaces are not unique unless b is positive definite or negative semidefinite, but among all the subspaces on which b is positive definite there must be some which have the largest dimension. Let W be such a subspace, $\{e_i\}$ an orthonormal basis which is numbered so that $a_1 = \cdots = a_k = 1$, $a_i \leq 0$ for $i > k$, and let X be the subspace spanned by $e_1, \ldots, e_k$. Then for any $v \in X$, $v = v^i e_i$, where $v^i = 0$ for $i > k$, and we have

$$b(v, v) = \sum_{i=1}^k (v^i)^2,$$

which is positive unless $v = 0$. Thus b is positive definite on X, and by the choice of W, $\dim W \geq \dim X = k$.

Now define a function $A : W \to X$ as follows. For $w \in W$, $w = w^i e_i$, let $Aw = \sum_{i=1}^k w^i e_i$. It is easily checked that A is linear. Suppose we have $Aw = 0$; that is, $w^i = 0$ for $i \leq k$. Then

$$b(w, w) = b(w^i e_i, w^j e_j)$$
$$= w^i w^j a_i \delta_{ij}$$
$$= \sum_{i=k+1}^d (w^i)^2 a_i$$
$$\leq 0$$

since $a_i \leq 0$ for $i > k$. Since b is positive definite on W, $w = 0$. Thus the only vector annihilated by A is 0, which proves that A is an isomorphism of W *into* X. Hence dim $W \leq$ dim X. Combining this with the previous inequality shows that dim $W = k$, as desired.

(c) The number of $a_i = -1$ is the dimension of a *maximal negative definite subspace* for b. The proof is the same as (b) except for obvious modifications. ∎

The number of $a_i = 0$, that is, the dimension of the null space N of b, is called the *nullity* of b. The number of $a_i = -1$ is called the *index* of b. If I is the index of b, $d - $ dim $N - 2I$ is called the *signature* of b. The signature is the difference between the number of $a_i = 1$ and the index.

In the proof by induction of the existence of an orthonormal basis, there is implicitly given a step-by-step construction of such a basis which may be actually carried out if the components of b with respect to some nonorthonormal basis $\{f_i\}$ are given. This construction is easier in the definite case since we do not encounter, at each step, the problem of finding some v such that $b(v, v) \neq 0$. If b is definite, any v will do, say $v = f_d$. However, we still must compute somehow the subspace $W = v^\perp$ at each stage; that is, we must find a basis for W. For this the formula $x - \alpha e_d \in W$ can be applied to $x = f_i$, $i < d$, to give a basis of W. This is essentially the *Gram-Schmidt orthonormalization process*, which in practice is carried out as follows, supposing that b is positive definite and $\{f_i\}$ is a basis of V.

Let

$$g_1 = f_1,$$
$$g_2 = f_2 - [b(f_2, g_1)/b(g_1, g_1)]g_1,$$
$$\vdots$$
$$g_i = f_i - \sum_{j=1}^{i-1} [b(f_i, g_j)/b(g_j, g_j)]g_j.$$

These g_i are mutually orthogonal and linearly independent since the f_i can be expressed in terms of them. The final step is simply to normalize them:

$$e_i = \alpha_i g_i, \qquad \text{where } 1/\alpha_i = \sqrt{b(g_i, g_i)}.$$

The advantage of waiting until the last step to normalize is that the taking of roots is delayed. Thus, if the $b(f_i, f_j)$ are all rational numbers, the whole process is carried out with rational numbers until the final step. For numerical computations with computers or desk calculators this is not much of an advantage, so that it is better to normalize at each step so as to make use of the simpler formula

$$g_i = f_i - \sum_{j=1}^{i-1} b(f_i, e_j)e_j.$$

Problem 2.21.3.   A subspace W of V is an *isotropy subspace* of b if $b(w, w) = 0$ for every $w \in W$.

(a) If W is an isotropy subspace, so is $W + N$.

(b) If s is the signature of b, then the dimension of a maximal isotropy subspace is $(d - |s| + \dim N)/2$.

(c) If $q(x, y, z) = x^2 + y^2 - z^2$, q a quadratic form on R^3, then the isotropy subspaces are the generators (lines through the vertex) of a cone. The maximal negative definite subspaces are the lines through the vertex of the cone and passing within the cone. The maximal positive definite subspaces are planes through the vertex which do not otherwise meet the cone.

Problem 2.21.4.   Reduce $q(x, y, z) = xy + yz + xz$ to a sum and difference of squares, finding an orthonormal basis, the index, the signature, and the nullity.

Problem 2.21.5.   Show that the index and nullity are a complete set of independent invariants for symmetric bilinear forms in the following sense. If b and c are symmetric bilinear forms on V having the same index and nullity, then there are bases $\{e_i\}$ and $\{f_i\}$ for which b and c have the same components, respectively; that is, $b(e_i, e_j) = c(f_i, f_j)$.

Problem 2.21.6.   Let b be a definite bilinear form on V and suppose that $v_1, \ldots, v_k$ are nonzero mutually orthogonal vectors. Show that $v_1, \ldots, v_k$ are linearly independent. Is this true if b is merely nondegenerate?

2.22.   Hodge Duality

We have noted in Section 2.18 that cross product of vectors in R^3 enjoys the same properties as the combination of wedge product and a correspondence between $\bigwedge^1 R^3$ and $\bigwedge^2 R^3$. In this section we shall show how to generalize this correspondence. As in the case of R^3, it will depend on an inner product.

The dimensions of the skew-symmetric tensor spaces over a vector space V of dimension d have a symmetry, $\binom{d}{p} = \binom{d}{d - p}$. The Hodge star operator is an isomorphism between the pairs of these spaces of equal dimension:

$$*: \bigwedge^p V \to \bigwedge^{d-p} V.$$

We assume that V is provided with a positive definite inner product b and an orientation.

An *orientation* of V is given by a nonzero element θ of $\bigwedge^d V$. If such an orientation is given, we divide the ordered bases of V into two classes, those in the orientation and those not. An ordered basis $(e_1, \ldots, e_d)$ is *in the orientation given by* θ if $e_1 \wedge \cdots \wedge e_d = \alpha\theta$, where $\alpha > 0$. Any positive multiple of θ will divide the bases in the same way, so that we say that θ and $\alpha\theta$ give the same orientation of V if $\alpha > 0$. The orientation itself is the collection of all ordered bases in the orientation. Any two such bases are related by a matrix having positive determinant, and if two bases are related by a matrix having positive determinant they are either both in the orientation or both not in it. There are clearly just two orientations.

If $(e_1, \ldots, e_d)$ is an ordered orthonormal basis in the orientation, then $e_1 \wedge \cdots \wedge e_d$ is the *volume element of the oriented vector space with inner product b*. We are justified in writing *the* volume element because it is unique. Indeed, if $(f_1, \ldots, f_d)$ is any other ordered orthonormal basis in the orientation, then

$$f_i = a_i^j e_j,$$
$$\begin{aligned} b(f_i, f_j) &= \delta_{ij} \\ &= b(a_i^h e_h, a_j^k e_k) \\ &= a_i^h a_j^k b(e_h, e_k) \\ &= a_i^h a_j^k \delta_{hk} \\ &= a_i^h a_j^h. \end{aligned}$$

Thus the inverse of the matrix (a_j^i) is its transpose (a_i^j); that is (a_j^i) is an *orthogonal matrix*. Since the determinant of the transpose of a matrix is equal to the determinant of the matrix,

$$\begin{aligned} \det (\delta_{ij}) &= 1 \\ &= \det (a_i^j) \det (a_j^i) \\ &= [\det (a_j^i)]^2, \end{aligned}$$

so that $\det (a_j^i) = 1$ or -1. But $(f_1, \ldots, f_d)$ and $(e_1, \ldots, e_d)$ are in the orientation, so $\det (a_j^i) > 0$. Hence $\det (a_j^i) = 1$ and

$$\begin{aligned} f_1 \wedge \cdots \wedge f_d &= \det (a_j^i) \, e_1 \wedge \cdots \wedge e_d \\ &= e_1 \wedge \cdots \wedge e_d. \end{aligned}$$

We define the *Hodge star operator* by specifying it first on the basis of $\bigwedge^p V$ obtained from an ordered orthonormal basis $(e_1, \ldots, e_d)$ in the orientation of V. (Actually there is one operator for each $p = 0, \ldots, d$.) We then show that it is independent of the choice of such basis. A typical basis element of $\bigwedge^p V$ is $e_{i_1} \wedge \cdots \wedge e_{i_p}$. Let $j_1, \ldots, j_{d-p}$ be chosen so that

$$(i_1, \ldots, i_p, j_1, \ldots, j_{d-p})$$

is an even permutation of $(1, \ldots, d)$. Then

$$*(e_{i_1} \wedge \cdots \wedge e_{i_p}) = e_{j_1} \wedge \cdots \wedge e_{j_{d-p}}. \tag{2.22.1}$$

An even permutation of $i_1, \ldots, i_p$ or of $j_1, \ldots, j_{d-p}$ will not effect either $e_{i_1} \wedge \cdots \wedge e_{i_p}$ or $e_{j_1} \wedge \cdots \wedge e_{j_{d-p}}$, so that the definition is independent of the choices of orders of the indices. We extend $*$ to be a linear transformation.

To show that $*$ does not depend upon the choice of $(e_1, \ldots, e_d)$ we decompose $*$ into the composition of maps which are independent of bases choices.

(a) Let $F: \bigwedge^p V \to (\bigwedge^{d-p} V)^*$ be defined by requiring that for $x \in \bigwedge^p V$, $y \in \bigwedge^{d-p} V$,

$$\langle y, Fx \rangle \theta = x \wedge y,$$

where $\theta = e_1 \wedge \cdots \wedge e_d$.

(b) We define $G: (\bigwedge^k V)^* \to \bigwedge^k (V^*)$, as follows. Let $\{e_i\}$ be any basis of V, $\{\varepsilon^i\}$ the dual basis. Then $\{\varepsilon^{i_1} \wedge \cdots \wedge \varepsilon^{i_k} \mid i_1 < \cdots < i_k\}$ is a basis of $\bigwedge^k (V^*)$; $\{e_{i_1} \wedge \cdots \wedge e_{i_k}\}$ is a basis of $\bigwedge^k V$ and so has a dual basis of $(\bigwedge^k V)^*$, $\{\varepsilon^{i_1 \cdots i_k}\}$. Let $G\varepsilon^{i_1 \cdots i_k} = \varepsilon^{i_1} \wedge \cdots \wedge \varepsilon^{i_k}$, and extend G by linearity. We show that G is independent of the choice of basis. Let

$$f_i = a_i^j e_j$$

be another basis and $\{\varphi^i\}$ the dual basis.

For any $A \in (\bigwedge^k V)^*$, $GA \in \bigwedge^k V^*$, so GA is a skew-symmetric k-linear function on $V^{**} = V$. In particular, for $A = \varepsilon^{i_1 \cdots i_k}$, if $i_1 < \cdots < i_k$ and $h_1 < \cdots < h_k$, then by Section 2.11 and (2.18.1),

$$GA(e_{h_1}, \ldots, e_{h_k}) = \varepsilon^{i_1} \wedge \cdots \wedge \varepsilon^{i_k}(e_{h_1}, \ldots, e_{h_k})$$

$$= \frac{1}{k!} \sum_\pi \operatorname{sgn} \pi \langle e_{h_{\pi 1}}, \varepsilon^{i_1} \rangle \cdots \langle e_{h_{\pi k}}, \varepsilon^{i_k} \rangle$$

$$= \frac{1}{k!} \delta_{h_1}^{i_1} \cdots \delta_{h_k}^{i_k}$$

$$= \frac{1}{k!} \langle e_{h_1} \wedge \cdots \wedge e_{h_k}, A \rangle.$$

Both sides of this equation are skew-symmetric in $h_1, \ldots, h_k$, so it follows that it is valid for any $A \in (\bigwedge^k V)^*$. Then we have for $A \in (\bigwedge^k V)^*$,

$$GA(f_{h_1}, \ldots, f_{h_k}) = GA(a_{h_1}^{j_1} e_{j_1}, \ldots, a_{h_k}^{j_k} e_{j_k})$$

$$= a_{h_1}^{j_1} \cdots a_{h_k}^{j_k} GA(e_{j_1}, \ldots, e_{j_k})$$

$$= \frac{1}{k!} a_{h_1}^{j_1} \cdots a_{h_k}^{j_k} \langle e_{j_1} \wedge \cdots \wedge e_{j_k}, A \rangle$$

$$= \frac{1}{k!} \langle a_{h_1}^{j_1} e_{j_1} \wedge \cdots \wedge a_{h_k}^{j_k} e_{j_k}, A \rangle$$

$$= \frac{1}{k!} \langle f_{h_1} \wedge \cdots \wedge f_{h_k}, A \rangle.$$

In particular, if $\{\varphi^{i_1\cdots i_k}\}$ is the dual basis to $\{f_{i_1}\wedge\cdots\wedge f_{i_k}\}$, this equation also holds for $A=\varphi^{i_1\cdots i_k}$. But we also have

$$\varphi^{i_1}\wedge\cdots\wedge\varphi^{i_k}(f_{h_1},\ldots,f_{h_k})=\frac{1}{k!}\delta^{i_1}_{h_1}\cdots\delta^{i_k}_{h_k}$$
$$=\frac{1}{k!}\langle f_{h_1}\wedge\cdots\wedge f_{h_k},\varphi^{i_1\cdots i_k}\rangle.$$

Since $G\varphi^{i_1\cdots i_k}$ and $\varphi^{i_1}\wedge\cdots\wedge\varphi^{i_k}$ coincide as multilinear functions, they are equal, which shows the desired independence of choice.

(c) Since b is a nondegenerate symmetric bilinear form on V, it may be reinterpreted as a nonsingular linear function $b_1\colon V\to V^*$. The inverse map b_1^{-1} has an extension to a homomorphism of the exterior algebras, giving us $B\colon\bigwedge^k V^*\to\bigwedge^k V$.

We now show that $*=B\circ G\circ F$. It is sufficient to check equality on any basis of $\bigwedge^p V$, in particular on $\{e_{i_1}\wedge\cdots\wedge e_{i_p}\}$, where (e_i) is an oriented orthonormal basis of V. Letting $k=d-p$, $i_1<\cdots<i_p$, and $j_1,\ldots,j_p$ be as in the definition of $*$,

$$\theta=\langle e_{j_1}\wedge\cdots\wedge e_{j_k},\varepsilon^{j_1\cdots j_k}\rangle\theta\quad\text{(not summed)}$$
$$=(e_{i_1}\wedge\cdots\wedge e_{i_p})\wedge e_{j_1}\wedge\cdots\wedge e_{j_k},$$

while for $h_1,\ldots,h_k$ not a permutation of $j_1,\ldots,j_k$,

$$\langle e_{h_1}\wedge\cdots\wedge e_{h_k},\varepsilon^{j_1\cdots j_k}\rangle\theta=0$$
$$=(e_{i_1}\wedge\cdots\wedge e_{i_p})\wedge e_{h_1}\wedge\cdots\wedge e_{h_k}.$$

Thus for

$$y=a^{h_1\cdots h_k}e_{h_1}\wedge\cdots\wedge e_{h_k}\in\bigwedge^k V,$$
$$\langle y,\varepsilon^{j_1\cdots j_k}\rangle\theta=e_{i_1}\wedge\cdots\wedge e_{i_p}\wedge y,$$

so we must have

$$Fe_{i_1}\wedge\cdots\wedge e_{i_p}=\varepsilon^{j_1\cdots j_k}.$$

By definition $G\varepsilon^{j_1\cdots j_k}=\varepsilon^{j_1}\wedge\cdots\wedge\varepsilon^{j_k}$.
Finally,

$$B\varepsilon^{j_1}\wedge\cdots\wedge\varepsilon^{j_k}=b_1^{-1}\varepsilon^{j_1}\wedge\cdots\wedge b_1^{-1}\varepsilon^{j_k}$$
$$=e_{j_1}\wedge\cdots\wedge e_{j_k}.$$

Thus $B\circ G\circ F$ coincides on a basis with $*$, so they are equal. ∎

The composition of $*$ with itself is a map which preserves degrees: $*\circ*\colon\bigwedge^p V\to\bigwedge^p V$. More than that, on $\bigwedge^p V$, $*\circ*$ is simply the identity or its negative, depending on p and d. For, if $(i_1,\ldots,i_p,j_1,\ldots,j_k)$ is an even permutation of $(1,\ldots,d)$, then $(j_1,\ldots,j_k,i_1,\ldots,i_p)$ is also a permutation of $(1,\ldots,d)$ which is even or odd depending on whether $pk=p(d-p)$ is even or odd, since each j must be transposed with each i to pass from permutation

$(i_1, \ldots, i_p, j_1, \ldots, j_k)$ to $(j_1, \ldots, j_k, i_1, \ldots, i_p)$, giving a total of pk transpositions. Thus if $*(e_{i_1} \wedge \cdots \wedge e_{i_p}) = e_{j_1} \wedge \cdots \wedge e_{j_k}$, then $*(e_{j_1} \wedge \cdots \wedge e_{j_k}) = (-1)^{pk} e_{i_1} \wedge \cdots \wedge e_{i_p}$. Thus we have proved

Theorem 2.22.1. *The composition of $*$ with itself, $* \circ *$, equals $(-1)^{p(d-p)} I_p$ on $\bigwedge^p V$, where I_p is the identity on $\bigwedge^p V$. In particular, if d is odd, $*$ is its own inverse. If d is even, $*$ is its own inverse on the spaces $\bigwedge^p V$ with even degree p, the negative of its inverse on spaces $\bigwedge^p V$ with odd degree p.*

Problem 2.22.1. At each point of E^3, euclidean space, (dx, dy, dz) is an oriented orthonormal basis of the dual space to the tangent space (covariant vectors). At points not on the z axis, spherical coordinates ρ, φ, θ are an admissible coordinate system (when suitably restricted), so that $(d\rho, d\varphi, d\theta)$ also is a basis of the covariant vector space. They are orthogonal but not normal. The normalizing factors are the lengths of the contravariant basis vectors $\partial/\partial\rho$, $\partial/\partial\varphi$, and $\partial/\partial\theta$, which can be found geometrically by letting each coordinate vary in turn at unit rate with the others fixed and observing the speed of motion. Find the normalizing factors and thus compute $*$ on $\bigwedge^1$ in terms of the bases $(d\rho, d\varphi, d\theta)$ of $\bigwedge^1$ and $(d\rho \wedge d\varphi, d\varphi \wedge d\theta, d\theta \wedge d\rho)$ of $\bigwedge^2$. Do the same for cylindrical coordinates r, θ, z.

On $\bigwedge^0 = R$, $*$ maps 1 into the volume element, $*1 = e_1 \wedge \cdots \wedge e_d$, and on $\bigwedge^d$, $*(e_1 \wedge \cdots \wedge e_d) = 1 \in \bigwedge^0$.

To obtain the $*$ operation for nondegenerate indefinite quadratic forms we obtain orthonormal bases by extending the scalar field to the complex numbers. Then if $(e_1, \ldots, e_d)$ is an orthonormal basis in the sense previously given, with $b(e_j, e_j) = -1$ for $1 \le j \le I$, where I is the index of b, $(ie_1, \ldots ie_I, e_{I+1}, \ldots, e_d)$, where $i^2 = -1$, will be an orthonormal basis having $b(ie_j, ie_j) = b(e_k, e_k) = 1$ for $1 \le j \le I$ and $k > I$. The definition of $*$ will then proceed as before. However, it may happen that $*$ maps real vectors into complex ones, as in the following.

Problem 2.22.2. The space-time continuum of special relativity is R^4 with coordinates x, y, z, t and a quadratic form on the covariant vector spaces having index 1—the one for which (dx, dy, dz, idt) is an orthonormal basis. The volume element is then $idx \wedge dy \wedge dz \wedge dt$. Compute $*$ on $\bigwedge^1, \bigwedge^2$, and $\bigwedge^3$ in terms of the real basis elements $dx, dy, dz, dt, dx \wedge dy$, etc.

2.23. Symplectic Forms

The *rank* of a skew-symmetric bilinear form is the minimum number of vectors in terms of which it can be expressed. We may think of a skew-symmetric bilinear form b on V as being in $\bigwedge^2 V^*$. If b can be written in terms of

$\varepsilon^1, \ldots, \varepsilon^r$, then we may discard any dependent ε^i's and extend to a basis, getting

$$b = b_{ij}\varepsilon^i \otimes \varepsilon^j, \qquad \text{where } b_{ij} = 0 \text{ unless } i, j \leq r,$$
$$= b_{ij}\varepsilon^i \wedge \varepsilon^j, \qquad \text{since } b_{ij} = -b_{ji},$$

so it does not matter, in the definition of rank, whether the mode of expressing b is in terms of tensor products or exterior products.

If we let W be the subspace spanned by $\varepsilon^1, \ldots, \varepsilon^r$, then $b \in \bigwedge^2 W$. If k is any integer such that $2k > r$, then the k-*fold exterior product* of b,

$$\bigwedge^k b = b \wedge \cdots \wedge b \in \bigwedge^{2k} W,$$

is zero, so we have proved

Proposition 2.23.1.   *If the rank of b is r and $2k > r$, then $\bigwedge^k b = 0$.*

If $\varepsilon^1, \ldots, \varepsilon^{2p}$ are linearly independent and we let

$$b = \varepsilon^1 \wedge \varepsilon^2 + \varepsilon^3 \wedge \varepsilon^4 + \cdots + \varepsilon^{2p-1} \wedge \varepsilon^{2p},$$

then the p-fold product of b is

$$\bigwedge^p b = p!\varepsilon^1 \wedge \varepsilon^2 \wedge \cdots \wedge \varepsilon^{2p} \neq 0. \tag{2.23.1}$$

By Proposition 2.23.1, b has rank $r \geq 2p$, but b is expressed in terms of $2p$ vectors, so $r \leq 2p$. Thus we have

Proposition 2.23.2.   *If $\varepsilon^1, \ldots, \varepsilon^{2p}$ are linearly independent, then the rank of* $b = \varepsilon^1 \wedge \varepsilon^2 + \varepsilon^3 \wedge \varepsilon^4 + \cdots + \varepsilon^{2p-1} \wedge \varepsilon^{2p}$ *is $2p$.*

Problem 2.23.1.   Prove formula (2.23.1).

Now suppose that r is the rank of b, and $\varepsilon^1, \ldots, \varepsilon^r$ are such that

$$b = \sum_{i<j} a_{ij}\varepsilon^i \wedge \varepsilon^j,$$

where $a = a_{12} \neq 0$. Then $b = \varepsilon^1 \wedge \sum_{1<j} a_{1j}\varepsilon^j + \varepsilon^2 \wedge \sum_{2<j} a_{2j}\varepsilon^j +$ terms involving $\varepsilon^3, \ldots, \varepsilon^r = \varepsilon^1 \wedge \varphi^1 + \varepsilon^2 \wedge \varphi^2 +$ terms involving $\varepsilon^3, \ldots, \varepsilon^r$, where φ^2 has only terms in $\varepsilon^3, \ldots, \varepsilon^r$. But $\varphi^1 = a\varepsilon^2 + \varphi^3$, where φ^3 has only terms in $\varepsilon^3, \ldots, \varepsilon^r$, so that

$$\varepsilon^2 = c\varphi^1 - c\varphi^3, \qquad \text{where } c = 1/a.$$

Thus

$$b = \varepsilon^1 \wedge \varphi^1 + c\varphi^1 \wedge \varphi^2 - c\varphi^3 \wedge \varphi^2 + \text{terms in } \varepsilon^3, \ldots, \varepsilon^r$$
$$= (\varepsilon^1 - c\varphi^2) \wedge \varphi^1 + \text{terms in } \varepsilon^3, \ldots, \varepsilon^r$$
$$= \alpha^1 \wedge \alpha^2 + b_1,$$

where b_1 has rank $r - 2$. Continuing in this way we obtain, for every k such that $2k \leq r$,

$$b = \alpha^1 \wedge \alpha^2 + \alpha^3 \wedge \alpha^4 + \cdots + \alpha^{2k-1} \wedge \alpha^{2k} + b_k,$$

where b_k has rank $r - 2k$. In particular, we can continue until $r - 2k = 0$ or 1. But it is not possible for the rank of $b_k \in \bigwedge^2 V^*$ to be 1, for the only thing expressible in terms of one $\alpha \in V^*$ is a multiple of $\alpha \wedge \alpha = 0$. We include this result in the following.

Theorem 2.23.1.   *If b is a skew-symmetric bilinear form of rank r, then*

(a) *r is even, $r = 2p$ for some p.*
(b) *There are linearly independent $\varepsilon^1, \ldots, \varepsilon^{2p}$ such that*

$$b = \varepsilon^1 \wedge \varepsilon^{p+1} + \varepsilon^2 \wedge \varepsilon^{p+2} + \cdots + \varepsilon^p \wedge \varepsilon^{2p}.$$

(c) $\bigwedge^{p+1} b = 0$ *and* $\bigwedge^p b \neq 0$.
(d) *The range of b when viewed as a linear function $b_1 : V \to V^*$ is 2p-dimensional and is spanned by $\varepsilon^1, \ldots, \varepsilon^{2p}$.*

Proof. All except (d) follow easily from the previous results. For (d) we extend $\{\varepsilon^i\}$ to a basis and let $\{e_i\}$ be the dual basis of V. Then

$$\begin{aligned} b_1 e_i &= (\varepsilon^1 \wedge \varepsilon^{p+1} + \cdots + \varepsilon^p \wedge \varepsilon^{2p})_1 e_i \\ &= \tfrac{1}{2}[\varepsilon^1 \otimes \varepsilon^{p+1} - \varepsilon^{p+1} \otimes \varepsilon^1 + \cdots]_1 e_i \\ &= \begin{cases} \tfrac{1}{2}\varepsilon^{i+p} & \text{if } i \leq p \\ -\tfrac{1}{2}\varepsilon^{i-p} & \text{if } i > p \\ 0 & \text{if } i > 2p. \end{cases} \end{aligned}$$

The space spanned by the values of b_1 on a basis is the range of b_1 and is clearly the span of $\varepsilon^1, \ldots, \varepsilon^{2p}$. ∎

Corollary.   *The only invariant of a skew-symmetric bilinear form under an arbitrary change of basis is its rank; that is, if b and c are two such forms, then there are bases $\{e_i\}$ and $\{f_i\}$ of V such that $b(e_i, e_j) = c(f_i, f_j)$ iff the ranks of b and c are equal.*

A *symplectic form* is a skew-symmetric bilinear form of maximal rank; thus the rank will be d if d is even, $d - 1$ if d is odd.

A *symplectic basis* for a symplectic form b is a basis $\{\varepsilon^i\}$ such that

$$b = \varepsilon^1 \wedge \varepsilon^{p+1} + \cdots + \varepsilon^p \wedge \varepsilon^{2p}.$$

The change of basis matrix between two symplectic bases must satisfy certain relations. This is analogous to the change of basis matrix between two orthonormal bases of a positive definite quadratic form, which must be an orthogonal matrix; that is, it satisfies the relation $AA^* = I$, where A^* is the transpose of the matrix A, obtained by interchanging the rows and columns of A, and I is the identity matrix.

To find the relations satisfied by a change of basis which leaves invariant the normal form of a skew-symmetric form $b = \sum_{i=1}^{p} \varepsilon^i \wedge \varepsilon^{i+p}$ we introduce a summation convention in which indices range over the following values:

$$i, j = 1, \ldots, p,$$
$$h, k = p + 1, \ldots, 2p,$$
$$m, n = 2p + 1, \ldots, d.$$
$$\alpha = 1, \ldots, d.$$

The change of basis matrix will be split into blocks corresponding to these ranges of indices:

$$\begin{pmatrix} A & B \\ C & D & G \\ E & F \end{pmatrix}, \qquad A = (a_i^j), \quad B = (b_i^k), \quad C = (c_h^j), \quad \text{etc.};$$

that is, the new basis is

$$\varphi^j = a_i^j \varepsilon^i + c_h^j \varepsilon^h + e_m^j \varepsilon^m,$$
$$\varphi^k = b_i^k \varepsilon^i + d_h^k \varepsilon^h + f_m^k \varepsilon^m,$$
$$\varphi^n = g_\alpha^n \varepsilon^\alpha.$$

Now if we are also to have $b = \sum_{u=1}^{p} \varphi^u \wedge \varphi^{u+p}$, then

$$b = \sum_{u=1}^{p} (a_i^u \varepsilon^i + c_h^u \varepsilon^h + e_m^u \varepsilon^m) \wedge (b_j^{u+p} \varepsilon^j + d_k^{u+p} \varepsilon^k + f_n^{u+p} \varepsilon^n)$$

$$= \sum_{u=1}^{p} [a_i^u b_j^{u+p} \varepsilon^i \wedge \varepsilon^j + c_h^u d_k^{u+p} \varepsilon^h \wedge \varepsilon^k + e_m^u f_n^{u+p} \varepsilon^m \wedge \varepsilon^n$$
$$+ (a_i^u d_h^{u+p} - b_i^{u+p} c_h^u)\varepsilon^i \wedge \varepsilon^h + (a_i^u f_m^{u+p} - b_i^{u+p} e_m^u)\varepsilon^i \wedge \varepsilon^m$$
$$+ (c_h^u f_m^{u+p} - d_h^{u+p} e_m^u)\varepsilon^h \wedge \varepsilon^m + e_m^u f_n^{u+p} \varepsilon^m \wedge \varepsilon^n]$$

$$= \sum_{u=1}^{p} \varepsilon^u \wedge \varepsilon^{u+p}.$$

Equating coefficients of $\varepsilon^i \wedge \varepsilon^j$, $i < j$, gives

$$\sum_{u=1}^{p} (a_i^u b_j^{u+p} - b_i^{u+p} a_j^u) = 0.$$

Since this trivialy holds for $i = j$ and for $j < i$ by skew-symmetry, we may write it in matrix terms as

$$AB^* - BA^* = 0.$$

Similarly, equating coefficients of $\varepsilon^i \wedge \varepsilon^h$, $\varepsilon^i \wedge \varepsilon^m$, and $\varepsilon^h \wedge \varepsilon^m$ gives matrix equations

$$AD^* - BC^* = I,$$
$$CD^* - DC^* = 0,$$
$$AF^* - BE^* = 0,$$
$$CF^* - DE^* = 0.$$

From these we deduce that [since $(RS)^* = S^*R^*$] the inverse of the block with A, B, C, and D can be expressed in terms of its own elements:

$$\begin{pmatrix} A & B \\ C & D \end{pmatrix}\begin{pmatrix} D^* & -B^* \\ -C^* & A^* \end{pmatrix} = \begin{pmatrix} AD^* - BC^* & -AB^* + BA^* \\ CD^* - DC^* & -CB^* + DA^* \end{pmatrix}$$

$$= \begin{pmatrix} I & 0 \\ 0 & I \end{pmatrix}.$$

(We have used the fact that we are justified in multiplying compatible block matrices using the rule for 2×2 matrices.) Moreover, if $H = \begin{pmatrix} A & B \\ C & D \end{pmatrix}$, then $H\begin{pmatrix} F^* \\ -E^* \end{pmatrix} = 0$, so that $\begin{pmatrix} F^* \\ -E^* \end{pmatrix} = H^{-1}H\begin{pmatrix} F^* \\ -E^* \end{pmatrix} = 0.$

A *symplectic matrix* is a $2p \times 2p$ matrix of the form $\begin{pmatrix} A & B \\ C & D \end{pmatrix}$ such that its inverse is $\begin{pmatrix} D & -C \\ -B & A \end{pmatrix}^*$, where A, B, C, and D are $p \times p$ matrices.

Theorem 2.23.2. *A change of basis matrix for which the change of basis leaves invariant the normal form* $b = \sum_{i=1}^{p} \varepsilon^i \wedge \varepsilon^{i+p}$ *of a rank* p *skew-symmetric bilinear form is a matrix of the form* $\begin{pmatrix} H & G \\ 0 & \end{pmatrix}$, *where* H *is a symplectic* $2p \times 2p$ *matrix.*

Problem 2.23.2. Show that the product and inverse of symplectic matrices are symplectic, both by actual computation and by an argument from change of basis considerations.

Problem 2.23.3. A complex matrix $U = A + iB$ is *unitary*, where A and B are real, if the inverse of U is $U^{-1} = U^* = A^* - iB^*$. Show that if $\begin{pmatrix} A & B \\ C & D \end{pmatrix}$ is both symplectic and orthogonal, then $A + iB$ is unitary, and, conversely, if $A + iB$ is unitary, then $\begin{pmatrix} A & B \\ -B & A \end{pmatrix}$ is both symplectic and orthogonal.

CHAPTER 3

Vector Analysis on Manifolds

3.1. Vector Fields

A *vector field* X on a subset E of a manifold M is a function which assigns to each $m \in E$ a vector $X(m)$ at m, so $X(m) \in M_m$. The domain of X is E and the range space is the tangent bundle TM of M, defined in Section 1.8 to be the collection of all tangents at all points of M.

If U is a coordinate neighborhood with coordinates x^i, then (by Theorem 1.7.1) at each $m \in U$, the $\partial_i(m)$ form a basis for M_m. Thus if $m \in E$, there are real numbers $X^i m$, the components of $X(m)$, such that $X(m) = (X^i m) \, \partial_i(m)$. As we let m vary through $E \cap U$, $m \to X^i m$ defines d real-valued functions X^i on $E \cap U$, the *components of X with respect to the coordinates x^i.*

If V is another coordinate neighborhood with coordinates y^i and Y^i are the components of X with respect to y^i, then on $E \cap U \cap V$, by the law-of-change formula (1.7.1),

$$Y^i = X^j \frac{\partial y^i}{\partial x^j}.$$

If f is a C^∞ real-valued function defined on an open set W of M, then Xf is the real-valued function defined on $W \cap E$ by

$$(Xf)m = X(m)f$$

for every $m \in W \cap E$. Just as single tangents were defined as operators on C^∞ functions to real numbers, vector fields could be defined directly as operators on C^∞ functions to real-valued functions which satisfy the linearity and product rules:

(a) $X(af + bg) = aXf + bXg$,
(b) $X(fg) = (Xf)g + fXg$.

Here, (b) is actually simpler because the right side consists of function products and sums, not products and sums of function values. However, Xf is not necessarily C^∞; indeed, E need not be an open set, so that differentiability of Xf would fail on that account.

On $U \cap W \cap E$ the expression for Xf is $Xf = X^i \, \partial_i f$.

A vector field X is C^∞ if its domain E is open and for every C^∞ function f, Xf is also a C^∞ function.

The components of X are $X^i = Xx^i$, so that if X is C^∞, its components are C^∞, since the x^i are C^∞. Conversely, if X has C^∞ components $X^i = Xx^i$ with respect to every coordinate system x^i, then X is C^∞. Indeed, if f is a C^∞ function with domain W, then for $m \in W$ there is a coordinate system x^i with domain U containing m. The expression for Xf in $U \cap W \cap E$ is $X^i \partial_i f$, which is a sum of products of C^∞ functions, hence is C^∞. This shows that Xf is C^∞ in a neighborhood of each point of its domain, so Xf is C^∞. We have proved

Proposition 3.1.1.   *A vector field X is C^∞ iff for every coordinate system x^i the components of X with respect to the x^i, $X^i = Xx^i$, are C^∞ functions.*

Our previous nomenclature, calling ∂_i coordinate vector fields, agrees with the present notation, since ∂_i are obviously C^∞ vector fields.

If t is a single tangent at m, we may choose coordinates at m, so $t = a^i \partial_i(m)$, where $a^i \in R$. Thus t is the value at m of the C^∞ vector field $X = a^i \partial_i$, where the a^i are regarded as being constant functions. Furthermore, if $g \colon M \to R$ is a C^∞ function such that $gm = 1$ and g vanishes identically outside a neighborhood W of m contained in the coordinate domain U [see Problem 1.6.2(**b**)], then we may define

$$Y = \begin{cases} gX & \text{on } W, \\ 0 & \text{outside } W. \end{cases}$$

Then Y is a C^∞ vector field on all of M such that $Y(m) = t$.

A C^∞ vector field Z is a C^∞ *extension* of $t \in M_m$ if Z is defined at m and $Z(m) = t$. The vector fields X and Y of the previous paragraph are C^∞ extensions of t.

Proposition 3.1.2.   *If $t \in M_m$, there is a C^∞ extension of t to all of M.*

Problem 3.1.1.   Let U be the domain of coordinates x^i, V the domain of coordinates y^i, and suppose that $\partial/\partial x^i = \partial/\partial y^i$, $i = 1, \dots, d$, on $U \cap V$. If $U \cap V$ is arcwise connected (see Section 0.2.7), show that $y^i = x^i - a^i$ on $U \cap V$, where the a^i are constant. On the other hand, show by examples of coordinates on the circle and the torus that if $U \cap V$ is not connected, then $y^i - x^i$ may have different values in the different connected components of $U \cap V$.

Problem 3.1.2. If X and Y are C^∞ vector fields and f is a C^∞ function, then Yf is a C^∞ function, to which X may be applied, getting C^∞ function XYf. Show that the operator $XY: f \rightarrow XYf$ has as its coordinate expression a second-order partial differentiation operator, $XY = F^{ij} \dfrac{\partial^2}{\partial x^i \, \partial x^j} + G^i \dfrac{\partial}{\partial x^i}.$ Express the coefficients F^{ij} and G^i in terms of the components $X^i = Xx^i$ and $Y^i = Yx^i$ of X and Y.

Problem 3.1.3. In contrast to single vectors, show that a C^∞ vector field X may not have a C^∞ extension to the whole manifold; that is, there may be no C^∞ vector field Z on all of M such that for every $m \in V$, the domain of X, $X(m) = Z(m)$. (*Hint:* Take U so that there is a point in M which can be approached from more than one part of U, and define X so that its limits on different approaches are different.)

3.2. Tensor Fields

For each type (r, s) of tensor and each $m \in M$, there is the corresponding tensor space M_{ms}^r over M_m. For fixed (r, s) the union of these tensor spaces as m varies is called the *bundle of tensors of type* (r, s) *over* M, denoted $T_s^r M$. Thus

$$T_s^r M = \bigcup_{m \in M} M_{ms}^r.$$

In particular, we have the tangent bundle $TM = T_0^1 M$, the *scalar bundle* $T_0^0 M$, and the *cotangent bundle* $T_1^0 M$. Other names for $T_1^0 M$ are the *bundle of differentials of* M (since it contains all the values at points of M of the differentials of real-valued C^∞ functions) and the *phase space* of M (this is customarily used when M is the configuration space of a mechanical system). The scalar bundle $T_0^0 M$ is the same as $M \times R$, since $V_0^0 = R$ for any vector space V.

A *tensor field* T *of type* (r, s) is a function $T: E \rightarrow T_s^r M$, where the domain E of T is a subset of M, such that for every $m \in E$ we have $T(m) \in M_{ms}^r$.

If $r = 1$, $s = 0$, then we again have vector fields; that is, a tensor field of type $(1, 0)$ is a vector field.

If $r = s = 0$, then T assigns a scalar to each $m \in E$, so a tensor field of type $(0, 0)$ is simply a real-valued function.

If f is a C^∞ function on $E \subset M$, then for every $m \in E$, $df_m \in M_m^* = M_{m1}^0$. Thus the differential of f, $df: E \rightarrow T_1^0 M$, is a tensor field of type $(0, 1)$.

We call a tensor field T *symmetric* if its value at every point m, $T(m)$, is a symmetric tensor. We define skew-symmetric tensor fields similarly.

If T is a tensor field of type (r, s), $\theta_1, \ldots, \theta_r$ tensor fields of type $(0, 1)$, and $X_1, \ldots, X_s$ vector fields, then we define a real-valued function on the intersection of all $r + s + 1$ domains (of T, the θ's, and the X's) by

$$T(\theta_1, \ldots, \theta_r, X_1, \ldots, X_s)m = T(m)(\theta_1(m), \ldots, \theta_r(m), X_1(m), \ldots, X_s(m)).$$

In particular, the *components of T with respect to coordinates x^i* are the d^{r+s} real-valued functions

$$T^{i_1 \ldots i_r}_{j_1 \ldots j_s} = T(dx^{i_1}, \ldots, dx^{i_r}, \partial_{j_1}, \ldots, \partial_{j_s}).$$

Turning the analogue of Proposition 3.1.1 into a definition, we say that tensor field T is C^∞ if its components are C^∞ functions. A tensor field of type $(0, 1)$ which is also C^∞ is called a *1–form (pfaffian form)*. The analogue of the definition of C^∞ for vector fields is the following, given without proof.

Proposition 3.2.1. *A tensor field T of type (r, s) is C^∞ iff for all 1–forms $\theta_1, \ldots, \theta_r$ and all C^∞ vector fields $X_1, \ldots, X_s$ the function $T(\theta_1, \ldots, \theta_r, X_1, \ldots, X_s)$ is C^∞.*

For the evaluation of 1–forms on vector fields we use the symmetric notation as with single vectors and covectors. That is, if X is a vector field and θ a 1–form, we write $\langle X, \theta \rangle$ for $\theta(X)$, a real-valued function on M.

If f is a C^∞ scalar field, then df is a 1–form. However, not every 1–form is of the form df for some C^∞ function f. In fact, if x^i are coordinates,

$$df = \partial_i f \, dx^i$$

is the coordinate expression for df, from which it follows that the components θ_i of df satisfy

$$\partial_j \theta_i = \partial_i \theta_j \tag{3.2.1}$$

no matter what coordinates are used. On the other hand, if U is a coordinate domain for the x^i, we define a 1–form on U by $\tau = x^1 \, dx^2$. The components of τ are $\tau_i = \delta_{i2} x^1$, so we have

$$\partial_2 \tau_1 = \partial_2 0 = 0,$$

but

$$\partial_1 \tau_2 = \partial_1 x^1 = 1.$$

It follows that there can be no function f such that $\tau = df$.

As a multilinear function of vector fields and 1–forms, a tensor field is linear in each variable with respect to multiplication by scalar fields:

$$T(\ldots, fX, \ldots) = fT(\ldots, X, \ldots). \tag{3.2.2}$$

Another fact, which may be derived as a consequence of (3.2.2), is that if one of the variables is zero at a point m, then so is the tensor field function of those variables:

$$\text{If } X(m) = 0, \text{ then } T(\ldots, X, \ldots)m = 0. \tag{3.2.3}$$

Indeed, in terms of coordinates $X = f^i \partial_i$, where $f^i m = 0$ for each i, so that $T(\ldots, X, \ldots)m = (f^i m)T(\ldots, \partial_i, \ldots)m = 0$. Furthermore, the values of a tensor field evaluated on vector fields and 1–forms depend only on the components of the vector fields and 1–forms, and not on the derivatives of these components. Or what is the same thing, if two sets of vector field and 1–form values are the same at m, then the values of T on them are the same at m: If $\theta_\alpha(m) = \tau_\alpha(m)$ and $X_\beta(m) = Y_\beta(m)$, then

$$T(\theta_1, \ldots, \theta_r, X_1, \ldots, X_s)m = T(\tau_1, \ldots, \tau_r, Y_1, \ldots, Y_s)m. \tag{3.2.4}$$

Problem 3.2.1. Let T be a function on r 1–forms and s C^∞ vector fields which assigns to them a C^∞ real-valued function such that **(a)** T is multilinear with respect to multiplication by constants: $T(\ldots, aX, \ldots) = aT(\ldots, X, \ldots)$, and **(b)** T is additive in each variable: $T(\ldots, X + Y, \ldots) = T(\ldots, X, \ldots) + T(\ldots, Y, \ldots)$. Show that if T satisfies any one of (3.2.2), (3.2.3), or (3.2.4), then T is a tensor field.

Problem 3.2.2. Let f be a fixed C^∞ function which is not constant. For C^∞ vector fields X and Y define $T(X, Y) = X Yf$. Show that T satisfies **(a)** and **(b)** of Problem 3.2.1 but that T is not a tensor field.

3.3.   Riemannian Metrics

A symmetric C^∞ tensor field of type $(0, 2)$ which is nondegenerate and has the same index at each point is called a *semi-riemannian metric*.* If the field is positive definite at each point, it is a *riemannian metric*. If the index is 1 or $d - 1$, it is called a *Lorentz metric*. A manifold which has one of these fields distinguished is called a *semi-riemannian, riemannian,* or *Lorentz manifold*, as the case may be.

If the manifold is connected, the condition that the index be constant is redundant. For, as we move along a continuous curve the index of a C^∞ symmetric tensor field of type $(0, 2)$ cannot jump from one value to another unless the form becomes degenerate at the jump point.

For a given manifold there are infinitely many different riemannian metrics. If g is a semi-riemannian metric and f is a positive C^∞ function, then fg is a semi-riemannian metric of the same index as g. Thus if there is one g, there are infinitely many.

On the other hand, the existence of a semi-riemannian metric of index

---

* Also called a *pseudo-riemannian metric*.

$k \neq 0$ or d depends on the topological structure of the manifold. For example, the only compact surfaces on which there is a Lorentz metric are the torus and the Klein bottle. In particular, the 2-sphere admits only definite (positive or negative) semi-riemannian metrics. Odd-dimensional manifolds always admit Lorentz metrics. In general the topological properties involved in the existence of metrics of a given index are difficult to study and not usually known in particular examples. It was only discovered in the 1950s what indices are possible for the spheres. For parallelizable manifolds (Appendix 3B) metrics of all indices from 0 to d exist.

3.4.   Integral Curves

If X is a vector field defined on $E \subset M$, a curve γ is an *integral curve of X* if the range of γ is contained in E and for every s in the domain of γ the tangent vector satisfies $\gamma_* s = X(\gamma s)$. If $\gamma 0 = m$, we say that γ *starts at m*. Note that the property of being an integral curve not only depends on the curve as a set of points (the range of γ) but also on the parametrization of γ. The allowable reparametrizations are rather restricted, as indicated by the following.

Proposition 3.4.1. *If γ and τ are integral curves of a nonzero vector field X which have the same range, then there is a constant c such that $\tau s = \gamma(s + c)$ for all s in the domain of τ. Conversely, if γ is an integral curve, then so is τ, $\tau s = \gamma(s + c)$, no matter what c is. In other words, a reparametrization of an integral curve is also an integral curve iff the reparametrization is a translation of the variable.*

Proof. Suppose $f: (a, b) \to (\alpha, \beta)$ is a reparametrizing function, so that $\tau = \gamma \circ f$; that is, $\tau s = \gamma(fs)$ for $a < s < b$. Then $\tau_* s = (f's)\gamma_*(fs)$, by the chain rule, so if $\tau_* s = X(\tau s)$ and $\gamma_* t = X(\gamma t)$ for $a < s < b$ and $\alpha < t < \beta$, then $f's = 1$. Thus $fs = s + c$ for some constant c. Conversely, if $fs = s + c$, then $f's = 1$ and $\tau_* s = \gamma_*(fs)$. ∎

Corollary. *The parametrization of an integral curve is entirely determined by specifying its value at one point.*

Proof. For the case of a nonzero vector field the result is evident from the theorem. However, the integral curve through a point where the vector field is zero is a constant curve, $\gamma t = m$ for all t, and a constant curve is unchanged by reparametrization and is certainly determined by its value at one point. ∎

In terms of coordinates the problem of finding integral curves reduces to a system of first-order differential equations. For coordinates x^i defined on U we have $X = X^i \partial_i$, where X^i are real-valued functions defined on $E \subset U$,

$$\gamma_* = \frac{d(x^i \circ \gamma)}{du} (\partial_i \circ \gamma),$$

$$X \circ \gamma = X^i \circ \gamma (\partial_i \circ \gamma).$$

Since ∂_i are a basis at each point, the condition for γ to be an integral curve is $d(x^i \circ \gamma)/du = X^i \circ \gamma$. The part of γ in U is determined by the functions $g^i = x^i \circ \gamma$. The components X^i determine their coordinate expressions, real-valued functions F^i defined on part of R^d such that $X^i = F^i(x^1, \ldots, x^d)$. Thus we have

Proposition 3.4.2. *A curve γ is an integral curve of X iff for every coordinate system the coordinate expressions g^i of γ and F^i of X satisfy the system of differential equations*

$$\frac{dg^i}{du} = F^i(g^1, \ldots, g^d). \tag{3.4.1}$$

Theorems on the existence and uniqueness of integral curves are based on corresponding theorems on the existence and uniqueness of solutions of such systems of ordinary differential equations in R^d; computations to find integral curves are based on techniques for solving such systems. For vector fields which are defined on a domain which is not included in a single coordinate domain, solutions are patched together (extended) from one system to the next. We state the following without proof.

Basic Existence and Uniqueness Theorem. Suppose F^i are C^∞ on the region determined by the inequality $\sum_i |u^i - a^i| \le b$, where $b > 0$, and let K be an upper bound for $\sum_i |F^i|$ on that region. Then there exist unique functions g^i defined and C^∞ on $|u - c| < b/K$ such that they satisfy the differential equations (3.4.1) and the initial conditions $g^i c = a^i$. (Reference: D. Greenspan, *Theory and Solution of Ordinary Differential Equations*, Macmillan, New York, 1960, p. 85, Theorem 5.5.)

Theorem 3.4.1. *Let X be a C^∞ vector field defined on $E \subset M$, $m \in E$, and $c \in R$. Then there is a positive number r and a unique integral curve γ of X defined on $|u - c| \le r$ such that $\gamma c = m$.*

By uniqueness we mean that if τ is an integral curve defined on $|u - c| \le r'$ and $\tau c = m$, then γ and τ coincide on the smaller of the two intervals.

Proof. An open set, such as E, will contain closed coordinate "cubes" of the sort mentioned in the basic theorem, with any given point m as center. The center is the point with coordinates a^i. For an upper bound of $\sum_i |F^i|$ on $\sum_i |u^i - a^i| \le b$ we can use its maximum, which exists since the sum is continuous and the cube is compact (see Proposition 0.2.8.3). ∎

Theorem 3.4.1. does not use the full strength of the basic existence and uniqueness theorem. It is not specific about the size of the interval on which the curve is defined. The number r may depend on the point m, and, in particular, there may be a sequence of points $\{m_i\}$ such that the limit of any corresponding sequence $\{r_i\}$ must be zero. By imposing a further condition, compactness, on the region through which the integral curve is to pass, we can make use of the specific estimate of the basic theorem to show that either an integral curve extends indefinitely or passes outside the compact set. The following lemma allows us to adapt the conditions of the basic theorem to a compact set.

Lemma 3.4.1.   *Let C be a compact set in a manifold M. Then there are a finite number of coordinate systems, each of which includes in its range the closed cube $\sum_i |u^i| \leq 2$, and such that every point of C is mapped into one or more of the open cubes $\sum_i |u^i| < 1$.*

Proof. At each $m \in C$ there is a coordinate system (x^i) such that $x^i m = 0$. The domain contains some closed cube of the form $\sum_i |x^i| \leq 2b$. Letting $y^i = x^i/b$, we obtain new coordinates such that the closed cube $\sum_i |y^i| \leq 2$ is in the y^i-domain. The open cubes $\sum_i |y^i| < 1$, one for each $m \in C$, form an open covering of C. Since C is compact, there are a finite number which cover C, as the lemma asserts.   ∎

Theorem 3.4.2.   *Let X be a C^∞ vector field, C a compact set contained in the domain of X, $m \in C$, and $c \in R$. Then there is an integral curve γ of X such that $\gamma c = m$ and*

   (a) *Either γ is defined on $[c, +\infty)$ or γ is defined on $[c, r]$, where $\gamma r \notin C$.*
   (b) *Either γ is defined on $(-\infty, c]$ or γ is defined on $[r', c]$, where $\gamma r' \notin C$.*

Proof. The domain of X is an open submanifold containing C, and we take this open submanifold to be the manifold of Lemma 3.4.1, so that we may assume that the larger closed cubes of Lemma 3.4.1 are contained in the domain of X. For each of these systems we have the coordinate expressions F^i for the components of X. Since $\sum_i |F^i|$ is continuous, it has a maximum on the closed cube $\sum_i |y^i| \leq 2$, and since there are a finite number of such cubes, there is a largest one of the maximums, which we call K.

If m' is any point in C, then m' is included in one of the coordinate cubes of size 1, say $y^i m' = a^i$, where $\sum_i |a^i| < 1$. By the triangle inequality the unit coordinate cube with center m' is contained in the larger closed cube; that is, from $\sum_i |y^i - a^i| \leq 1$ and $\sum_i |a^i| < 1$ we conclude $\sum_i |y^i| \leq 2$. Thus the maximum of $\sum_i |F^i|$ on $\sum_i |y^i - a^i| \leq 1$ is no greater than K. It follows from the basic theorem that the integral curve of X through m' is defined on an interval of length at least $2/K$, with m' corresponding to the center of the interval. The importance of this is that we may always extend by the fixed amount $1/K$ as long as we start from a point of C. It should now be clear that we can start at m and extend step by step in both directions either until the endpoint of some step falls outside C or until we get beyond any given parameter value.   ∎

The following is an important corollary.

Theorem 3.4.3. *Suppose X is a C^∞ vector field defined on all of a compact manifold M. Then every integral curve may be extended to all of R.*

A vector field is said to be *complete* if all its integral curves may be extended to all of R. Thus a globally defined C^∞ vector field on a compact manifold is complete.

Examples. Let $M = R^2$ with cartesian coordinates x and y, and corresponding coordinate vector fields ∂_x and ∂_y. Let $X = x\partial_x + y\partial_y$ and $Y = -y\partial_x + x\partial_y$. Then with the customary disregard for the distinction between x and $x \circ \gamma$, etc., the equations for the integral curves of X are $dx/du = x$ and $dy/du = y$. The general solution of these equations is $x = ae^u$, $y = be^u$. The unique integral curve γ such that $\gamma 0 = (a, b)$ is thus given by $\gamma = (ae^u, be^u)$. It is defined on all of R, so X is complete. The integral curve starting at $(0, 0)$ is the constant curve $\gamma = (0, 0)$. The other curves, as sets of points, are the open half-lines with origin $(0, 0)$.

The equations for the integral curves of Y are $dx/du = -y$ and $dy/du = x$. The general solution (use $d^2x/du^2 = -dy/du = -x$, etc.) is $x = a \cos u - b \sin u$, $y = b \cos u + a \sin u$. The integral curve γ such that $\gamma 0 = (a, b)$ is given by $\gamma = (a \cos u - b \sin u, b \cos u + a \sin u)$. When $a = b = 0$, it is the constant curve $(0, 0)$; otherwise it is a circle traversed uniformly in a counterclockwise direction so that a change in u by 2π gives one revolution. Y is complete.

Problem 3.4.1. Let $X = F\partial_x + G\partial_y$ be a C^∞ vector field defined on all of R^2 and suppose that there is a constant K such that $|F| + |G| \le K$. Show that X is complete. Is this a necessary condition for completeness?

Problem 3.4.2. Show that X is complete iff there is $r > 0$ such that for every m the integral curve of X starting at m is defined on $(-r, r)$.

Problem 3.4.3. Let X be a C^∞ vector field on M, and let $f: M \to R$ be a positive C^∞ function. Show that for every integral curve γ of X there is a function g having positive derivative and there is an integral curve τ of fX such that γ is the reparametrization $\tau \circ g$ of τ. Find the relation between g, γ, and f.

Problem 3.4.4. Find the integral curves of $X = \partial_x + e^{-y}\partial_y$. Is X complete?

Problem 3.4.5. Find the integral curves (in R^3) of $\partial_x + x^2\partial_y + (3y - x^3)\partial_z$.

3.5. Flows

If a vector field represents the velocity field of a flowing fluid, then the path traced by a particle parametrized by time is an integral curve. However, there is another significant viewpoint: We can ask where the fluid occupying a certain region has moved to after a fixed elapsed time. This viewpoint leads us to a purely mathematical notion associated with a vector field—its flow.

The *flow* of a vector field X is the collection of maps $\{\mu_s \colon E_s \to M \mid s \in R\}$, such that $\mu_s m = \gamma_m s$ for each $m \in E_s$, where γ_m is the integral curve of X starting at m. Thus m and $\mu_s m$ are always on the same integral curve of X and the difference in parameter values at $\mu_s m$ and m is s; in other words, μ_s is the map which pushes each point along the integral curve by an amount equal to the parameter change s. The domain of μ_s, E_s, consists of those points m such that γ_m is defined at s. Thus if $0 < s < t$ or if $t < s < 0$, then $E_t \subset E_s$, since if γ_m is defined at t, then it is defined at every point between 0 and t. If X is complete, then $E_s = E$, the domain of X, for every s. If X is C^∞, then there is an integral curve γ_m for every $m \in E$, and since γ_m is defined at 0, $E_0 = E$. Moreover, it is evident that $\mu_0 \colon E_0 \to M$ is the identity map on E_0, since $\gamma_m 0 = m$. The flow of a vector field X conveys no more information than the totality of integral curves of X.

For a C^∞ vector field the domains E_s are all open. More specifically, we have:

Proposition 3.5.1. *If X is a C^∞ vector field with domain $E \subset M$, then for every $m \in E$ there is a neighborhood U of m and an interval $(-r, r)$ such that μ_s is defined on U for every $s \in (-r, r)$.*

The proof requires little more than a translation of the information given by the basic existence and uniqueness theorem and so is omitted.

In the past, vector fields have been called *infinitesimal transformations* and they were thought of as generating finite transformations, that is, their flows.

A *one-parameter group* is a collection of objects $\{\mu_s \mid s \in R\}$, provided with an operation $\circ$ which is related to the parametrization by the rule $\mu_s \circ \mu_t = \mu_{s+t}$ such that there is $c > 0$, for which the μ_s with $-c < s < c$ are all distinct. For examples, the real numbers themselves, $\mu_s = s$ with $\circ = +$, and the circle of unit complex numbers, $\mu_s = e^{is}$ with $\circ =$ multiplication, are one-parameter groups. In fact, "up to isomorphism" these are the only one-parameter groups.

Proposition 3.5.2. *The flow $\{\mu_s\}$ of a complete C^∞ vector field X which is not identically 0 is a one-parameter group under the operation of composition.*

Proof. There are two things to verify:

(a) For every s, t, $\mu_s \circ \mu_t = \mu_{s+t}$.
(b) There is $c > 0$ such that the μ_s, $-c < s < c$, are all distinct.

Part (a) may be viewed as a restatement of Proposition 3.4.1. For if $m \in E$, and γ_m is the integral curve of X starting at m, then τ defined by $\tau s = \gamma_m(s + t)$, for t fixed, is also an integral curve of X; indeed, $\tau 0 = \gamma_m(0 + t) = \gamma_m t$, so $\tau = \gamma_n$, where $n = \gamma_m t = \mu_t m$. But then $\mu_s n = \mu_s \circ \mu_t m = \gamma_n s = \tau s = \gamma_m(s + t) = \mu_{s+t} m$, which proves $\mu_s \circ \mu_t = \mu_{s+t}$.

To prove **(b)** we let m be a point such that $X(m) \neq 0$. Choose coordinates x^i at m such that $X^1 m \neq 0$; because X^1 is continuous we may restrict to a smaller coordinate domain on which $X^1 \neq 0$. By reversing the sign of x^1 if necessary, we also can obtain $X^1 > 0$. By the first differential equation for integral curves, $dg^1/du = F^1(g^1, \ldots, g^d)$, we then have that $dg^1/du > 0$, so that g^1 is a strictly increasing function along any integral curve within the x^i-domain. Thus for the r of Theorem 3.4.1 for γ_m and an x^i-coordinate cube, we have that $g^1 s = x^1 \gamma_m s$ is strictly increasing for $-r \leq s \leq r$, so that the points $\gamma_m s = \mu_s m$, $-r < s < r$ are all distinct. Hence the μ_s for $-r < s < r$ are all distinct. ∎

A *local one-parameter group* is a collection of objects $\{\mu_s\}$ parametrized by an interval (possibly unbounded) of real numbers $\{s\}$ containing 0, provided with an operation $(\mu_s, \mu_t) \to \mu_s \circ \mu_t$ which is defined at least for all pairs s, t in some interval about 0 and satisfies $\mu_s \circ \mu_t = \mu_{s+t}$ whenever defined, and such that the μ_s are all distinct for $-c < s < c$, for some $c > 0$.

By a slight abuse of language we can claim that the flow of a C^∞ vector field is a one-parameter group, but more precisely what we have is the following.

Proposition 3.5.3. *Let X be a C^∞ vector field. Then for each m in the domain of X such that X is not identically 0 in some neighborhood of m, there is a neighborhood U of m such that the collection of restrictions $\{\mu_s|_U\}$ of the flow of X to U is a local one-parameter group.*

The proof is very much like that of Proposition 3.5.2 except for automatic modifications needed to fit the local definition, so it is left as an exercise.

Examples. **(a)** Let $M = R^d$ and let the vector field be ∂_1, where the coordinates are the cartesian coordinates on R^d. The integral curves of ∂_1 are given by the differential equations $du^1/du = 1$, $du^i/du = 0$, $i > 1$, so that if $m = (c^1, \ldots, c^d)$, $\gamma_m s = \mu_s m = (c^1 + s, c^2, \ldots, c^d)$. Thus μ_s is translation by amount s in the u^1 direction.

(b) If $X = x\partial_x + y\partial_y$ on R^2, then the integral curves are $\gamma_{(a,b)} s = (ae^s, be^s)$. Thus $\mu_s(a, b) = e^s(a, b)$ and μ_s is a magnification by factor e^s and center 0.

(c) If $\gamma = -y\partial_x + x\partial_y$ on R^2, then the integral curves are $\gamma_{(a,b)} s = (a \cos s - b \sin s, a \sin s + b \cos s)$, and μ_s is a rotation by angle s with center 0.

(d) If X is the "unit" radial field on $M = R^2 - \{0\}$, in polar coordinates $X = \partial_r$ and μ_s is a translation by s in the r-direction, which is given in cartesian coordinates by $\mu_s(a, b) = [(r + s)/r](a, b)$, where $r = (a^2 + b^2)^{1/2}$. It is defined whenever $-s < r$. For $s \geq 0$, μ_s is defined on all of M, $E_s = M$. For $s < 0$, μ_s is defined outside a disk of radius $-s$; that is,

$$E_s = \{(a, b) \mid (a^2 + b^2)^{1/2} \geq -s\}.$$

For some applications of the flow of a C^∞ vector field we need the smoothness properties stated in the following theorem.

Proposition 3.5.4. *Let $\{\mu_s\}$ be the flow of a C^∞ vector field X. Then the function F, defined on an open submanifold of $M \times R$ by $F(m, s) = \mu_s m$, is C^∞. In other words, $\mu_s m$ is a C^∞ function of both m and s.*

The proof is too technical to give here. It involves an initial reduction to coordinate expressions, which we have already seen, and then a proof that solutions of systems of C^∞ differential equations have a C^∞ dependence on initial conditions. Theorems of this sort are found in more advanced treatises on differential equations, for example, E. Coddington and N. Levinson, *Theory of Ordinary Differential Equations*, McGraw-Hill, New York, 1955, p. 22, Theorem 7.1.

If we are given a collection of maps which behave like the flow of a vector field, it is possible to obtain a vector field for which they are the flow. Specifically, we have

Proposition 3.5.5. *Let $\{\mu_s\}$ be a local one-parameter group of C^∞ maps, $\mu_s: E_s \to M$, such that the group operation is composition and the function F given by $F(m, s) = \mu_s m$ is a C^∞ function on an open subset of $M \times R$. Then there is a C^∞ vector field such that $\{\mu_s\}$ is a restriction of its flow.*

Proof. For $m \in E_0 = E$ and $f: M \to R$ any C^∞ function, define $X(m)f = \partial f \circ F/\partial s(m, 0)$. The linearity and derivation properties of X are easily verified, so X is a vector field on E. Moreover, Xf is the composition of the C^∞ function $\partial f \circ F/\partial s$ with the C^∞ injection $m \to (m, 0)$, so X is a C^∞ vector field.

It remains to show that $\{\mu_s\}$ coincides with the flow of X on its domain. For this it suffices to show that the curves $\tau_m: s \to \mu_s m$ are integral curves of X. The derivative of $f: M \to R$ along τ_m at $s = t$ is

$$\begin{aligned}
\tau_{m*}(t)f &= \frac{d}{ds}(t)f(\mu_s m) \\
&= \frac{d}{ds}(t)f(\mu_{s-t}\mu_t m) \\
&= \frac{d}{ds}(0)f(\mu_s\mu_t m) \\
&= \frac{\partial f \circ F}{\partial s}(\mu_t m, 0) \\
&= X(\mu_t m)f \\
&= X(\tau_m t)f.
\end{aligned}$$

Thus $\tau_{m*}(t) = X(\tau_m t)$ and τ_m is an integral curve of X. ■

As an application of Proposition 3.5.4 we obtain a local canonical form for a nonzero C^∞ vector field.

Theorem 3.5.1. *If X is a C^∞ vector field and $X(m) \neq 0$, then there are coordinates x^i at m such that in the coordinate neighborhood $X = \partial_1$. In other words, every nonzero vector field is locally a coordinate vector field.*

Proof. Choose coordinates y^i at m such that $y^i m = 0$ and $X(m)$, $\partial/\partial y^2(m)$, ..., $\partial/\partial y^d(m)$ is a basis of M_m. Define a map F of a neighborhood of the origin in R^d into M by

$$F(s, a^2, \ldots, a^d) = \mu_s \theta^{-1}(0, a^2, \ldots, a^d),$$

where $\{\mu_s\}$ is the flow of X and $\theta = (y^1, \ldots, y^d)$ is the y^i coordinate map. It is clear that F is C^∞.

As s varies $\mu_s m'$ moves along an integral curve of X for every m'. Thus the integral curves of X correspond under F to the u^1 coordinate curves in R^d. If F is the inverse of a coordinate map, then the first coordinate curves must be the integral curves of X and hence X the first coordinate vector field.

To show that F is the inverse of a coordinate map at m we employ the inverse function theorem. This requires that we show that F_* is nonsingular at the origin 0 in R^d, since the matrix of F_* with respect to some coordinate vector basis is the jacobian matrix. The values of F_* on the basis $\partial/\partial u^i(0)$ of R_0^d can be found by mapping the curves to which they are tangent. We have already done this for $\partial/\partial u^1(0)$, so we know that $F_*(\partial/\partial u^1(0)) = X(m)$. The other coordinate curves through 0 have $s = 0$, so that F coincides with θ^{-1} on them. It follows that $F_*(\partial/\partial u^i(0)) = \partial/\partial y^i(m)$, $i > 1$. Since F_* maps a basis into a basis, it is nonsingular. By the inverse function theorem there is a neighborhood U of m such that $F^{-1} = (x^1, \ldots, x^d)$ is defined and C^∞ on U. ■

Problem 3.5.1.   **(a)** If $\{\mu_s\}$ is the flow of ∂_1 and $\{\theta_t\}$ is the flow of ∂_2, show that $\mu_s \circ \theta_t = \theta_t \circ \mu_s$ for all s and t.

  (b) Suppose X and Y are C^∞ vector fields with flows $\{\mu_s\}$ and $\{\theta_t\}$, respectively, such that $\mu_s \circ \theta_t = \theta_t \circ \mu_s$ for all s and t. If $X(m)$ and $Y(m)$ are linearly independent, show that there are coordinates at m such that $X = \partial_1$ and $Y = \partial_2$ in the coordinate domain.

  (c) Generalize **(a)** and **(b)** to k vector fields, where $k \leq d$.

3.6.   Lie Derivatives

If X is a C^∞ vector field, then X operates on C^∞ scalar fields to give C^∞ scalar fields. The Lie derivation with respect to X is an extension of this operation to an operator L_X on all C^∞ tensor fields which preserves type of tensor fields.

Let $\{\mu_s\}$ be the flow of X and let m be in the domain of X. It follows from the definition of μ_s that it has an inverse, μ_{-s}, and since both are C^∞, μ_s is a diffeomorphism. Hence for each s for which $\mu_s m$ is defined, μ_{s*} is an isomorphism $M_m \to M_{\gamma s}$, where $\gamma s = \mu_s m$. If $\{e_i\}$ is a basis of M_m, then $\{\mu_{s*} e_i\}$ is a basis of $M_{\gamma s}$. The vectors $E_i(s) = \mu_{s*} e_i$ form curves in TM lying "above" the integral curve γ of X starting at m and giving a basis of the tangent space at each point γs. We say that the E_i form a *moving frame* along γ.

If T is a C^∞ tensor field defined in a neighborhood of m, then the components of $T(\gamma s)$ with respect to the basis $E_i(s)$ are C^∞ functions T^α of s, $\alpha = 1, \ldots, d^p$, where we have numbered the components in some definite way. The derivatives $U^\alpha = dT^\alpha / ds$ are the components of tensors U along γ; that is, U is a function of s such that $U(s)$ is a tensor over $M_{\gamma s}$. The tensor-valued function U is independent of the choice of initial basis $\{e_i\}$ of M_m. For if we take a new basis $\{f_i\}$ and let $F_i(s) = \mu_{s*} f_i$, then, letting $f_i = a_i^j e_j$, we have $F_i = a_i^j E_j$, since the μ_{s*} are linear. Here the a_i^j are constants, not functions of s. Thus the components of T with respect to the e and f bases are related by constant functions of s, $T^{f,\alpha} = A_\beta^\alpha T^{e,\beta}$, where β is summed from 1 to d^p, so that the s-derivatives are related in the same way and are therefore the components of the same tensor-valued function U with respect to the different bases. By varying m we obtain values for a tensor field at points other than those on a single curve.

The tensor field derived from T in the above way by differentiating with respect to the parameters of the integral curves of X is called the *Lie derivative of T with respect to X* and is denoted $L_X T$.

In the following proposition we list some of the elementary properties of Lie derivatives.

Proposition 3.6.1.   (a) *If T is a C^∞ tensor field, then $L_X T$ is a C^∞ tensor field of the same type as T, defined on the intersection of the T and X domains.*

(b) *$L_X T$ has the same symmetry or skew-symmetry properties as T does.*

(c) *L_X is additive: $L_X(S + T) = L_X S + L_X T$.*

(d) *L_X satisfies a product rule, so it is a derivation:*

$$L_X(S \otimes T) = (L_X S) \otimes T + S \otimes L_X T.$$

(e) *In the case of a scalar field f, $L_X f = Xf$.*

(f) *If $X = \partial_1$, the first coordinate vector field of some coordinate system, then the components U^α of $L_X T$ with respect to the basis ∂_i are $\partial_1 T^\alpha$, the first coordinate derivatives of the components of T.*

Proof. The first five are simple consequences of the definition. For (f), if $X = \partial_1$, then we have seen that μ_s, in the coordinate domain, is the translation of the first coordinate by amount s. This translation takes coordinate curves into coordinate curves and hence also takes coordinate vector fields into themselves; that is, $\mu_{s*} \partial_i = \partial_i$. Thus if we let $e_i = \partial_i(m)$, then $E_i = \partial_i$ and the components of T with respect to the E_i are just the coordinate components of T. Differentiation with respect to s is clearly the same as applying ∂_1. ∎

It is an interesting computation to verify directly that the process of (**f**) does not depend on the coordinate system used for which $X = \partial_1$. That is, if we suppose that $X = \partial/\partial x^1 = \partial/\partial y^1$, then the same result is obtained by differentiating the x^i-components of T with respect to x^1 as by differentiating the y^i-components of T with respect to y^1. Indeed, we have $T^{y,\alpha} = A^\alpha_\beta T^{x,\beta}$, where the A^α_β are products of the $\partial y^i/\partial x^j$ and the $\partial x^i/\partial y^j$. But we have

$$X \frac{\partial y^i}{\partial x^j} = \frac{\partial^2 y^i}{\partial x^1 \partial x^j} = \frac{\partial}{\partial x^j}\frac{\partial y^i}{\partial x^1} = \frac{\partial}{\partial x^j}\frac{\partial y^i}{\partial y^1} = \frac{\partial}{\partial x^j}\delta^i_1 = 0$$

and, similarly, $X(\partial x^h/\partial y^k) = 0$. It follows that $XA^\alpha_\beta = 0$, so $XT^{y,\alpha} = A^\alpha_\beta XT^{x,\beta}$. Thus $XT^{y,\alpha}$ and $XT^{x,\beta}$ are the components of the same tensor with respect to the different bases.

Remark. The x^i coordinate components of $L_X T$ can be expressed algebraically in terms of the x^i components of $L_X \partial_i$, $L_X dx^i$, and XT^α, where the T^α are the components of T. This follows from (**c**), (**d**), and (**e**), since $T = T^\alpha P_\alpha(\partial_i, dx^j)$, where $P_\alpha(\partial_i, dx^j)$ is a tensor product of the ∂_i and dx^j.

Problem 3.6.1. Prove that $\mu_{s*}X(m) = X(\mu_s m)$ and that $L_X X = 0$.

Theorem 3.6.1. *Let X be a C^∞ vector field, T a C^∞ tensor field, x^i coordinates with coordinate vector fields ∂_i, $X^i = Xx^i$ the components of X, and $T^{i_1\cdots i_r}_{j_1\cdots j_s}$ those of T. Then the components of $L_X T$ are*

$$(L_X T)^{i_1\cdots i_r}_{j_1\cdots j_s} = XT^{i_1\cdots i_r}_{j_1\cdots j_s} - \sum_{\alpha=1}^{r} T^{i_1\cdots i_{\alpha-1}h i_{\alpha+1}\cdots i_r}_{j_1\cdots j_s}\partial_h X^{i_\alpha}$$

$$+ \sum_{\alpha=1}^{s} T^{i_1\cdots i_r}_{j_1\cdots j_{\alpha-1}h j_{\alpha+1}\cdots j_s}\partial_{j_\alpha}X^h. \qquad (3.6.1)$$

Proof. We obtain the validity of the formula at points for which $X \neq 0$ by using Proposition 3.6.1(**f**) and the transformation law. The zeros of X will be handled as special cases. We suppose T has type $(1, 1)$ and it will be evident what modifications in the proof are necessary for other types.

If $X(m) \neq 0$, then by Theorem 3.5.1 there are coordinates y^i at m such that $X = \partial/\partial y^1$. By (**f**), if $U = L_X T$ we have $U^{y,i}_j = XT^{y,i}_j$, and the x^i components of X are $X^i = Xx^i = \partial x^i/\partial y^1$. Thus

$$U^{x,i}_j = U^{y,h}_k \frac{\partial x^i}{\partial y^h}\frac{\partial y^k}{\partial x^j}$$

$$= (XT^{y,h}_k)\frac{\partial x^i}{\partial y^h}\frac{\partial y^k}{\partial x^j}$$

$$= X\Big(T^{x,p}_q \frac{\partial y^h}{\partial x^p}\frac{\partial x^q}{\partial y^k}\Big)\frac{\partial x^i}{\partial y^h}\frac{\partial y^k}{\partial x^j}$$

$$= XT^{x,i}_j + T^{x,p}_j\Big(X\frac{\partial y^h}{\partial x^p}\Big)\frac{\partial x^i}{\partial y^h} + T^{x,i}_q\Big(X\frac{\partial x^q}{\partial y^k}\Big)\frac{\partial y^k}{\partial x^j}.$$

Now we would like to eliminate the y^i coordinates:

$$\left(X\frac{\partial y^h}{\partial x^p}\right)\frac{\partial x^i}{\partial y^h} = X\left(\frac{\partial y^h}{\partial x^p}\frac{\partial x^i}{\partial y^h}\right) - \frac{\partial y^h}{\partial x^p}X\frac{\partial x^i}{\partial y^h}$$

$$= X\delta^i_p - \frac{\partial y^h}{\partial x^p}\frac{\partial}{\partial y^1}\frac{\partial x^i}{\partial y^h}$$

$$= 0 - \frac{\partial y^h}{\partial x^p}\frac{\partial}{\partial y^h}\frac{\partial x^i}{\partial y^1}$$

$$= -\frac{\partial X^i}{\partial x^p},$$

$$\frac{\partial y^k}{\partial x^j}\frac{\partial}{\partial y^k}(Xx^q) = \frac{\partial X^q}{\partial x^j}.$$

These may now be substituted above to give (3.6.1).

For those m such that $X(m) = 0$ we have two cases:

(a) If there is a neighborhood of m such that $X = 0$ in that neighborhood, then μ_s is the identity map on that neighborhood, so that the components of T with respect to $\mu_{s*}e_i = e_i$ are simply the components of T with respect to e_i. The s-derivatives vanish, so $(L_X T)m = 0$. But the X^i are identically 0 in the neighborhood so $(\partial_h X^i)m = 0$, and finally $(XT^i_j)m = X(m)T^i_j = 0$. Thus both sides of (3.6.1) are 0.

(b) Otherwise m is a limit point of a sequence on which X is nonzero. The formula (3.6.1) is valid on the sequence and both sides are continuous, so (3.6.1) is proved valid at m by taking limits. ∎

Corollary. $L_{X+Y} = L_X + L_Y$.

Problem 3.6.2.   For a scalar field f prove that $L_X\,df = d(Xf)$.

Problem 3.6.3.   Prove that L_X commutes with contractions (see Section 2.14); that is, if C is the operator which assigns to a tensor its contraction on the pth covariant index and qth contravariant index, then $L_X(CT) = C(L_X T)$.

Problem 3.6.4.   For C^∞ vector fields X and Y and 1–form θ prove that $X\langle Y, \theta\rangle = \langle L_X Y, \theta\rangle + \langle Y, L_X\theta\rangle$.

Problem 3.6.5.   For C^∞ vector fields X and Y and scalar field f prove that $(L_X Y)f = XYf - YXf$ and hence that $XY - YX$ is a vector field and $L_X Y = -L_Y X$.

Problem 3.6.6.   The Lie derivation L_X commutes with the symmetrizing and alternating operators $\mathscr{S}$ and $\mathscr{A}$, and therefore L_X is a derivation with respect to symmetric and exterior products of symmetric and skew-symmetric tensors of unmixed types. That is, for symmetric tensor fields T and U, $L_X(TU) = (L_XT)U + TL_XU$, and for skew-symmetric tensor fields T and U,

$$L_X(T \wedge U) = (L_XT) \wedge U + T \wedge L_XU.$$

Problem 3.6.7. Show that

$$L_XL_Y - L_YL_X = L_{L_XY}. \tag{3.6.2}$$

Problem 3.6.8. **(a)** If A is a tensor field of type $(1, 1)$, we may view A as a field of linear functions, $A(m): M_m \to M_m$. Show that there is a unique extension D_A of A to a *derivation* of tensors such that for vector fields X we have $D_AX = AX$ and

    (1) For each tensor field T, D_AT is a tensor field of the same type as T.

    (2) For tensor fields T and U we have the product rule, $D_A(T \otimes U) = (D_AT) \otimes U + T \otimes D_AU$.

    (3) D_A is additive: $D_A(T + U) = D_AT + D_AU$.

    (4) D_A commutes with contractions, $C(D_AT) = D_A(CT)$.

Show that for scalar fields f, $D_Af = 0$.

    (b) In the notation of Section 2.12 the linear function on M_m given by $A(m)$ should be written $A(m)_2: M_m \to M_m$. Show that in the same notation the restriction of D_A to covariant vectors is $D_A|_{M_m{}^*} = -A(m)_1: M_m{}^* \to M_m{}^*$.

    (c) Prove that $L_{fX} = fL_X - D_{X \otimes df}$, where X is a vector field and f is a scalar field. (*Hint:* Since both sides are derivations which commute with contractions the identity needs to be verified only on scalar and vector fields.)

    (d) If T is of type $(2, 1)$, then the component formula for D_AT is $(D_AT)^{ij}_k = T^{pj}_k A^i_p + T^{ip}_k A^j_p - T^{ij}_p A^p_k$. Generalize this formula to other types. (In particular, D_A is entirely algebraic, requiring no derivatives of the A or T components.)

Problem 3.6.9. Suppose that X and Y are C^∞ vector fields, that $X(m)$ and $Y(m)$ are linearly independent, and $L_XY = 0$. Let $\{\mu_s\}$ be the flow of X and suppose that for some number b the domain of Y includes all the points $\mu_s m$ such that s is between 0 and b. Prove that $X(\mu_b m)$ and $Y(\mu_b m)$ are linearly independent.

Problem 3.6.10. Suppose that $L_XY = 0$ and $Y(m) = \alpha X(m)$. For a number b as in Problem 3.6.9 prove that $Y(\mu_b m) = \alpha X(\mu_b m)$.

Problem 3.6.11. For the vector field $X = (ax - by)\partial_x + (bx + ay)\partial_y$ on E^2, where a and b are constants, show that $L_Xg = 2ag$, where $g = dx \otimes dx + dy \otimes dy$ is the euclidean metric of E^2.

3.7. Bracket

An important special case of a Lie derivative is the *Lie bracket* of two C^∞ vector fields, $[X, Y] = L_X Y$. In Problem 3.6.5 we have given another formula for bracket, and it is this which we take as our working definition, to be used for further development: $[X, Y] = XY - YX$, where the product XY is to be understood as the composition of the operators Y and X on scalar fields. With this definition it is not a priori clear that it is a vector field, since XY and YX are second-order partial differential operators, but not usually vector fields. It is clearly additive but the product rule needs verification:

$$(XY - YX)(fg) = X[(Yf)g + fYg] - Y[(Xf)g + fXg]$$
$$= (XYf)g + (Yf)Xg + (Xf)Yg + fXYg - (YXf)g$$
$$- (Xf)Yg - (Yf)Xg - fYXg$$
$$= [(XY - YX)f]g + f(XY - YX)g.$$

Of course, this verification is unnecessary if we adhere to the Lie derivative approach, as in Problem 3.6.5.

The bracket of two coordinate vector fields from the same coordinate system is 0 because second partial derivatives are the same in either order on C^∞ functions: $\partial_i\partial_j f - \partial_j\partial_i f = [\partial_i, \partial_j]f = 0$. However, for two coordinate systems x^i and y^i it is *not* generally true that $\partial/\partial x^i$ and $\partial/\partial y^j$ commute. For example, on R^2 we have the two coordinate systems x, y and r, θ and it is easily computed that $[\partial_x, \partial_r] = (-\sin\theta/r^2)\partial_\theta$.

If X^i and Y^i are the x^i coordinate components of vector fields X and Y, respectively, then the components of $[X, Y]$ are

$$[X, Y]^i = [X, Y]x^i$$
$$= XY^i - YX^i$$
$$= X^j\partial_j Y^i - Y^j\partial_j X^i.$$

This formula makes it obvious that the bracket operation $(X, Y) \to [X, Y]$ is not some interpretation of a tensor field. Indeed, a tensor field only deals with the components, not the derivatives of the components of the variables. Some tensor properties are valid for bracket: It is *additive* in each variable:

$$[X + Y, Z] = [X, Z] + [Y, Z],$$
$$[X, Y + Z] = [X, Y] + [X, Z],$$

and skew-symmetric:

$$[X, Y] = -[Y, X].$$

The other linearity property fails and we have instead

$$[fX, gY] = (fX)(gY) - (gY)(fX)$$
$$= f(Xg)Y + fgXY - g(Yf)X - gfYX$$
$$= fg[X, Y] + f(Xg)Y - g(Yf)X.$$

Of course, when f and g are constants the last two terms vanish.
Another property is the *Jacobi identity*:

$$[[X, Y], Z] + [[Y, Z], X] + [[Z, X], Y] = 0.$$

The proof is automatic.

There are other interpretations of the Jacobi identity, one of which is the formula (3.6.2) of Problem 3.6.7 as applied to vector fields. If we define the bracket of any operators A, B to be $[A, B] = AB - BA$, then (3.6.2) may be written $[L_X, L_Y] = L_{[X,Y]}$. This may be thought of as telling us that $L: X \to L_X$ is a Lie algebra *homomorphism*; a *Lie algebra* is a vector space provided with an internal product, called the bracket operation, which is skew-symmetric, bilinear, and satisfies the Jacobi identity. The Lie algebras connected by the homomorphism L are infinite-dimensional vector spaces—the space of C^∞ vector fields and the space of derivations of tensor fields. Another interpretation of the Jacobi identity is that L_X is a derivation with respect to bracket multiplication:

$$L_X[Y, Z] = [L_X Y, Z] + [Y, L_X Z].$$

Problem 3.7.1. Let X, Y, Z be the vector fields on R^3 with components $(0, z, -y)$, $(-z, 0, x)$, $(y, -x, 0)$, respectively, with respect to cartesian coordinates x, y, z. Show that the correspondence

$$\mu: aX + bY + cZ \to a\mathbf{i} + b\mathbf{j} + c\mathbf{k}$$

is not only a linear isomorphism but that under μ brackets go into cross products: $\mu[U, V] = (\mu U) \times (\mu V)$. Consequently, ordinary three-dimensional vector algebra with cross product multiplication is a Lie algebra, and in particular, the cross product satisfies the Jacobi identity.

Problem 3.7.2. For $U = aX + bY + cZ$ as in Problem 3.7.1, show that the flow of U is a rotation of R^3 about an axis through 0, with angular velocity $-\mu U$.

Now we extend Theorem 3.5.1 to the case of two vector fields.

Theorem 3.7.1. *Let X and Y be C^∞ vector fields such that $[X, Y] = 0$ and suppose m is a point for which $X(m)$ and $Y(m)$ are linearly independent. Then there are coordinates at m such that $X = \partial_1$ and $Y = \partial_2$ in the coordinate domain. For s and t sufficiently close to 0 and on a neighborhood of m, $\mu_s \circ \theta_t = \theta_t \circ \mu_s$, where $\{\mu_s\}$ and $\{\theta_t\}$ are the one-parameter groups of X and Y.*

Proof. The second statement follows from the first, since coordinate translations commute. The proof of the first is similar to the proof of Theorem 3.5.1. Choose coordinates y^i in a neighborhood of m such that $y^i m = 0$ and

$X(m)$, $Y(m)$, $\partial/\partial y^3(m), \ldots, \partial/\partial y^d(m)$ are a basis of M_m. Define $F: V \to M$, where V is a neighborhood of 0 in R^d, by

$$F(s, t, a^3, \ldots, a^d) = \mu_s \theta_t \varphi^{-1}(0, 0, a^3, \ldots, a^d),$$

where $\varphi = (y^1, \ldots, y^d)$. Just as in Theorem 3.5.1. we prove that F_* is non-singular on R_0^d, so that $F^{-1} = (x^1, \ldots, x^d)$ exists and is a coordinate map in a neighborhood of m. Moreover, $X = \partial_1$ is proved as before and if we restrict to points where $x^1 = s = 0$, we have $Y = \partial_2$.

The x^i components of Y are Yx^i. At points of the slice $x^1 = 0$ we have just seen that $Yx^i = \delta_2^i$. Now we show that if we move crossways to the slice on the x^1 coordinate curves Yx^i does not change. In fact, since $[X, Y] = 0$, we have $X(Yx^i) = Y(Xx^i) = Y\delta_1^i = 0$. Thus $Yx^i = \delta_2^i$ on all points which can be reached on an x^1 coordinate curve starting at a point where $x^1 = 0$. Since such points fill a neighborhood of m we are done. ∎

It requires no additional technique to extend Theorem 3.7.1 to any number, up to d, of commuting, linearly independent vector fields. We state the result without further proof.

Theorem 3.7.2.   *Let $X_1, \ldots, X_k$ be C^∞ vector fields such that $[X_i, X_j] = 0$ for all i, j, and let m be such that $X_1(m), \ldots, X_k(m)$ are linearly independent. Then there is a coordinate system at m such that $X_i = \partial_i$, $i = 1, \ldots, k$, on the co-ordinate domain.*

3.8.   Geometric Interpretation of Bracket

We have seen in Theorem 3.7.1 that the flows of X and Y commute if $[X, Y] = 0$. The commutativity of μ_s and θ_t on m may be written

$$\theta_{-t}\mu_{-s}\theta_t\mu_s m = m.$$

In general, the effect of applying $\theta_{-t}\mu_{-s}\theta_t\mu_s$ to m is to push m along the sides of a "parallelogram" whose sides are integral curves of X, Y, $-X$, and $-Y$,

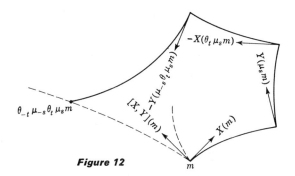

Figure 12

in that order (see Figure 12). We shall now see that these "parallelograms" are not usually closed curves, but that the gap between the first and last point is approximately $st[X, Y](m)$. Thus the bracket is a measure of how much such parallelograms fail to close.

If we replace Y by $(t/s)Y = Z$, then the same parallelogram is the parallelogram for X and Z but for which the parameter changes along each side are all equal to s. Thus without loss of generality we may assume $s = t$.

We give two formulations of the same result, one in terms of coordinates and one intrinsic.

Theorem 3.8.1.   (a) *Let x^i be coordinates at m with $x^i m = 0$ and let $[X, Y](m) = c^i \partial_i(m)$. Then the Taylor expansion of $\gamma s = \theta_{-s}\mu_{-s}\theta_s\mu_s m$ has the form*

$$x^i(\gamma s) = c^i s^2 + g^i(s)s^3, \qquad \text{where } g^i \text{ is } C^\infty.$$

(b) *With γ as in (a), $\gamma_* 0 = 0$ and $\gamma_{**} 0 = [X, Y](m)$.*
(For the definition of the second-order tangent $\gamma_{**} 0$, see Problem 1.6.1.)

Proof. It is easily seen that part (a) is simply the coordinate form of part (b), so it suffices to prove part (a).

If $X(m) = Y(m) = 0$, then $[X, Y](m) = 0$ (this is left as an exercise) and μ_s and θ_s leave m fixed for all s, so $\gamma s = m$, $x^i(\gamma s) = 0s^2$, and $c^i = 0$. This proves the result in this case.

Thus we may assume one of $X(m)$ and $Y(m)$ is nonzero. We shall do the case $X(m) \neq 0$, leaving the case $X(m) = 0$ and $Y(m) \neq 0$ as an exercise. We use this proof as an illustration of two important techniques:

(a) Choosing the coordinates to fit the problem. For this problem we can simplify μ_s by choosing coordinates for which $X = \partial_1$. Then μ_s is a translation by s in the x^1 direction.

(b) Using finite Taylor expansions. In using these expansions we guess what order will suffice and retain only those terms in computations which do not exceed that order. If too few terms are retained, the computation must be started over with more terms. If too many terms are retained, the procedure is more laborious than necessary but otherwise no harm is done.†

We suppose that coordinates x^i have been chosen such that $X = \partial_1$, $x^i m = 0$, and the components of Y, $Y^i = Yx^i$ have Taylor expansions of the second order,

$$Y^i = b^i + b^i_j x^j + g^i_{jk} x^j x^k,$$

where the g^i_{jk} are C^∞ functions on M and b^i and b^i_j are constants.

---

† The use of infinite Taylor series would display similar technique but requires the assumption that the series converge, not generally valid for C^∞ functions.

First we relate the components of $[X, Y](m)$ to the expansion of Y^i:

$$c^i = ([X, Y]x^i)m$$
$$= (\partial_1 Y x^i - Y\partial_1 x^i)m$$
$$= (\partial_1 Y^i)m$$
$$= b^i_1.$$

Now we compute the expansions of the coordinates of the corners of the parallelogram in turn, in powers of s. For this we must also compute the equations for the integral curves, which we parametrize by u. In these equations we shall use expansions of the third order in s and u for coordinate functions and second order for their derivatives. Instead of using specific notation for the remainder terms, we shall merely indicate them by $O(3)$ or $O(2)$. Thus $O(2)$ will stand for a number of different functions of s and u, all having the form $O(2) = \alpha(s, u)s^2 + \beta(s, u)su + \gamma(s, u)u^2$, where α, β, γ are C^∞ functions. Similarly, $O(3)$ will denote something in the form of a homogeneous cubic polynomial in s and u with C^∞ coefficients.

For the first corner after m we have

$$x^i\mu_s m = \delta^i_1 s, \tag{3.8.1}$$

since μ_s is translation by s in the x_1-direction and m is the origin.

Now we let $g^i = x^i\theta_u\mu_s m$. As a function of u these are the coordinates of the integral curve of Y starting at $\mu_s m$, so the terms not dependent on u are given by (3.8.1), and g^i has the form

$$g^i = \delta^i_1 s + a^i_1 u + a^i_2 su + a^i_3 u^2 + O(3).$$

The differential equations for an integral curve of Y must be satisfied by these g^i, so we have

$$\frac{\partial g^i}{\partial u} = a^i_1 + a^i_2 s + 2a^i_3 u + O(2)$$
$$= b^i + b^i_j g^j + O(2)$$
$$= b^i + b^i_1 s + b^i_j a^j_1 u + O(2).$$

Equating coefficients we get

$$a^i_1 = b^i, \qquad a^i_2 = b^i_1, \qquad a^i_3 = b^i_j b^j/2,$$

so the next corner of the parallelogram has coordinates

$$g^i(s, s) = (\delta^i_1 + b^i)s + (b^i_1 + b^i_j b^j/2)s^2 + O(3).$$

Translating the first coordinate by $-s$ gives us the third corner,

$$x^i\mu_{-s}\theta_s\mu_s m = b^i s + (b^i_1 + b^i_j b^j/2)s^2 + O(3).$$

These are used as initial conditions for the integral curve forming the fourth side of the parallelogram.

Letting $h^i = x^i \theta_{-u}\mu_{-s}\theta_s\mu_s m$, we have

$$h^i = b^i s + a^i_4 u + (b^i_1 + b^i_j b^j/2)s^2 + a^i_5 su + a^i_6 u^2 + O(3).$$

The differential equations for an integral curve of $-Y$ are

$$\frac{\partial h^i}{\partial u} = a^i_4 + a^i_5 s + 2a^i_6 u + O(2)$$
$$= -b^i - b^i_j h^j + O(2)$$
$$= -b^i - b^i_j b^j s - b^i_j a^j_4 u + O(2).$$

Comparing the two expansions gives

$$a^i_4 = -b^i, \qquad a^i_5 = -b^i_j b^j, \qquad a^i_6 = -b^i_j a^j_4/2 = b^i_j b^j/2.$$

Now we have the desired expansion of γ:

$$x^i(\gamma s) = h^i(s, s)$$
$$= (b^i - b^i)s + (b^i_1 + b^i_j b^j/2 - b^i_j b^j + b^i_j b^j/2)s^2 + O(3)$$
$$= b^i_1 s^2 + O(3)$$
$$= c^i s^2 + O(3). \quad \blacksquare$$

Problem 3.8.1. Compute the curve $\gamma s = \theta_{-s}\mu_{-s}\theta_s\mu_s m$ directly in the following instances, verifying that its Taylor expansion has the form specified in Theorem 3.8.1:

(a) $M = R$, $X = d/du$, $Y = uX$.

(b) $M = R^2$, $X = \partial_x$, $Y = x\partial_y$. Sketch some of the parallelograms in this case.

3.9.　Action of Maps

If $\varphi: M \to N$ is a C^∞ map we have seen that there are corresponding maps $\varphi_{*m}: M_m \to N_{\varphi m}$, which map individual tangent vectors to M. However, it is not generally possible to map vector fields into vector fields via φ_*, since φ can map two points m and m' into the same $n \in N$ and it may happen that $\varphi_* X(m) \neq \varphi_* X(m')$ for a given vector field X on M. Thus we may not be able to assign a unique value to $(\varphi_* X)m$. Even if we were able to assign unique values there is no assurance in general that the result is C^∞ if X is C^∞. For example, the image φM may not be an open set, and a continuous extension of $\varphi_* X$ to a larger set which is open may be impossible.

We say that vector fields X and Y on M and N, respectively, are φ-*related* if for every m in the domain of X, $\varphi_* X(m) = Y(\varphi m)$. Equivalently, we have that X and Y are φ-related iff for every C^∞ function $f: N \to R$, $(Yf) \circ \varphi = X(f \circ \varphi)$. Indeed, the values at $m \in$ domain of X of both sides of this equation are $(Yf) \circ \varphi m = Y(\varphi m)f$ and $X(f \circ \varphi)m = X(m)(f \circ \varphi) = (\varphi_* X(m))f$, so equality at m is the same as $\varphi_* X(m) = Y(\varphi m)$.

Proposition 3.9.1. *If X_i is φ-related to Y_i, $i = 1, 2$, then $[X_1, X_2]$ is φ-related to $[Y_1, Y_2]$.*

Proof. For all $f: N \to R$ we have

$$([Y_1, Y_2]f) \circ \varphi = (Y_1 Y_2 f - Y_2 Y_1 f) \circ \varphi$$
$$= (Y_1(Y_2 f)) \circ \varphi - (Y_2(Y_1 f)) \circ \varphi$$
$$= X_1((Y_2 f) \circ \varphi) - X_2((Y_1 f) \circ \varphi)$$
$$= X_1(X_2(f \circ \varphi)) - X_2(X_1(f \circ \varphi))$$
$$= [X_1, X_2](f \circ \varphi). \quad \blacksquare$$

A C^∞ map $\varphi: M \to N$ is *regular* if for every $m \in M$, φ_{*m} is 1–1 on M_m.

Lemma 3.9.1. *If φ is regular, then for every $m \in M$ there are coordinates y^α at φm, $\alpha = 1, \ldots, e$, such that $x^i = y^i \circ \varphi$, $i = 1, \ldots, d$, are coordinates at m.*

Proof. The fact that φ_{*m} is 1–1 can be expressed in terms of a matrix for φ_{*m} by the condition that some $d \times d$ submatrix, obtained by omitting $e - d$ rows, be nonsingular. If we choose coordinates z^i at m, then any coordinates y^α at φm can be numbered so that the first d rows of the matrix $(\partial y^\alpha \circ \varphi / \partial z^i(m))$ of φ_{*m}, with respect to the bases $\partial/\partial z^i(m)$ and $\partial/\partial y^\alpha(\varphi m)$, is a nonsingular submatrix. But this simply means that the functions $x^i = y^i \circ \varphi$, $i = 1, \ldots, d$, are related by the nonzero jacobian determinant, $\det(\partial x^i/\partial z^j(m))$, to the coordinates z^j, so the x^i are a coordinate system at m. $\quad \blacksquare$

Proposition 3.9.2. *Let φ be regular and let Y be a C^∞ vector field on N such that for every $m \in \varphi^{-1}$ (domain of Y), $Y(\varphi m) \in \varphi_* M_m$. Then there is a unique C^∞ vector field X defined on φ^{-1} (domain of Y) which is φ-related to Y.*

Proof. It is clear that X is unique, for $X(m) = \varphi_{*m}^{-1}(Y(\varphi m))$, which makes sense since $\varphi_{*m}: M_m \to N_{\varphi m}$ is 1–1 and contains $Y(\varphi m)$ in its range.

To show that X is C^∞ we compute its components with respect to coordinates $x^i = y^i \circ \varphi$ as in Lemma 3.9.1. Indeed, $X^i = Xx^i = X(y^i \circ \varphi) = (Yy^i) \circ \varphi = Y^i \circ \varphi$, where Y^α are the components of Y with respect to the y^α coordinates. Thus X^i is C^∞, since Y^i and φ are C^∞. $\quad \blacksquare$

The notion of φ-relatedness is easily extended to contravariant tensor fields. The linear function φ_{*m} has an extension to a *homomorphism* of the algebra of contravariant tensors over M_m, which means that the extension is linear and commutes with tensor product formation: $\varphi_{*m}(A \otimes B) = (\varphi_{*m}A) \otimes (\varphi_{*m}B)$, for all contravariant tensors A and B over M_m. Then we define contravariant tensor field S on M to be φ-related to contravariant tensor field T on N if for every $m \in$ (domain of S), $\varphi_{*m}S(m) = T(\varphi m)$. The following lemma is not difficult to prove and has the generalization of Proposition 3.9.1 as an almost immediate consequence.

Lemma 3.9.2.   *If X is φ-related to Y, {μ$_s$} is the flow of X, and {θ$_s$} the flow of Y, then for every s, φ ∘ μ$_s$ = θ$_s$ ∘ φ; that is, the following diagram is commutative:*

(To prove this show that φ maps an integral curve of X into an integral curve of Y.)

Proposition 3.9.3.   *If C^∞ contravariant tensor fields S and T are φ-related and C^∞ vector fields X and Y are φ-related, then $L_X S$ is φ-related to $L_Y T$.*
(Proof omitted.)

If φ is regular, then Proposition 3.9.2 also extends to contravariant tensor fields with very little effort.

The situation for covariant tensor fields is quite different. First of all, the direction in which they are mapped is opposite to that of contravariant tensors, since the dual of $\varphi_{*m}: M_m \to N_{\varphi m}$ is the "transpose" $\varphi^*_m: N^*_{\varphi m} \to M^*_m$, defined by the relation $\langle \varphi_{*m} v, \tau \rangle = \langle v, \varphi^*_m \tau \rangle$ for all $v \in M_m$ and $\tau \in N_{\varphi m}$. For this reason the classical names "contravariant" and "covariant" are backwards from a mapping-oriented viewpoint, since the tangent vectors go "with" the map φ and the dual vectors go "against" φ. From this viewpoint names such as "tangential tensor fields" and "cotangential tensor fields" would seem more appropriate.

Second, there is no problem of existence for φ-related covariant tensor fields, whereas with contravariant fields we only have results when we *assume* the φ-related fields are given and we have proved existence only under the restrictive assumption that φ is regular. Even then the existence result was in a peculiar direction, reverse to the direction in which individual contravariant tensors map. The following result for covariant fields is more natural and inclusive.

Proposition 3.9.4.   *Let T be a C^∞ covariant tensor field on N. Then there is a unique C^∞ covariant tensor field $S = \varphi^* T$ defined on $E = \varphi^{-1}$ (domain of T) such that for every $m \in E$, $\varphi^*_m T(\varphi m) = S(m)$. (Here φ^*_m has been extended as a homomorphism.) An alternative definition is that for $v_1, \ldots, v_q \in M_m$, $S(v_1, \ldots, v_q) = T(\varphi_{*m} v_1, \ldots, \varphi_{*m} v_q)$, where $(0, q)$ is the type of T.*

Symmetry and skew-symmetry are preserved by φ^*. Moreover, φ^* commutes with the symmetrizing and alternating operators, so is a homomorphism of the covariant symmetric and Grassmann algebras.

Proof. To show that the two definitions are equivalent we compute as follows on a typical term of T, where $\tau_1, \ldots, \tau_q$ are chosen from a local basis of 1–forms dy^α, and f is a C^∞ scalar field on N:

$$
\begin{aligned}
(\varphi_m^* f \tau_1 \otimes \cdots \otimes \tau_q)(v_1, \ldots, v_q) &= f(\varphi m)\varphi_m^* \tau_1(v_1) \cdots \varphi_m^* \tau_q(v_q) \\
&= f(\varphi m)\langle v_1, \varphi_m^* \tau_1 \rangle \cdots \langle v_q, \varphi_m^* \tau_q \rangle \\
&= f(\varphi m)\langle \varphi_{*m} v_1, \tau_1 \rangle \cdots \langle \varphi_{*m} v_q, \tau_q \rangle \\
&= f(\varphi m)\tau_1 \otimes \cdots \otimes \tau_q(\varphi_{*m} v_1, \ldots, \varphi_{*m} v_q).
\end{aligned}
$$

To show that $\varphi^* T$ is C^∞ it suffices to prove it for C^∞ scalar fields and basis 1–forms dy^α, since in general $\varphi^* T$ is a sum of products of $\varphi^* f$'s and $\varphi^* dy^\alpha$'s.

For a scalar field f we have $(\varphi^* f)m = f(\varphi m) = (f \circ \varphi)m$, that is, $\varphi^* f = f \circ \varphi$, which is C^∞ if f is C^∞.

If x^i are coordinates at m and y^α coordinates at φm, then the matrix of φ_{*p} with respect to bases $\partial/\partial x^i(p)$ and $\partial/\partial y^\alpha(\varphi p)$ is $(\partial(y^\alpha \circ \varphi)/\partial x^i(p))$, where α is the row index and i is the column index; p is any point in the x^i coordinate domain. The transpose, with the same entries but with i as the row index and α as the column index, is the matrix of φ_p^*. Thus we have $\varphi^* \, dy^\alpha = [\partial(y^\alpha \circ \varphi)/\partial x^i] \, dx^i = d(y^\alpha \circ \varphi)$, which is C^∞ since y^α and φ are C^∞.

The fact that φ^* preserves symmetry and skew-symmetry and commutes with the operators is evident from the second definition. ∎

In terms of coordinates the operation of φ^* on a covariant tensor field T amounts to substituting the coordinate equations for φ into the coordinate expression for T. More explicitly, suppose that the equations for φ are

$$
y^\alpha = F^\alpha(x^1, \ldots, x^d) = F^\alpha(x), \qquad \alpha = 1, \ldots, e,
$$

and that T is of type $(0, 2)$, with expression in the y^α coordinates

$$
T = T_{\alpha\beta}(y^1, \ldots, y^e) \, dy^\alpha \otimes dy^\beta.
$$

Then the x^i coordinate expression for $\varphi^* T$ is

$$
\varphi^* T = T_{\alpha\beta}(F^1(x), \ldots, F^e(x))(\partial F^\alpha/\partial x^i)(\partial F^\beta/\partial x^j) \, dx^i \otimes dx^j.
$$

Remark. It has been noted in the proof that for coordinate function y^α, $\varphi^* \, dy^\alpha = d(\varphi^* y^\alpha)$. More generally, φ^* and d commute on any function:

$$
\begin{aligned}
(\varphi^* \, df)v &= df(\varphi_* v) \\
&= (\varphi_* v)f \\
&= v(f \circ \varphi) \\
&= v(\varphi^* f) \\
&= (d\varphi^* f)v,
\end{aligned}
$$

so we have $\varphi^* \, df = d\varphi^* f$. In Chapter 4 we extend the operator d to act on skew-symmetric tensor fields (differential forms), and the property of commutation with φ^* will also be extended.

Problem 3.9.1. If g is a riemannian metric on N and $\varphi: M \to N$ is regular, show that $\varphi^* g$ is a riemannian metric on M. Show by examples that if g is a semi-riemannian metric, then $\varphi^* g$ is not necessarily a semi-riemannian metric, and it might also be a semi-riemannian metric of different index than g. Show that if φ is not regular and g is a riemannian metric, then $\varphi^* g$ is not a riemannian metric.

Problem 3.9.2. Let M be the hyperboloid of revolution in R^3 given by the equation $x^2 + y^2 - z^2 = -1$ and inequality $z > 0$, let g be the *Minkowski metric* on R^3, that is, $g = (dx)^2 + (dy)^2 - (dz)^2$ (since semi-riemannian metrics are symmetric it is customary to use symmetric product notation in writing them), and let $\varphi: M \to R^3$ be the inclusion map. Show that $h = \varphi^* g$ is a riemannian metric on M. The riemannian manifold (M, h) is called the *hyperbolic plane*. [It has constant curvature -1 (see Section 5.14) and is a negative dual of the euclidean sphere $S^2: x^2 + y^2 + z^2 = 1$, which has constant curvature 1. Many of the properties of M are similar to those for a sphere. For example, the geodesics (shortest paths on the surface; see Section 5.13) are the intersections of M with planes through the origin of R^3, just as the geodesics of S^2 (great circles) are intersections of S^2 with planes through the origin.]

Problem 3.9.3. If T is a skew-symmetric tensor field of type $(0, q)$ on N and $q > \dim M$, show that $\varphi^* T = 0$.

3.10. Critical Point Theory

A *critical point* of a C^∞ scalar field $f: M \to R$ is a point m such that $df_m = 0$. In terms of coordinates this means that all the partial derivatives of f are 0 at m, $\partial_i f(m) = 0$. A point m is a *relative maximum (minimum)* point of f if there is a neighborhood U of m such that for every $n \in U$, $fm \geq fn$ ($fm \leq fn$). If m is a critical point, relative maximum point, or relative minimum point of f, we say that fm is a critical value, relative maximum value, or relative minimum value of f, respectively.

Proposition 3.10.1. *If m is a relative maximum or minimum point of f, then m is a critical point of f.*

Proof. If $v \in M_m$, let γ be a curve such that $\gamma_* 0 = v$. Then $f \circ \gamma$ is a real-valued function of one variable such that 0 is a relative maximum or minimum point. Hence $d(f \circ \gamma)/du(0) = 0 = (\gamma_* 0)f = vf = df_m(v)$, so that $df_m = 0$; that is, m is a critical point of f. ∎

A point m is a *maximum (minimum) point* of f if for every $n \in M$, $fm \geq fn$ ($fm \leq fn$). A maximum (minimum) point is clearly a relative maximum (minimum) point, and hence a critical point.

If m is a critical point of f, we define the *hessian of f at m* to be the bilinear form H_f on M_m defined as follows. If $v, w \in M_m$, let W be any extension of w to a C^∞ vector field. Then

$$H_f(v, w) = v(Wf).$$

It is not immediately clear from this definition that H_f is well defined, since there may be a dependance on the choice of extension W. Thus we need

Proposition 3.10.2.   *The hessian of f at m is well defined and symmetric. The components of H_f with respect to a coordinate basis $\partial_i m$ are $(\partial_i \partial_j f)m$.*

Proof. Let V be a C^∞ extension of v. Then we have $[V, W](m)f = 0$, since $df_m = 0$. Hence $vWf = (VWf)m = (WVf)m = wVf$. In the equation $vWf = wVf$, vWf does not depend on which extension V of v is used, wVf does not depend on which extension W of w is used, and it is clear that the common value $H_f(v, w)$ is symmetric in v and w.

One extension of $\partial_i m$ is ∂_i. Thus the coordinate components of H_f are

$$H_f(\partial_i m, \partial_j m) = \partial_i(m)\partial_j f = (\partial_i \partial_j f)m. \quad \blacksquare$$

A critical point m of f is *nondegenerate* if H_f is nondegenerate. This is equivalent to $\det((\partial_i \partial_j f)m) \neq 0$; if we let $y^i = \partial_i f$, then $((\partial_i \partial_j f)m) = ((\partial_i y^j)m)$ is the jacobian matrix of the functions y^i with respect to the coordinates. Thus we have

Proposition 3.10.3.   *A critical point m of f is nondegenerate iff the partial derivatives $\partial_i f = y^i$ form a coordinate system at m.*

A function f is *nondegenerate* if all its critical points are nondegenerate.

By choosing a basis of M_m which is orthonormal with respect to H_f, and then choosing coordinates x^i such that $x^i m = 0$ and $\partial_i m$ is the orthonormal basis for H_f, we have $(\partial_i \partial_j f)m = \delta_{ij} \varepsilon_i$ (i not summed), where $\varepsilon_i = 1, -1$, or 0. We number so that the -1's are first, $\varepsilon_i = -1$, $i = 1, \ldots, I$, and the 0's last. The second-order Taylor expansion for f at m has the form $f = fm + f_{ij} x^i x^j$, where the f_{ij} are C^∞ functions such that $f_{ij} m = \delta_{ij} \varepsilon_i$ and $f_{ij} = f_{ji}$. Now, assuming not all $\varepsilon_i = 0$, we may proceed in a manner similar to the process for diagonalizing a quadratic function, obtaining coordinates for which f has as simple an expression as possible near m. Namely, we let

$$y^1 = (-f_{11} x^1 - f_{12} x^2 - \cdots - f_{1d} x^d)/(-f_{11})^{1/2},$$

and we find that the jacobian matrix of $y^1, x^2, \ldots, x^d$ with respect to $x^1, \ldots, x^d$ is nonsingular at m, so $y^1, x^2, \ldots, x^d$ is a coordinate system at m. Furthermore,

$$f = fm - (y^1)^2 + \sum_{i,j \geq 2} g_{ij} x^i x^j, \qquad \text{where } g_{ij} m = \delta_{ij} \varepsilon_i.$$

This is the first step of a recursive procedure for which the continuation should now be obvious. In the steps for which $i > I$ the formula for y^i resembles that for y^1 above except that all the signs are changed:

$$y^i = (k_{ii}x^i + \ldots + k_{id}x^d)/k_{ii}^{1/2}.$$

The procedure ends when we have generated r new coordinates, $y^1, \ldots, y^r$, which together with $x^{r+1}, \ldots, x^d$ form a coordinate system at m, where $d - r$ is the rank of H_f. In terms of these new coordinates the expression for f has the form

$$f = fm - \sum_{i=1}^{I} (y^i)^2 + \sum_{i=I+1}^{r} (y^i)^2 + \sum_{i,j \geq r+1} h_{ij}x^i x^j.$$

When $r = d$ the formula has no annoying remainder with h_{ij}'s. The formula obtained thus says that f is a quadratic form plus a constant. This is known as the *Morse Lemma.* A more trivial step is the case where the differential of the function is not zero at the point and hence the function may be taken as the first coordinate function. These two steps are the substance of the following proposition.

Proposition 3.10.4. (a) *If m is not a critical point of f, then there are coordinates y^i such that $f = y^1$ in a neighborhood of m.*

(b) (**Morse Lemma.**) *If m is a nondegenerate critical point of f, then there are coordinates y^i at m such that*

$$f = fm - \sum_{i=1}^{I} (y^i)^2 + \sum_{i=I+1}^{d} (y^i)^2,$$

where I is the index of H_f.

Corollary. *If m is a nondegenerate critical point of f, then m is an isolated critical point of f; that is, there are no other critical points in some neighborhood of m.*

Proof. It is permissible to search for critical points as we do in advanced calculus, by equating all the partial derivatives to 0. In terms of the "Morse coordinates" this process is very easy, since the equations simply read $y^i = 0$ for all i. As long as we stay within the coordinate neighborhood, m is the only solution. ∎

We shall call a critical point m of f a *quadratic critical point* if there are coordinates y^i at m for which the coordinate expression for f is a quadratic function of the y^i. In particular, a nondegenerate critical point is quadratic. However, if the nullity of H_f is not zero, then a quadratic critical point will not be isolated. In fact, the equations for critical points show that all of the points in a coordinate slice $y^1 = 0, \ldots, y^r = 0$, are again quadratic critical points with hessians of the same index and nullity. Thus, under the assumption that a function has only quadratic critical points the set of all critical points has a nice structure, since these coordinate expressions show that it consists of a union of closed, nonoverlapping submanifolds.

The classification of critical points without such special assumptions is a subject of intense current research. There have been remarkable developments since the first edition of this book was published. For those who wish to pursue the subject we point out that the book by V. Guillemin and M. Golubitsky, *Stable Mappings and Their Singularities*, NY, Springer, 1974, is a thorough introduction. However, it is not easy unless one is familiar with commutative ring theory.

Problem 3.10.1.   (a) If $f = xy(x + y): R^2 \to R$, show that f has an isolated critical point for which $H_f = 0$. (The shape of the surface $f = 0$ at the critical point is called a *monkey saddle*. Why?)

   (b) Show that $f = x^3: R^2 \to R$ has a submanifold of nonisolated critical points, for all of which $H_f = 0$.

   (c) What are the submanifolds of critical points of $f = x^2 y^2: R^2 \to R$ for which $H_f \neq 0$? Are these submanifolds closed? Are there any critical points such that $H_f = 0$?

   (d) Show that the critical points of $f = x^3 y^3: R^2 \to R$ at which $H_f = 0$ do not form a submanifold.

Problem 3.10.2.   If f has only quadratic critical points, show that the critical points at which H_f has fixed index I and fixed nullity n form a submanifold M_I^n of dimension n. Moreover, M_I^n is a closed submanifold and f is constant on each connected component of M_I^n.

Problem 3.10.3.   Let M be the usual doughnut-shaped torus contained in R^3 with center at the origin and the z axis as the axis of revolution. Find the critical submanifolds M_I^n in the cases $f = r|_M$, where r is the spherical radial coordinate on R^3 (fm = the distance from m to 0 in R^3), and $f = z|_M$, the *height* function on M. What happens if M is pushed slightly off center or tilted?

Problem 3.10.4.   Show that if f is nondegenerate and M is compact, then f has only a finite number, at least two, of critical points.

   The problem of finding the maximum or minimum of a function on a manifold frequently can be solved by employing Proposition 3.10.1 and, in the more difficult cases, the other results above. The first step is always to solve for the critical points of the function. This usually involves only a finite number of sets of equations $\{\partial_i f = 0\}$, since most manifolds arising in applications can be covered by finitely many coordinate systems. Then one compares the values of f on these critical points (or submanifolds of critical points in the more difficult cases) and sorts out the greatest or least.

   Sometimes the manifold M is a hypersurface or the intersection of hypersurfaces of another manifold, and f is the restriction to M of a function (still called f) on the larger manifold N. A hypersurface is determined, at least locally, as a level hypersurface $g = c$ of a function g on N, such that $dg \neq 0$ on M. We can express the condition for a critical point of f on the hypersurface $g = c$ by the method of *lagrangian multipliers*. The rule is that we solve the equations $df(n) = \lambda\, dg(n)$, $gn = c$ simultaneously for λ and n, and then n is a critical point of $f|_M$. This rule is proved easily by using the fact that there are coordinates on N in any neighborhood of $n \in M$ which have the form $x^1 = g, x^2, \ldots, x^d$. Then $y^2 = x^2|_M, \ldots, y^d = x^d|_M$ are coordinates on M and if $df = \lambda_i\, dx^i$ on N, the restriction to M is $d(f|_M) = \sum_{i>1} \lambda_i\, dy^i$. Thus the condition that $f|_M$ have a critical point at $n \in N$ is that $n \in M$ ($gn = c$) and that $\lambda_i n = 0, i > 1$; that is, $df(n) = (\lambda_1 n)\, dg(n)$.

If M is the intersection of hypersurfaces $g_1 = c_1, \ldots, g_k = c_k$, then it is easy to generalize the rule as follows:

Solve $df(n) = \sum_{\alpha=1}^k \lambda_\alpha\, dg_\alpha(n)$ and $g_\alpha n = c_\alpha, \alpha = 1, \ldots, k$, simultaneously for $\lambda_1, \ldots, \lambda_k$ and n. In applying this rule, the g_α and f may be expressed in terms of any convenient coordinates z^i on N, of course.

Example.   Suppose we wish to find the maximum of $f = xy + yz - xz$ on the sphere $S: x^2 + y^2 + z^2 = 1$. In the direct method we would choose several coordinate systems which cover S (say, x, y, z in pairs restricted to various hemispheres) and express f in terms of them [substitute $z = (1 - x^2 - y^2)^{1/2}$ in f, etc.], and solve for the zeros of the partial derivatives. In this case the method of langrangian multipliers is simpler. We need to solve $df = \lambda\, dg$; that is,

$$(y - z)\, dx + (x + z)\, dy + (y - x)\, dz = \lambda(2x\, dx + 2y\, dy + 2z\, dz)$$

and $x^2 + y^2 + z^2 = 1$ for x, y, z, and λ. That is,

$$y - z = 2\lambda x,$$
$$x + z = 2\lambda y,$$
$$y - x = 2\lambda z,$$

or in matrix form,

$$\begin{pmatrix} -2\lambda & 1 & -1 \\ 1 & -2\lambda & 1 \\ -1 & 1 & -2\lambda \end{pmatrix} \begin{pmatrix} x \\ y \\ z \end{pmatrix} = 0.$$

Since $(x, y, z) \neq (0, 0, 0)$, the matrix must be singular, so its determinant $-8\lambda^3 + 6\lambda - 2 = 0$. The roots of this are $\lambda = -1, 1/2, 1/2$.

If $\lambda = -1$, we get $y = -x$ and $z = x$ from the linear equations. Then from $g = 1$ we get

(1) $(x, y, z) = (1/\sqrt{3}, -1/\sqrt{3}, 1/\sqrt{3})$

or

(2) $(x, y, z) = (-1/\sqrt{3}, 1/\sqrt{3}, -1/\sqrt{3})$.

If $\lambda = 1/2$, we get only $z = y - x$ from the linear equations. Since this is a plane through the origin, the intersection with M is the great circle in that plane, a critical submanifold of dimension 1 which is connected. Hence $f|_M$ is constant on this submanifold and we need to check the value only at one point, say at

$$(3) \quad (x, y, z) = (1/\sqrt{2}, 1/\sqrt{2}, 0).$$

Since f is quadratic in x, y, z, it has the same values on opposite points, $fp = f(-p)$, so the values on the first two points (1) and (2) are the same,

$$f(1/\sqrt{3}, -1/\sqrt{3}, 1/\sqrt{3}) = -1/3 - 1/3 - 1/3 = -1.$$

On the critical submanifold f has value $f(1/\sqrt{2}, 1/\sqrt{2}, 0) = 1/2$.

We conclude that f has two minimum points (1) and (2) and a great circle of maximum points: $z = y - x$, $x^2 + y^2 + z^2 = 1$.

An elaborate theory (*Morse theory*) has been developed by Marston Morse which relates the number and types of critical points to certain topological invariants of M called *Betti numbers* (see Section 4.6). This theory can be used in either direction; that is, a knowledge of these invariants can be used to assert the existence of critical points of certain types, and in many cases the Betti numbers can be computed by a judicious choice of a function for which the critical point structure is easily calculated.

The theory is applicable to any C^∞ function on a compact manifold. For functions on noncompact manifolds it is assumed that the *level sets*, $f^c = \{m \mid fm \le c\}$, are compact, at least until c is large enough so that f^c contains all the critical points. It is easier to apply the theory in the case when f is nondegenerate, since then it only involves counting the number of critical points, M_I, for which H_f has index I. We call M_I the Ith *Morse number* of f. To show that the easiest case of his theory is quite general, Morse has shown that any C^∞ function can be perturbed slightly so that it becomes nondegenerate (cf. Problem 3.10.3, where a slight displacement of the torus causes the height and radial functions to become nondegenerate).

We illustrate Morse theory by giving a direct plausibility argument for the case of a function having only quadratic critical points on the sphere S^2. Since there are only two connected one-dimensional manifolds, R and the circle S^1, and the components of the one-dimensional critical submanifolds must be closed, hence compact (Problem 3.10.2), the only critical submanifolds consist of isolated points (zero-dimensional) and circles (one-dimensional). (The only two-dimensional submanifolds of S^2 are open submanifolds, and if one such is also closed, then by the connectedness of S^2 it must be all of S^2. Hence if f has a two-dimensional critical submanifold, f is constant, contradicting the hypothesis that the critical points are of the quadratic type.) We further classify the connected critical submanifolds by the index, using the following descriptive terms, arrived at by thinking of f as being an "elevation function" on an earthly surface. The connected critical submanifolds are isolated (they are contained in an open set having no other critical points) and hence finite in number (by the compactness of S^2).

$M_0^0 = $ those critical points at which H_f is positive definite

     $= $ a finite number P_0 of points, the *pits* (local minima).

$M_1^0 = $ those critical points at which H_f is nondegenerate and indefinite

     $= $ a finite number P_1 of points, the *passes* (saddle points).

$M_2^0 = $ a finite number P_2 of points, the *peaks* (local maxima, H_f negative definite).

$M_0^1 = $ those points m at which f has a local coordinate expression $fm + x^2$,
     x and y coordinates, nonisolated local minima

     $= $ a finite number R_0 of circles, *circular valleys*.

$M_1^1 = $ a finite number R_1 of circles, *circular ridges*, which are nonisolated local maxima.

We reiterate that our assumption of quadratic critical points forces the components of the set of critical points to be compact submanifolds, hence circles or points; there can be no segment-type ridges, since at the end of such a ridge there would be a critical point which is not quadratic.

We shall obtain the Morse relation $P_2 - P_1 + P_0 = 2$ by the following device. (The integer 2 is a topological invariant of S^2, called its *Euler-Poincaré characteristic*.) View the surface as an initially bone dry earth on which there is about to fall a deluge which ultimately covers the highest peak. We count the number of lakes and connected land masses formed and destroyed in this rainstorm to obtain the result.

For each pit there will be one lake formed.

For each pass there will be either two lakes joined (there are P_{11} of this type), or a single lake doubling back on itself and disconnecting one land mass from another (there are P_{12} of this type).

For each peak a land mass will be eliminated.

For each circular valley a lake and a land mass will be formed.

For each circular ridge two lakes will be joined and a land mass inundated.

Thus we have

number of lakes formed $= P_0 - P_{11} + R_0 - R_1$,

number of land masses formed $= P_{12} - P_2 + R_0 - R_1$,

initial situation: one land mass,

final situation: one lake,

lake count: $0 + P_0 - P_{11} + R_0 - R_1 = 1$,

land count: $1 + P_{12} - P_2 + R_0 - R_1 = 0$.

Subtracting the last two equations and using $P_1 = P_{11} + P_{12}$ gives

$$P_2 - P_1 + P_0 = 2,$$

as desired.

Problem 3.10.5. Modify the above procedure to obtain the corresponding result for a function on a toroidal earth: $P_2 - P_1 + P_0 = 0$. (The Euler-Poincaré characteristic of the torus is 0.) Note that twice two lakes will join in each direction around the torus without disconnecting any land mass.

Problem 3.10.6. Construct a function on the torus which has only three critical points. Why must at least one be nonquadratic? [*Hint:* View the torus as an identified square. Divide it into two triangles by a diagonal, put a maximum in the inside of one triangle, a minimum inside the other, and a monkey saddle at the vertex (the identified corners) so that the edges of the triangles all have one f-value.]

Problem 3.10.7. Let f be a C^∞ function on the plane R^2 such that f^c is compact for every c and such that f has only quadratic critical points. Show that the connected critical submanifolds are either points (zero-dimensional) or circles (one-dimensional). If there are only a finite number of them, show that $P_2 - P_1 + P_0 = 1$ and $P_1 + R_1 \geq P_0 - 1$. (The notation is the same as in the above example. The Euler-Poincaré characteristic of R^2 is 1. The inequality is another Morse relation, which is trivial in the case of compact manifolds because for them it follows from the existence of both a maximum and a minimum.)

3.11.   First-Order Partial Differential Equations

In this section we are concerned with partial differential equations of the simplest sort, linear homogeneous first-order partial differential equations in one unknown. If f is the unknown, these have the form

$$X^i \partial_i f = 0.$$

We shall also treat systems of such equations, that is, the problem of finding an f which simultaneously satisfies

$$X_1^i \partial_i f = 0,$$
$$X_2^i \partial_i f = 0,$$
$$\vdots$$
$$X_k^i \partial_i f = 0.$$

By *linear* we mean that if f and g are solutions and α is any real number, then αf and $f + g$ are solutions. Thus the solutions form a vector space. By *homogeneous* we mean that the right sides of the equations are zeros.

One of our goals is to generalize the formulation of the problem from a search for a function f on a coordinate neighborhood to a search for a function on a manifold.

The reason we can treat these problems here is that they are not, in a sense, partial differential equations at all, since their solutions, when possible, are obtained by means of ordinary differential equations and the use to which they have been put in the study of flows of vector fields.

As a first step we simplify our notation for the problem by writing $X_\alpha = X_\alpha^i \partial_i$, $\alpha = 1, \ldots, k$, so what we are looking for are functions annihilated by the k vector fields $X_1, \ldots, X_k$; that is,

$$X_\alpha f = 0, \qquad \alpha = 1, \ldots, k.$$

We assume that the maximum number of linearly independent $X_\alpha(m)$ is constant as a function of m. It is not that the case where this number is non-constant is uninteresting, but it is more difficult and would require nonuniform techniques from point to point. Thus if the number of linearly independent $X_\alpha(m)$ varies as a function of m, we call the problem *degenerate*. With the assumption of nondegeneracy, if, say, $X_1(m), \ldots, X_h(m)$ are a maximum number of linearly independent $X_\alpha(m)$ at m, then by continuity $X_1(n), \ldots, X_h(n)$ are linearly independent for all n in some neighborhood U of m. It follows that we have $X_\alpha(n) = \sum_{\beta=1}^h F_\alpha^\beta(n) X_\beta(n)$ for each $n \in U$, $\alpha = h+1, \ldots, k$. Thus if $X_\beta f = 0$, then $X_\alpha f = 0$, so we can always reduce locally to the linearly independent number h of equations.

We call $X_1, \ldots, X_h$ a *local basis* of the system in the neighborhood U, and h is called the *dimension* of the system. If we move to another point p outside U, then $X_1, \ldots, X_h$ may become linearly dependent, but in some neighborhood V of p, some other h of the X_α's will be a local basis. In the intersection $U \cap V$ we have two or more local bases, and in general many local bases. In fact, if we have $Y_\alpha = \sum_{\beta=1}^h G_\alpha^\beta X_\beta$, $\alpha = 1, \ldots, h$, where the matrix (G_α^β) of C^∞ functions on U has nonzero determinant at each point, then the equations $Y_\alpha f = 0$ have the same solutions on U as $X_\alpha f = 0$, so the Y_α should be considered a local basis also. In fact, what we do to solve such systems is, in a sense, to choose a local basis Y_α in as simple a fashion as possible. We illustrate this first in the case $h = 1$.

If $h = 1$ then, say, $X_1(m) \neq 0$. By Theorem 3.5.1 there are coordinates x^i at m such that $X_1 = \partial_1$. Our equations, in terms of x^i coordinates, become simply $\partial_1 f = 0$. Hence a solution is given by any function not dependent on x^1, that is, a function of $x^2, \ldots, x^d$. In other words, f is any function which is constant along each of the integral curves (*trajectories*) of X_1. Of course, this latter fact is quite evident from the original equation $X_1 f = 0$.

The step from $h = 1$ to $h = 2$ is difficult due to the following fact: If $X_\alpha f = 0$ for $\alpha = 1, \ldots, h$, then $[X_\alpha, X_\beta]f = 0$ for $\alpha, \beta = 1, \ldots, h$. This is trivial since $[X_\alpha, X_\beta]f = X_\alpha(X_\beta f) - X_\beta(X_\alpha f) = X_\alpha 0 - X_\beta 0 = 0$. As a consequence, if, say, $h = 2$ and $[X_1, X_2]$ is linearly independent of the local basis

X_1 and X_2, then the system $X_1 f = 0$, $X_2 f = 0$ for which $h = 2$, does not have more solutions than the system $X_1 f = 0$, $X_2 f = 0$, $[X_1, X_2]f = 0$ for which $h = 3$. Thus the number of variables on which f depends is determined not only by h but also by the relation of the X_α to each other.

In the following we shall use Greek letters α, β, and γ as summation indices running from 1 to h.

To generalize the concept of a linear homogeneous system to manifolds we fix our attention on the subspaces spanned by the $X_\alpha(m)$. Thus we have assigned to every m an h-dimensional subspace, $D(m)$, of M_m. If $X_\alpha f = 0$ for every α, then for every $t \in D(m)$, t is a linear combination of the $X_\alpha(m)$ with coefficients, say, c^α, so that $tf = c^\alpha X_\alpha(m)f = (c^\alpha X_\alpha f)m = 0$. Conversely, if $tf = 0$ for every $t \in D(m)$ and for every m, then $(X_\alpha f)m = X_\alpha(m)f = 0$ for all α and m, since $X_\alpha(m) \in D(m)$. Hence the problem of finding a function annihilated by all vectors in $D(m)$ for every m is equivalent to the solution of the system of partial differential equations under discussion.

A function D which assigns to each $m \in M$ an h-dimensional subspace $D(m)$ of M_m is called an *h-dimensional distribution*† on M. An h-dimensional distribution D is C^∞ if for every $m \in M$ there is a neighborhood U of m and C^∞ vector fields $X_1, \ldots, X_h$ defined on U such that for every $n \in U$, $X_1(n), \ldots,$ $X_h(n)$ is a basis of $D(n)$. Such $X_1, \ldots, X_h$ are then called a *local basis* for D at m. (An h-dimensional distribution is also called a *differential system of h-planes* or simply a *field of h-planes*. If $h = 1$, we say we have a *field of line elements*.)

A vector field X *belongs to* D, written $X \in D$, if for every m in the domain of X, $X(m) \in D(m)$. A C^∞ distribution D is *involutive* if for all $X, Y \in D$ we have $[X, Y] \in D$.

Proposition 3.11.1. *A C^∞ distribution D is involutive iff for every local basis $X_1, \ldots, X_h$ the brackets $[X_\alpha, X_\beta]$ are linear combinations of the X_γ; that is, there are C^∞ functions $F^\gamma_{\alpha\beta}$ such that $[X_\alpha, X_\beta] = F^\gamma_{\alpha\beta} X_\gamma$.*

Proof. If D is involutive, then $[X_\alpha, X_\beta] \in D$ and hence $[X_\alpha, X_\beta]$ can be expressed as a linear combination of the local basis $X_1, \ldots, X_h$. The fact that the coefficients of these linear combinations are C^∞ is left as an exercise.

If $[X_\alpha, X_\beta] = F^\gamma_{\alpha\beta} X_\gamma$, then for $X, Y \in D$ we may write $X = G^\alpha X_\alpha$, $Y = H^\alpha X_\alpha$, where the G^α and H^α are C^∞ functions (same exercise!). Then

$$[X, Y] = [G^\alpha X_\alpha, H^\beta X_\beta]$$
$$= G^\alpha(X_\alpha H^\beta)X_\beta - H^\beta(X_\beta G^\alpha)X_\alpha + G^\alpha H^\beta F^\gamma_{\alpha\beta} X_\gamma,$$

which clearly belongs to D. ∎

---

† There is no connection with Schwartz distributions, that is, generalized functions such as the Dirac delta function. A more reasonable name would be *tangent subbundle*.

Remark. The equations $[X_\alpha, X_\beta] = F^\gamma_{\alpha\beta} X_\gamma$, usually written in coordinate form

$$X^i_\alpha \partial_i X^j_\beta - X^i_\beta \partial_i X^j_\alpha = F^\gamma_{\alpha\beta} X^j_\gamma,$$

are called the *integrability conditions* of the system of equations $X^i \partial_i f = 0$. They are the classical hypotheses for the local complete integrability theorem of Frobenius stated in Section 3.12.

Examples.   (a) Let $Z = y\partial_x - x\partial_y$, $X = z\partial_y - y\partial_z$, and $Y = x\partial_z - z\partial_x$, restricted to $M = R^3 - \{0\}$. Then at any $m \in M$, X, Y, and Z span a two-dimensional subspace $D(m)$ of M_m. We may describe D directly by the fact that $D(m)$ is the subspace of M_m normal to the line in E^3 through 0 and m. (E^3 is R^3 with the usual euclidean metric.) Since $[X, Y] = Z$, $[Y, Z] = X$, and $[Z, X] = Y$, the distribution is involutive.

(b) The distribution on R^d with local basis $\partial_1, \ldots, \partial_h$ is involutive since $[\partial_\alpha, \partial_\beta] = 0 \in D$. One way of stating Frobenius' theorem is that locally every involutive distribution has this form; that is, for an involutive distribution there exist coordinates at each point such that $\partial_1, \ldots, \partial_h$ is a local basis of D.

An *integral submanifold* of D is a submanifold N of M such that for every $n \in N$ the tangent space of N at n is contained in $D(n)$; that is, $N_n \subset D(n)$.

If $X \in D$ and $X(m) \neq 0$, then the range of an integral curve γ of X is a one-dimensional integral submanifold if γ is defined on an open interval. Locally γ can be inverted, so that the parameter of γ becomes a coordinate on the one-dimensional manifold. The parameter can be used as a single global coordinate provided γ is 1–1. If γ is not 1–1, then it is periodic and the submanifold is diffeomorphic to a circle; the parameter may be restricted in different ways to become a local coordinate.

In Example (a), any euclidean sphere with center 0 is an integral submanifold. Other integral submanifolds consist of any open subset of such a sphere and unions of such open subsets contained in countably many such spheres. (The countability is required so that the submanifold will be separable.)

An h-dimensional distribution is *completely integrable* if there is an h-dimensional integral submanifold through each $m \in M$. The one-dimensional C^∞ distributions are completely integrable, since the local basis field will always have integral curves. The "spherical" two-dimensional distribution of Example (a) is completely integrable, since there is a central sphere through each point. Not every two-dimensional C^∞ distribution is integrable, since, for example, the vector fields ∂_x and $\partial_y + x\partial_z$ on R^3 span a two-dimensional distribution but $[\partial_x, \partial_y + x\partial_z] = \partial_z$ does not belong to the distribution. The following proposition then tells us that this particular distribution and many others are not completely integrable. It is the converse of Frobenius' theorem.

Proposition 3.11.2. *A completely integrable C^∞ distribution is involutive.*

Proof. Suppose D is completely integrable and that $X, Y \in D$. Let $m \in$ domain of $[X, Y]$ and let N be an h-dimensional integral submanifold of D through m. Then the inclusion map $i: N \to M$ is regular and for every $n \in N \cap$ (domain of $[X, Y]$) we have $X(n) \in D(n) = i_*(N_n)$ and $Y(n) \in i_*(N_n)$. By Proposition 3.9.2 there are unique C^∞ vector fields, called $X|_N$ and $Y|_N$, which are i-related to X and Y, respectively. By Proposition 3.9.1, $[X|_N, Y|_N]$ is i-related to $[X, Y]$, so in particular $i_*[X|_N, Y|_N](m) = [X, Y](m) \in N_m = D(m)$. Thus we have proved that $[X, Y] \in D$; that is, D is involutive. ∎

A *solution function*, that is, a *first integral* of D, is a C^∞ function f such that for every $m \in$ (domain of f) and every $t \in D(m)$, $tf = 0$; that is, $D(m)$ annihilates f, or, df annihilates $D(m)$. Of course, constants are solution functions, but are rather useless in studying D. If f is a solution function such that $df_m \neq 0$, that is, m is not a critical point of f, then the level hypersurface $f = c$, where $c = fm$, is a $(d-1)$-dimensional submanifold M_1 in a neighborhood of m on which $df \neq 0$. The tangent spaces of M_1 are the subspaces of the tangent spaces of M on which $df = 0$, and since $df(D(p)) = 0$ for every $p \in M_1$, $D(p) \subset (M_1)_p$. Thus D also defines an h-dimensional distribution D_1 on M_1. If X_α is a local basis of D_1, then $X_\alpha|_{M_1}$, defined and proved to be C^∞ as in the proof of Proposition 3.11.2, is a local basis of D_1. Thus D_1 is C^∞. Finding a first integral reduces the complexity of the problem by one dimension. If we can find $d - h$ functionally independent first integrals, then we have a complete local analysis of D.

Proposition 3.11.3. *Let D be a C^∞ h-dimensional distribution. Suppose that $f_1, \ldots, f_{d-h}$ are solution functions such that the df_i are linearly independent at some $m \in M$. Then there are coordinates x^i at m such that $x^{h+i} = f_i$, $i = 1, \ldots, d - h$. For any such coordinates $\partial_1, \ldots, \partial_h$ is a local basis for D, and the coordinate slices $f_i = c^i$, $i = 1, \ldots, d - h$, are h-dimensional integral submanifolds of D. Finally, if D is restricted to such a coordinate neighborhood, it is involutive.*

(We shall omit the proof since much of what is stated just reiterates what we have said before, and that which is new requires only routine applications of the inverse function theorem.)

For the spherical distribution of Example **(a)** it is easily verified that $f = r$ is a first integral, where $r^2 = x^2 + y^2 + z^2$ and $r > 0$. Since $d - h = 1$, any coordinate system of the form x^1, x^2, r gives ∂_1 and ∂_2 as a local basis for D. In particular, this is true for spherical polar coordinates. The level surfaces $r = c$ are the central spheres of R^3, which are integral submanifolds, as we have seen. Any function of r, such as $r^2 - 3r + 2$, is also a first integral and, conversely, every first integral is a function of r. Note that $r^2 - 3r + 2 = 0$ consists of two spheres, $r = 1$ and $r = 2$.

A *maximal connected integral submanifold* of D is an h-dimensional connected integral submanifold which is not contained in any larger connected integral submanifold. In Example **(a)** the maximal connected integral submanifolds are the single whole spheres. In Example **(b)** they are the h-dimensional coordinate planes of R^d on which the last $d - h$ cartesian coordinates are constant. In contrast to integral manifolds in general, the maximal connected one containing a given point is unique if it exists.

Theorem 3.11.1.   *Let D be a C^∞ h-dimensional distribution on a manifold M. For each $m \in M$ there is at most one maximal connected integral submanifold N of D through m. It exists if there is any h-dimensional integral submanifold through m, in which case N is the union of all connected h-dimensional integral submanifolds through m. In particular, every connected h-dimensional integral submanifold containing m is an open submanifold of N.*

The proof consists in showing that N, the union of the connected h-dimensional integral submanifolds containing m, is actually an integral submanifold. The local theory, showing that N looks like an h-dimensional submanifold in the neighborhood of any point, is essentially covered in the next theorem. The difficult part is to show that N has a countable basis of neighborhoods, and these topological details are too technical to be given here.

Corollary.   *If D is a C^∞ completely integral distribution on M, then for each $m \in M$ there is a unique maximal connected integral submanifold through m.*

The h-dimensional integral submanifolds of a distribution D can be parametrized in terms of the flows of a local basis of D. As a consequence they have a local uniqueness not available for lower-dimensional integral submanifolds. Moreover, the method allows us to construct solution functions by using these flows, and hence by solving ordinary differential equations.

Theorem 3.11.2.   *Let $X_1, \ldots, X_h$ be a local basis at m of the C^∞ distribution D and let $\{{}^\alpha\mu_s\}$ be the flow of X_α. If there is an h-dimensional integral submanifold N through m, then a neighborhood of m in N coincides with part of the range of the map F defined on a neighborhood of 0 in R^h by*

$$F(s^1, \ldots, s^h) = {}^1\mu_{s^1} \cdots {}^h\mu_{s^h} m.$$

Proof. As in Proposition 3.11.2, the restrictions $X_\alpha|_N$ are C^∞ vector fields on N and their flows are the restriction $\{{}^\alpha\mu_s|_N\}$. Thus F can be entirely defined as a map into N. Since F_* is 1–1 at 0, because $F_*\partial/\partial u^\alpha(0) = X_\alpha(m)$, F is 1–1 on a neighborhood of 0 and its inverse is a coordinate map on N. Therefore, the range of F fills a neighborhood of m in N. ∎

Remark. Another map which could be used just as well as F defined above to generate h-dimensional integral submanifolds may be described as follows. It uses a "radial" method instead of the "step-by-step" method of F. For each $s = (s^1, \ldots, s^h)$, let $X_s = s^\alpha X_\alpha$ and let $\{^s\mu_t\}$ be the flow of X_s. We define $Gs = {}^s\mu_1 m$. The proof that G is C^∞ is based on the C^∞ dependence of solutions of systems of ordinary differential equations on parameters entering the functions defining the system in a C^∞ manner. Here we have the system

$$\frac{dx^i}{du} = s^\alpha X_\alpha^i = F^i(x^1, \ldots, x^d, s^1, \ldots, s^h),$$

where the X_α^i are the components of X_α and the F^i are clearly C^∞ functions of both the x^i and the s^α.

Finally, we can construct our original objective, a solution function of a system of partial differential equations, by letting values of f vary arbitrarily (but C^∞) in directions transverse to D, but constant on the integral manifolds. Specifically we have

Theorem 3.11.3. *Let D be a C^∞ completely integrable distribution, $X_1, \ldots, X_h$ a local basis of D at m, and $\{^\alpha\mu_s\}$ the flow of X_α. Let x^i be coordinates at m such that $X_1(m), \ldots, X_h(m), \partial_{h+1}(m), \ldots, \partial_d(m)$ are a basis of M_m. Let g be any C^∞ function on R^{d-h} and define f on a neighborhood of m by*

$$f(^1\mu_{s^1} \cdots {}^d\mu_{s^d} m) = g(s^{h+1}, \ldots, s^d),$$

where $\{^i\mu_s\}$ is the flow (translation!) of ∂_i, $i > h$. Then f is a solution function of D. Every solution function of D in a neighborhood of m is given in this way by some function g.

(The proof is left as an exercise.)

Problem 3.11.1. Let D be the spherical distribution on $R^3 - \{0\}$, $X_1 = -Z$, $X_2 = -Y$, where Y and Z are as in Example (a), and let $m = (1, 0, 0)$. Show that the parametrization F of Theorem 3.11.2 is almost the usual spherical angle parametrization of the unit sphere.

3.12. Frobenius' Theorem

Suppose that D is a two-dimensional involutive distribution. Let X, Y be a local basis at m of D. Let us try to choose a new local basis which has vanishing brackets. As a first step we can choose coordinates x^i such that $x^i m = 0$,

$Y = \partial_2$, and $X(m)$, $Y(m)$, $\partial_3(m), \ldots, \partial_d(m)$ is a basis of M_m. Let $X_1 = X - (Xx^2)\partial_2 = X^1\partial_1 + X^3\partial_3 + \cdots + X^d\partial_d$. Then X_1, ∂_2 are a new local basis for D and $[X_1, \partial_2] = -(\partial_2 X^1)\partial_1 - (\partial_2 X^3)\partial_3 - \cdots - (\partial_2 X^d)\partial_d$ is a linear combination of X_1 and ∂_2, say, $fX_1 + g\partial_2$. Since the components of ∂_2 must match in the coordinate expression for $[X_1, \partial_2]$, we must have $g = 0$ and $[X_1, Y] = fX_1$.

Now, let $\theta = (x^1, \ldots, x^d)$ be the coordinate map, $\{\mu_s\}$ the flow of X_1, and define

$$F(s, a^2, \ldots, a^d) = \mu_s \theta^{-1}(0, a^2, \ldots, a^d).$$

Then F_* is nonsingular at m, so F^{-1} exists and is a coordinate map in a neighborhood of m, $F^{-1} = (y^1, \ldots, y^d)$. When $y^1 = 0$ it is clear that θ and F^{-1} coincide, so at such points $Yy^i = \partial y^i/\partial x^2 = \partial x^i/\partial x^2 = \delta_2^i$. When y^1 varies we are moved along integral curves of X_1 by μ, so $X_1 = \partial/\partial y^1$. If we move along these y^1 curves from a point at which $y^1 = 0$, the derivative of Yy^i, $i \geq 2$, is

$$\begin{aligned}
X_1 Yy^i &= YX_1 y^i + X_1 Yy^i - YX_1 y^i \\
&= Y\delta_1^i + [X_1, Y]y^i \\
&= 0 + fX_1 y^i \\
&= 0,
\end{aligned}$$

so Yy^i is constant along such curves, hence everywhere, $i \geq 2$. This gives $Y = (Yy^i)\,\partial/\partial y^i = (Yy^1)\,\partial/\partial y^1 + \partial/\partial y^2 = (Yy^1)X_1 + \partial/\partial y^2$, so $\partial/\partial y^2 = Y - (Yy^1)X_1 \in D$ and $\partial/\partial y^1$, $\partial/\partial y^2$ is a new local basis for D. It follows [cf. Example **(b)**] that D is completely integrable, having integral submanifolds in the y^i coordinate neighborhood which consist of the coordinate slices $y^i = c^i$, $i > 2$.

The above pattern can be extended to the case of an h-dimensional involutive distribution. The first step is to modify a local basis so as to produce a local basis $X_1, \ldots, X_h$ for which bracket multiplication is *diagonal*: If $\alpha < \beta$, then $[X_\alpha, X_\beta] = \sum_{\gamma=1}^{\beta-1} f_{\alpha\beta}^\gamma X_\gamma$. ($X_1$ and Y correspond to this basis in the case $h = 2$ above.) Then we define a map F in terms of the flows of the X_α and an auxiliarly coordinate system, and invert F to get a new coordinate system for which ∂_α are a local basis of D. The result is known as the *complete integrability theorem of Frobenius* and is the converse of Proposition 3.11.2.

Theorem 3.12.1. *A C^∞ involutive distribution is completely integrable; locally there are coordinates x^i such that $\partial_1, \ldots, \partial_h$ are a local basis, the coordinate slices $x^i = c^i$, $i > h$, are integral submanifolds, and the solution functions are the C^∞ functions of $x^{h+1}, \ldots, x^d$.*

The details of the proof are omitted.

If a C^∞ distribution D is not involutive, then its study is more difficult.

There are two viewpoints which can be taken. First, if we assume that solution functions are the goal, we try to include the distribution in a higher-dimensional distribution which is involutive by throwing in the brackets of vector fields which belong to D and the brackets of the brackets, etc., until we obtain a system $\bar{D}$ for which further brackets will not increase the dimension. This procedure may fail because the larger system $\bar{D}$ can be degenerate; that is, the dimension of the subspaces $\bar{D}(m)$ of M_m may vary as a function of m. If $\bar{D}$ is a nondegenerate system, hence a distribution, then it will be involutive and the solution functions of $\bar{D}$ will coincide with those of D. Of course, it may happen that $\bar{D}(m) = M_m$, in which case the only solution functions would be constant.

Second, we may desire to obtain integral submanifolds of lower dimension than that of D, but whose dimensions are as large as possible. Of course, we can obtain one-dimensional integral submanifolds from the integral curves of vector fields, but generally the structure of the maximal dimensional integral submanifolds is quite difficult to determine. The work that has been done on this problem has used the *dual formulation* in terms of 1–forms (*pfaffian systems*), which will be discussed briefly in Chapter 4. This work has not been very successful except when more smoothness assumptions are imposed, that is, the objects involved are assumed to be real-analytic (that is, expressible in terms of convergent power series in several variables) instead of C^∞.

Problem 3.12.1. (a) Show that the system of partial differential equations $Xf = 0$, $Yf = 0$, where $X = 9y\partial_x - 4x\partial_y$, $Y = x\partial_x + y\partial_y + 2(z + 1)\partial_z$, on $R^3 - \{0\}$, has nonconstant solutions.

(b) For a nonconstant solution f, find parametric equations of the level surface $f = c$ which passes through the point $(3, 0, 0)$.

Problem 3.12.2. Show that the only solutions on R^3 of $(\partial_x + x\partial_y)f = 0$, $(\partial_y + y\partial_z)f = 0$ are $f = $ constant.

Problem 3.12.3. Show that the system on R^4, $(\partial_y + x\partial_z)f = 0$, $(\partial_x + y\partial_w)f = 0$, where x, y, z, w are cartesian coordinates, has just one functionally independent solution.

Problem 3.12.4. Show that the distribution on R^4 spanned by $\partial_y + x\partial_z$ and $\partial_x + y\partial_w$ has no two-dimensional integral submanifolds.

Appendix to Chapter 3

3A. Tensor Bundles

It is natural to make $T^r_s M$ into a manifold. Since a tensor in M^r_{ms} can vary in d^{r+s} independent directions within M^r_{ms} and m can vary on M in d independent directions, the dimension of $T^r_s M$ is $d + d^{r+s}$. The manifold structure is defined as in Section 1.2(**f**) by patching together coordinate neighborhoods. We realize a coordinate neighborhood in $T^r_s M$ as the set of all tensors based at the points of a coordinate neighborhood U of M. For coordinates we take the d coordinates on U plus the d^{r+s} components of the tensors with respect to the coordinates on U. We shall give the details only in the case of the tangent bundle.

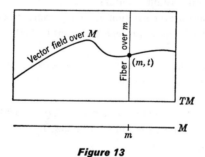

Figure 13

It is convenient (see Figure 13) to denote the points of TM by pairs (m, t), where $m \in M$ and $t \in M_m$; the "m" in this pair is redundant, of course, but it avoids naming the base point of t all the time. Let x^i be coordinates on U and let $V = \{(m, t) \mid m \in U\}$; that is, $V = TU$. We define $2d$ coordinates y^i, y^{i+d} on V by the following formulas:

$$y^i(m, t) = x^i m,$$
$$y^{i+d}(m, t) = dx^i(t).$$

The map $\mu = (y^1, \ldots, y^{2d})$ is clearly 1–1 on V.

At each $m \in U$ the components of a tangent may be specified to be an arbitrary member of R^d. Thus the range of μ is $W \times R^d$, where $W \subset R^d$ is the range of $(x^1, \ldots, x^d)$. The manifold structure is defined as in Section 1.2(f) by patching together these coordinate neighborhoods V. Thus V is homeomorphic via μ to $W \times R^d$. Subsets of TM which are not entirely within such a V, are open iff the intersection with each such V is open.

We show that TM is a Hausdorff space. If we have two points of the form (m, s) and (m, t), where $s \neq t$, we may suppose that U is a coordinate neighborhood of m. Since R^d is Hausdorff, there are open sets G and H containing $(y^{1+d}s, \ldots, y^{2d}s)$ and $(y^{1+d}t, \ldots, y^{2d}t)$, respectively, such that G and H do not intersect. Then $\mu^{-1}(W \times G)$ and $\mu^{-1}(W \times H)$ are nonintersecting open sets containing (m, s) and (m, t), respectively. If we have two points of the form (m, s) and (n, t), where $m \neq n$, we may include m and n in nonintersecting co-ordinate neighborhoods U and U_1, respectively, and then TU and TU_1 are nonintersecting open sets containing (m, s) and (n, t), respectively.

If $\{U_\alpha \mid \alpha = 1, 2, 3, \ldots\}$ is a countable basis of neighborhoods for M, we may assume they are coordinate neighborhoods, with coordinate maps $\{\varphi_\alpha\}$, and corresponding coordinate maps $\{\mu_\alpha\}$ on $\{V_\alpha = TU_\alpha\}$. Let $\{G_\beta \mid \beta = 1, 2, 3, \ldots\}$ be a countable basis of neighborhoods for R^d, and let $W_\alpha = \varphi_\alpha U_\alpha$. Then $\{\mu_\alpha^{-1}(W_\alpha \times G_\beta) \mid \alpha, \beta = 1, 2, 3, \ldots\}$ is a countable basis for TM, so TM is separable.

Finally, it is necessary to show that the coordinate systems are C^∞ related. Let x^i be coordinates on U, z^i coordinates on U_1, y^i and y^{i+d} the corresponding coordinates on $V = TU$, and w^i and w^{i+d} those on $V_1 = TU_1$. The x^i and z^i are related on $U \cap U_1$ by C^∞ expressions $x^i = f^i(z^1, \ldots, z^d)$.

For the first d of the coordinates on $T(U \cap U_1) = V \cap V_1$ we have, for $(m, t) \in V \cap V_1$,

$$
\begin{aligned}
y^i(m, t) &= x^i m \\
&= f^i(z^1 m, \ldots, z^d m) \\
&= f^i(w^1(m, t), \ldots, w^d(m, t)).
\end{aligned}
$$

Thus the first d are related in the same way as the x^i and the z^i. For the rest we have

$$
\begin{aligned}
y^{i+d}(m, t) &= dx^i(t) \\
&= \frac{\partial x^i}{\partial z^j}(m)\, dz^j(t) \\
&= f^i_j(z^1 m, \ldots, z^d m) w^{j+d}(m, t) \qquad (\text{sum on } j) \\
&= f^i_j(w^1(m, t), \ldots, w^d(m, t)) w^{j+d}(m, t),
\end{aligned}
$$

where the f^i_j are the partial derivatives of the f^i. Thus the last d relations are

$$
y^{i+d} = w^{j+d} f^i_j(w^1, \ldots, w^d),
$$

which are clearly C^∞.

We define the *projection map* $\pi\colon TM \to M$ by $\pi(m, t) = m$. It is easy to show that π is C^∞. Indeed, its coordinate expression in terms of the special coordinates x^i and y^i, y^{i+d} above is $(x^1, \ldots, x^d) \circ \pi = (y^1, \ldots, y^d)$, which follows directly from the definition of the y^i by applying both sides to (m, t).

A vector field is a map $X\colon E \to TM$, where $E \subset M$. However, this is not an arbitrary map but must satisfy the further condition that $X(m) \in M_m$, which is the same as saying $\pi X(m) = m$. That is, $\pi \circ X$ is the identity map on E. The converse is clearly true, and in fact we have

Proposition 3.A.1. *A map $X\colon E \to TM$ is a vector field iff $\pi \circ X$ is the identity on E. Moreover, X is a C^∞ vector field iff X is C^∞ as a map.*

Proof. If $\pi \circ X$ is the identity on E, then the coordinate expressions for X are

$$y^i \circ X = x^i,$$
$$y^{i+d} \circ X = Xx^i,$$

since $y^i X(m) = y^i(m, X(m)) = x^i m$ and $y^{i+d} X(m) = y^{i+d}(m, X(m)) = X(m)x^i = (Xx^i)m$. Here we have used the redundant m or not as we please. The first d of these coordinate expressions are always C^∞ if the domain of X is open. The last d are the components of X and are C^∞ iff X is a C^∞ vector field. ∎

Problem 3.A.1. Generalize the projection map π to a projection of the tensor bundle $T_s^r M$ into M and prove the analogue of Proposition 3.A.1.

3B.   Parallelizable Manifolds

The special coordinate neighborhoods V in TM are diffeomorphic to the product manifolds $U \times R^d$. Thus if M is covered by a single coordinate system, TM is diffeomorphic to the product manifold $M \times R^d$.

This is not the only case where TM is diffeomorphic to $M \times R^d$. A manifold is called *parallelizable* if there are C^∞ vector fields $X_1, \ldots, X_d$ defined on all of M such that for every $m \in M$, $\{X_1(m), \ldots, X_d(m)\}$ is a basis of M_m. The vector fields X_i are then called a *parallelization* of M. An equivalent formulation of this property is given in the following proposition, which is stated without a complete proof.

Proposition 3.B.1. *M is parallelizable iff there is a diffeomorphism $\mu\colon TM \to M \times R^d$ such that the first factor of μ is $\pi\colon TM \to M$ and for each m the second factor of μ restricted to M_m is a linear function $M_m \to R^d$.*

Outline of Proof. If M is parallelizable by vector fields $X_1, \ldots, X_d$, let $\tau^1, \ldots, \tau^d$ be the dual basis of 1–forms. The map μ is then defined by

$$\mu(m, t) = (m, \langle t, \tau^1(m) \rangle, \ldots, \langle t, \tau^d(m) \rangle).$$

It is clear that the first factor of μ is π and that the second factor is linear on each M_m. It is left as an exercise to prove that μ and μ^{-1} are C^∞.

Conversely, suppose $\mu: TM \to M \times R^d$ is a diffeomorphism of the type required. Let $\delta_i = (\delta_{1i}, \ldots, \delta_{di}) \in R^d$, that is, the natural basis for R^d, and define $X_i: M \to TM$ by $X_i(m) = \mu^{-1}(m, \delta_i)$. Then it may be shown that $X_1, \ldots, X_d$ is a parallelization of M. ∎

Problem 3.B.1.   If X_i is a parallelization of M and (f_j^i) is a matrix of real-valued C^∞ functions which has nonzero determinant at every point of M, then $Y_j = f_j^i X_i$ is a parallelization of M. Conversely, any two parallelizations are related by such a matrix.

If M and N are manifolds and X is a vector field on M, then we can think of X as a vector field on $M \times N$. In terms of product coordinates the components of X are independent of the coordinates on N and the last e ($e = dim\ N$) components of X vanish. Formally, if $j_n: M \to M \times N$ is the injection, $j_n m = (m, n)$, then the values of X as a vector field on $M \times N$ are given by $X(m, n) = j_{n*}X(m)$. Moreover, if $p: M \times N \to M$ and $q: M \times N \to N$ are the projections, then $p_* X(m, n) = X(m)$ and $q_* X(m, n) = 0$, and these facts also determine X uniquely as a vector field on $M \times N$. Similarly, a vector field Y on N determines a vector field, also called Y, on $M \times N$, such that $p_* Y = 0$ and $q_* Y = Y$.

If X_i is a parallelization of M and Y_α is a parallelization of N, then X_i, Y_α is a parallelization of $M \times N$. Thus we have

Proposition 3.B.2.   *The product of parallelizable manifolds is parallelizable.*

As an example which is not an open submanifold of R^d we note that the circle is parallelizable. Indeed, we need only take $X = d/d\theta$, where θ is any restriction of the polar angle to an interval of length 2π. This does actually define an X on all the circle because any two determinations of θ are related locally by a translation of amount $2n\pi$ for some integer n, and hence give the same coordinate vector field.

By Proposition 3.B.2 it now follows that the torus is parallelizable since it is the product of two circles.

To obtain examples of manifolds which are not parallelizable we need only have a nonorientable manifold (see Appendix 3.C). Indeed, if M has parallelization X_i, then those coordinate systems x^i such that $X_i = f_i^j \partial_j$, where $\det f_i^j > 0$, form a consistently oriented atlas. Thus if M is parallelizable, then M is orientable.

An important class of parallelizable manifolds are those which carry a *Lie group structure*. On a *Lie group* G there is a group operation which is a C^∞ map $G \times G \to G$. For each fixed $g \in G$, multiplication on the left by g is a diffeomorphism $L_g: G \to G$, $L_g h = gh$. If we take a basis $\{t_1, \ldots, t_d\}$ of G_e, $e =$ the identity of G, then $X_i(g) = L_{g*}t_i$ defines C^∞ vector fields X_i which form a parallelization of G. The X_i are *left invariant* in that $L_{g*}X_i = X_i$ for every $g \in G$. The collection of left invariant vector fields form a d-dimensional vector space spanned by the X_i, the *Lie algebra* of G. The Lie algebra is closed under bracket; that is, the bracket of two left invariant vector fields is again left invariant. The properties of a Lie group are largely determined by those of its Lie algebra. In particular, the matrix groups are Lie groups, so the orthogonal, unitary, and symplectic groups should be studied mostly in terms of their Lie algebras.

3C. Orientability

A pair of coordinate systems x^i and y^i is *consistently oriented* if the jacobian determinant $\det(\partial x^i/\partial y^j)$ is positive wherever defined. A manifold M is *orientable* if there is an atlas such that every pair of coordinate systems in the atlas is consistently oriented. Such an atlas is said to be *consistently oriented*. It determines an *orientation* on M and M is said to be *oriented* by such an atlas. Two atlases such that every coordinate system of one is related by negative jacobian determinant to every coordinate system of the other are said to determine *opposite orientations*. If an atlas $\{\mu_\alpha = (x_\alpha^i) \mid \alpha \in A\}$ is consistently oriented, then we can obtain an oppositely oriented atlas by reversing the sign of each x_α^1; that is, the atlas $\{\varphi_\alpha = (-x_\alpha^1, x_\alpha^2, \ldots, x_\alpha^d) \mid \alpha \in A\}$ determines the opposite orientation. An odd permutation of the coordinates also reverses orientation.

An open submanifold of R^d is orientable, since it has an atlas consisting of one coordinate system, which is, of course, consistently oriented with itself.

A connected orientable manifold has just two orientations, and every coordinate system with connected domain is consistent with either one or the other. If the domain of a coordinate system is not connected, then the coordinate system may be split into its restrictions to the various connected components of its domain, and these parts may agree or disagree independently with a given orientation of M.

An orientable manifold with k connected components has 2^k orientations, since the orientation on each component has two possibilities independent of the choice of orientation on the other components. If just one component is nonorientable, the whole manifold is nonorientable.

A surface in R^3, that is, a two-dimensional submanifold of R^3, is orientable iff there is a continuous nonzero field of vectors normal to the surface. If the surface is orientable, then for a consistently oriented atlas, the cross-product of coordinate vectors can be divided by their lengths to produce a unit normal vector field which is consistent, hence continuous, in passing from one coordinate system to another. Conversely, if a continuous normal field exists, then those coordinate systems for which the cross-product of the coordinate vectors is a positive multiple of the normal field form a consistently oriented atlas. Note that the definition of cross-product requires an orientation of R^3 in addition to the euclidean structure to make "normal" meaningful. However, this euclidean structure should be regarded as a convenient tool to expedite the proof; the result is still true if we use "nontangent" fields instead of "normal" fields, and the concept of nontangency does not require a euclidean structure. The result can be extended to say that a hypersurface (d-dimensional submanifold) of a ($d + 1$)-dimensional orientable manifold is orientable iff there is a continuous nontangent field defined on it. Orientability of surfaces is also described as *two-sidedness*.

For manifolds with a finite atlas it is possible to either construct a consistently oriented atlas or demonstrate nonorientability by a recursive process. As a first step, if the coordinate domains are not all connected, split them into their restrictions to the connected components. Consequently, if we alter the orientation of the coordinate system at one point we must alter it throughout. Choosing one coordinate system we alter all those intersecting it so that they are consistently oriented with the first one and with each other. This may be impossible, in that the intersection of two coordinate domains might be disconnected, with the coordinates consistently oriented in one component and not so in another, or two which are altered to match the first may be inconsistent in some part of their intersection not meeting the first one. In these cases the manifold is nonorientable. Otherwise we obtain a second collection of altered coordinate systems which are consistently oriented. (The first collection consisted of the initial coordinate system alone.) We try to alter those adjacent to this second collection to produce a larger consistently ordered third collection, etc.

To illustrate this procedure consider the d-dimensional projective space P^d. This may be realized as proportionality classes $[a_0 : a_1 : \cdots : a_d]$ of nonzero elements of R^{d+1}. If u^α are the standard cartesian coordinates on R^{d+1}, then the ratios u^α/u^β are well-defined functions on the open subset of P^d for which $u^\beta \neq 0$. There are $d + 1$ coordinate systems $\{(x_\alpha^i) \mid \alpha = 0, \ldots, d, \alpha \neq i\}$ forming an atlas on P^d, defined by $x_\alpha^i = u^i/u^\alpha$. The range of each coordinate system is all of R^d. The coordinate transformations are $x_\alpha^i = x_\beta^i/x_\beta^\alpha$, where we let $x_\beta^\beta = 1$ for the case $i = \beta$. The intersection of the two coordinate domains of (x_α^i) and (x_β^i) has two connected components which are mapped into the two half-spaces $x_\beta^\alpha > 0$ and $x_\beta^\alpha < 0$ of R^d by the β-coordinate map. If we order the coordinates x_0^i from $i = 1$ to $i = d$ and order the coordinates x_α^i in the order $i = 1, 2, \ldots, \alpha - 1, 0, \alpha + 1, \ldots, d$, then the jacobian matrix $(\partial x_\alpha^i/\partial x_0^j)$ is a diagonal matrix with diagonal entries $-1/x_0^\alpha$ except for the $j = \alpha$ column, which has diagonal entry $\partial x_\alpha^0/\partial x_0^\alpha = -1/(x_0^\alpha)^2$. The determinant is thus $(-1)^d/(x_0^\alpha)^{d+1}$. This determinant has consistently negative value if $d + 1$ is even, but has opposite signs in the two components if $d + 1$ is odd. If d is odd we can alter the 0-coordinate system so as to make all 0-α-jacobian determinants positive simultaneously. Since the 0-α-β intersection meets both components of the α-β intersection, positivity of the 0-α and 0-β determinants implies positivity of their quotient, the α-β determinant, at some points of each component of its domain, hence everywhere. Thus P^d is orientable if d is odd and nonorientable if d is even.

Problem 3.C.1.   Prove that the following are orientable: the cartesian product of orientable manifolds, every one-dimensional manifold, the torus, the d-sphere, and the tangent bundle TM of any manifold M.

Problem 3.C.2.   Prove that the following are nonorientable: the cartesian product of any nonorientable manifold and any other manifold, the Klein bottle, and the Möbius strip.

Problem 3.C.3.   Let M be a nonorientable manifold and $\{(\mu_\alpha, U_\alpha)\}$ an atlas for M. For each α and β, $U_\alpha \cap U_\beta = V_{\alpha\beta}^+ \cup V_{\alpha\beta}^-$, where μ_α and μ_β are related by positive jacobian determinant on $V_{\alpha\beta}^+$ and by negative jacobian determinant on $V_{\alpha\beta}^-$. Specify a new manifold ^0M by patching together coordinate domains as follows. The atlas of ^0M will have twice as many members as that of M, designated by $\{(\mu_\alpha^+, U_\alpha^+), (\mu_\alpha^-, U_\alpha^-)\}$. The range of μ_α^+ and μ_α^- is the same as that of μ_α. For all four possible choices of signs $(a, b) = (+, +), (+, -)$, $(-, +)$, or $(-, -)$, the coordinate transformation $\mu_\alpha^a \circ (\mu_\beta^b)^{-1}$ equals the restriction of $\mu_\alpha \circ \mu_\beta^{-1}$ to $V_{\alpha\beta}^{ab}$, where $+ + = +$, $+ - = -$, $- + = -$, and $- - = +$. Show that

(a) These coordinate domains and transformations do give a well-defined manifold ^0M by means of the patching-together process in Section 1.2(f).

(b) ^0M is orientable.

(c) If M is connected, so is ^0M.

(d) The mappings $\varphi_\alpha^a = \mu_\alpha^{-1} \circ \mu_\alpha^a : U_\alpha^a \to U_\alpha$ are consistent on the intersections, that is, $\varphi_\alpha^a|_{U_\alpha^a \cap U_\beta^b} = \varphi_\beta^b|_{U_\alpha^a \cap U_\beta^b}$, and so there is a unique well-defined map $\varphi: {}^0M \to M$ such that $\varphi|_{U_\alpha^a} = \varphi_\alpha^a$.

(e) For every α, $\varphi^{-1}(U_\alpha) = U_\alpha^+ \cup U_\alpha^-$ and φ_α^a is a diffeomorphism of U_α^a onto U_α. Moreover, $U_\alpha^+ \cap U_\alpha^-$ is empty.

Property (e) is described by saying that ^0M is a *twofold covering* of M with covering map φ. This property and the fact that ^0M is orientable determine ^0M uniquely (up to diffeomorphism). We call ^0M the *twofold orientable covering* of M. Results on orientable manifolds sometimes can be extended to nonorientable manifolds by considering the relation between ^0M and M.

CHAPTER **4**

Integration Theory

4.1. Introduction

Of all the types of tensor fields, the skew-symmetric covariant ones, that is, differential forms, seem to be the most frequently encountered and to have the widest applications. Electromagnetic theory (Maxwell's equations) can be given a neat and concise formulation in terms of them, a formulation which does not suffer when we pass to the space-time of relativity. Differential forms have been used by de Rham to express a deep relation between the topological structure of a manifold and certain aspects of vector analysis on a manifold. In the work of the famous French geometer E. Cartan, he uses differential forms almost exclusively to formulate and develop his results on differential systems and riemannian geometry. The generalization of Stokes' theorem and the divergence theorem to higher dimensions and more general spaces is very clumsy unless one employs a systematic development of the calculus of differential forms. It is this calculus and its use in formulating integration theory and the dual method of the study of distributions which are the topics taken up in this chapter.

Sections 4.2 through 4.5 deal with the calculus of differential forms. This consists of an algebraic part which has already been discussed in Section 2.18; we modify the notation of this algebra and introduce the interior product operator in Section 4.4; and an analytic part, in which a differential operator is defined and its properties developed. This differential operator generalizes and unifies the vector-analysis operators of gradient, curl, and divergence. Moreover, it replaces the bracket operation in the dual formulation of distributions.

We then turn to a description of the objects on which integration takes place. A "set calculus" is introduced in which a new operator called the boundary operator plays a fundamental part (see Section 4.6).

A review of the basic facts about multiple integration of functions on R^d is provided in Section 4.7.

The material of the previous sections is combined to give a theory of integration of differential forms on oriented parametrized regions of a manifold, culminating in the generalized Stokes' theorem (see Section 4.9).

In Section 4.10 we return to the material of Sections 3.11 and 3.12, showing how the concept of an involutive distribution has a dual formulation.

4.2. Differential Forms

A (*differential*) *p–form* is a C^∞ skew-symmetric covariant tensor field of degree p [type $(0, p)$]. Thus a 0–form is a real-valued C^∞ function. This definition of a 1–form agrees with that given in Section 3.2. There are no p–forms when $p > d$, where d is the dimension of the manifold.

If x^i are coordinates, then the dx^i are a local basis for 1–forms, in that any 1–form can be expressed locally as $f_i\, dx^i$, where the f_i are C^∞ functions. By exterior products the dx^i generate local bases for forms of higher orders. Thus $\{dx^i \wedge dx^j \mid i < j\}$ is a local basis for 2–forms; $dx^1 \wedge \cdots \wedge dx^d$ is a local basis for d–forms.

Since we are concerned in this chapter exclusively with wedge products of forms, not with symmetric products, we can simplify our notation slightly. Thus we shall omit the wedges between coordinate differentials, writing $dx^i\, dx^j$ instead of $dx^i \wedge dx^j$. Moreover, since a local basis for p–forms consists of $\binom{d}{p}$ coordinate p–forms $dx^{i_1} \cdots dx^{i_p}$ where $i_1 < \cdots < i_p$, it is convenient to have a summation convention which gives us sums running through the *increasing* sets of indices. We indicate this alternative type of sum by placing the string of indices to which it is to apply in parentheses in one of its occurrences in the formula, thus: $(i_1 \cdots i_p)$. For example, if $d = 3$,

$$a_{(i_1 i_2)}\, dx^{i_1}\, dx^{i_2} = a_{12}\, dx^1\, dx^2 + a_{13}\, dx^1\, dx^3 + a_{23}\, dx^2\, dx^3.$$

This convention does not prevent us from multiplying coordinate differentials in nonincreasing order and we have not suspended the previous summation convention.

Finally, by the *components* of a p–form we mean its components with respect to the increasing-index basis $\{dx^{(i_1} \cdots dx^{i_p)}\}$, not with respect to the tensor product basis $\{dx^{i_1} \otimes \cdots \otimes dx^{i_p}\}$ as in Chapter 2. Thus we now say that the components of $a_{(i_1 i_2)}\, dx^{i_1}\, dx^{i_2}$ above are a_{12}, a_{13}, and a_{23}, whereas in Chapter 2, since

$$dx^{i_1}\, dx^{i_2} = \tfrac{1}{2}(dx^{i_1} \otimes dx^{i_2} - dx^{i_2} \otimes dx^{i_1}),$$

the components would have been said to be, say, $b_{11} = 0$, $b_{12} = \tfrac{1}{2}a_{12}$, $b_{13} = \tfrac{1}{2}a_{13}$, $b_{21} = -b_{12}$, etc. However, it is also useful to define other scalars which are not all components, by using skew-symmetry for the nonincreasing indices: $a_{11} = 0$, $a_{21} = -a_{12}$, etc.

Problem 4.2.1. (a) Show that the rule for evaluating basis forms on basis vector fields is

$$dx^{i_1}\cdots dx^{i_p}(\partial_{j_1}, \ldots, \partial_{j_p}) = \frac{1}{p!}\, \delta^{i_1}_{j_1}\cdots \delta^{i_p}_{j_p},$$

where $i_1\cdots i_p$ and $j_1\cdots j_p$ are both increasing index sets.

(b) If $\theta_{i_1\ldots i_p}$ are the components of a p–form θ, show that $\theta(\partial_{j_1}, \ldots, \partial_{j_p}) = \frac{1}{p!}\,\theta_{j_1\cdots j_p}$.

4.3.   Exterior Derivatives

The *exterior derivative* of a p–form θ is a $(p + 1)$–form which we denote by $d\theta$. We have already defined $d\theta$ in the case $p = 0$ [see equation (1.8.5)]. There are several approaches to its definition, each of which gives important information about the operator d.

(a) In terms of coordinates d merely operates on the component functions:

$$d\theta = (d\theta_{(i_1\cdots i_p)}) \wedge dx^{i_1}\cdots dx^{i_p}. \tag{4.3.1}$$

It is not immediately clear that this defines anything at all, since the right side might depend on the choice of coordinates x^i. However, it is easily verified that this formula satisfies the axioms for d given below. Since the axioms are coordinate free and determine d, it is a consequence that (4.3.1) is invariant under change of coordinates.

In the case of $M = R^3$ and cartesian coordinates x, y, z the formula bears a strong, nonaccidental resemblance to grad, curl, and div:

$$df = f_x\, dx + f_y\, dy + f_z\, dz,$$
$$\begin{aligned}
d(f\, dx + g\, dy + h\, dz) &= df \wedge dx + dg \wedge dy + dh \wedge dz \\
&= (f_x\, dx + f_y\, dy + f_z\, dz) \wedge dx \\
&\quad + (g_x\, dx + g_y\, dy + g_z\, dz) \wedge dy \\
&\quad + (h_x\, dx + h_y\, dy + h_z\, dz) \wedge dz \\
&= (h_y - g_z)\, dy\, dz + (f_z - h_x)\, dz\, dx \\
&\quad + (g_x - f_y)\, dx\, dy,
\end{aligned}$$
$$\begin{aligned}
d(f\, dy\, dz + g\, dz\, dx + h\, dx\, dy) &= df \wedge dy\, dz + dg \wedge dz\, dx \\
&\quad + dh \wedge dx\, dy \\
&= (f_x + g_y + h_z)\, dx\, dy\, dz.
\end{aligned}$$

(We have indicated partial derivatives by subscripts.) The discrepancies from the usual formulas for grad, curl, and div can be erased by introducing the euclidean inner product on R^3, for which dx, dy, dz is an orthonormal basis at

each point. This gives us an isomorphism between contravariant and covariant
vectors, $a\mathbf{i} + b\mathbf{j} + c\mathbf{k} = a\partial_x + b\partial_y + c\partial_z \leftrightarrow a\,dx + b\,dy + c\,dz$; we shall
ignore this isomorphism and deal with only the covariant vectors. If we also
impose the orientation given by $dx\,dy\,dz$, then we get the Hodge star operator
(2.22): $*dx = dy\,dz$, $*dy = dz\,dx$, $*dz = dx\,dy$, $*(dx\,dy\,dz) = 1$, and for the
other cases we can use $** = $ the identity. Then we have

$$*d(f\,dx + g\,dy + h\,dz) = (h_y - g_z)\,dx + (f_z - h_x)\,dy + (g_x - f_y)\,dz,$$
$$*d*(f\,dx + g\,dy + h\,dz) = f_x + g_y + h_z.$$

These formulas show that a more precise version of the resemblance between
curl and div and d on 1–forms and 2–forms, respectively, is that the covariant
forms of curl and div are the operators curl $= *d$ and div $= *d*$, both operating on 1–forms. The covariant form of grad is grad $= d$, the exterior derivative
on 0–forms.

(b) There are a few important properties of d which are also sufficient to
determine d completely, that is, *axioms* for d:

(1) If f is a 0–form, then df coincides with the previous definition; that is,
$df(X) = Xf$ for every vector field X.

(2) There is a wedge-product rule which d satisfies; as a memory device, we
think of d as having degree 1, so a factor of $(-1)^p$ is produced when d commutes with a p–form: If θ is a p–form and τ a q–form, then

$$d(\theta \wedge \tau) = d\theta \wedge \tau + (-1)^p \theta \wedge d\tau;$$

that is, d is a *derivation*.

(3) When d is applied twice the result is 0, written $d^2 = 0$: $d(d\theta) = 0$ for
every p–form θ. [As an axiom for the determination of d it would suffice to
assume $d(df) = 0$ only for 0–forms f, but the more general result **(3)** is a
theorem which we need.]

(4) The operator d is linear. Only the additivity need be assumed, because
commutation with constant scalar multiplication is a consequence of **(1)** and
(2): If θ and τ are p–forms, then $d(\theta + \tau) = d\theta + d\tau$.

The coordinate definition (4.3.1) is an easy consequence of these axioms,
because by **(2)** and **(3)**,

$$d(dx^{i_1}\cdots dx^{i_p}) = (d^2 x^{i_1}) \wedge dx^{i_2}\cdots dx^{i_p} - dx^{i_1} \wedge (d^2 x^{i_2}) \wedge dx^{i_3}\cdots dx^{i_p}$$
$$+ \cdots + (-1)^{p-1} dx^{i_1}\cdots dx^{i_{p-1}} \wedge d^2 x^{i_p} = 0.$$

Thus we have

$$d(f\,dx^{i_1}\cdots dx^{i_p}) = df \wedge dx^{i_1}\cdots dx^{i_p} + f d(dx^{i_1}\cdots dx^{i_p})$$
$$= df \wedge dx^{i_1}\cdots dx^{i_p},$$

which, with additivity **(4)**, gives (4.3.1).

The converse, that formula (4.3.1) satisfies the axioms, is a little harder. Of course, (1) and (4) are trivial. To prove (2) we need the product rule for functions: $d(fg) = (df)g + f\,dg$. The components of $\theta \wedge \tau$ are sums of products of the components of θ and τ. Applying the product rule for functions gives two indexed sums, which we want to factor to get (2), and this is done by shifting the components of τ and their differentials over the coordinate differentials corresponding to θ, which in the second case requires a sign $(-1)^p$:

$$d(\theta \wedge \tau) = \frac{1}{p!q!}\, d(\theta_{i_1 \cdots i_p} \tau_{j_1 \cdots j_q}) \wedge dx^{i_1} \cdots dx^{i_p}\, dx^{j_1} \cdots dx^{j_q}$$

$$= \frac{1}{p!q!}\, [d\theta_{i_1 \cdots i_p} \wedge dx^{i_1} \cdots dx^{i_p} \tau_{j_1 \cdots j_q}\, dx^{j_1} \cdots dx^{j_q}$$
$$+ \theta_{i_1 \cdots i_p}(-1)^p\, dx^{i_1} \cdots dx^{i_p} \wedge d\tau_{j_1 \cdots j_q} \wedge dx^{j_1} \cdots dx^{j_q}].$$

(The factor $1/p!q!$ is inserted because we are unable to keep $i_1 \cdots i_p j_1 \cdots j_q$ in increasing order when we are only given $i_1 \cdots i_p$ and $j_1 \cdots j_q$ in increasing order, so we have switched to the full sum and consequent duplication of terms, $p!$ for θ and $q!$ for τ.)

Axiom (3) is known as the *Poincaré lemma*, although there is some confusion historically, so that in some places the converse, "if $d\theta = 0$, then there is some τ such that $\theta = d\tau$," is referred to as the Poincaré lemma. The converse is true only locally (Section 4.5). The proof that (4.3.1) satisfies (3), $d^2 = 0$, uses the equality of mixed derivatives of functions in either order, a symmetry property, which combines with the skew-symmetry of wedge products to give 0.

(c) There is an intrinsic formula for d in terms of values of forms on arbitrary vector fields. This formula involves bracket and shows that the ability to form an intrinsic derivative of p–forms is related to the ability to form an intrinsic bracket of two vector fields. We only give the formula in the low-degree cases for which it has the greatest use.

f a 0–form: $df(X) = Xf$.

θ a 1–form: $d\theta(X, Y) = \frac{1}{2}\{X\theta(Y) - Y\theta(X) - \theta[X, Y]\}$
$$= \tfrac{1}{2}(X\langle Y, \theta\rangle - Y\langle X, \theta\rangle - \langle [X, Y], \theta\rangle).$$

θ a 2–form: $d\theta(X, Y, Z) = \frac{1}{3}\{X\theta(Y, Z) + Y\theta(Z, X) + Z\theta(X, Y)$
$$- \theta([X, Y], Z) - \theta([Y, Z], X)$$
$$- \theta([Z, X], Y)\}.$$

[The annoying factors $\frac{1}{2}, \frac{1}{3}, \ldots$ can be eliminated by using another definition of wedge products. This alternative definition, which does not alter the essential properties of wedge product, is obtained by magnifying our present wedge product of a p–form and a q–form by the factor $(p + q)!/p!q!$. Both products are in common use and we shall continue with our original definition.]

Problem 4.3.1. Show that axiom (2) unifies the following formulas of vector analysis:

(a) grad $(fg) = g$ grad $f + f$ grad g.
(b) curl $(f\theta) = $ grad $f \times \theta + f$ curl θ.
(c) div $(f\theta) = $ grad $f \cdot \theta + f$ div θ.
(d) div $(\theta \times \tau) = $ curl $\theta \cdot \tau - \theta \cdot$ curl τ.

Hint: Use the following expressions for cross and dot product in terms of $\wedge$ and $*$:

$$\theta \times \tau = *(\theta \wedge \tau).$$
$$\theta \cdot \tau = *(*\theta \wedge \tau).$$

Problem 4.3.2. Show that axiom (3) gives:

(a) curl grad $f = 0$.
(b) div curl $f = 0$.

Problem 4.3.3. (a) Show that the laplacian operator on functions is div grad $= *d*d$.

(b) The cylindrical coordinate vectors $\partial_r, \partial_\theta, \partial_z$ are orthogonal and have lengths $1, r, 1$. Hence $dr, r\, d\theta, dz$ is an orthonormal coherently-oriented covariant basis, so the cylindrical coordinate formulas for $*$ are

$$*dr = r\, d\theta\, dz, \qquad *d\theta = \frac{1}{r}\, dz\, dr, \qquad *dz = r\, dr\, d\theta, \qquad *(dr\, d\theta\, dz) = \frac{1}{r}.$$

Use these to obtain the cylindrical coordinate formula for the laplacian $*d*d$.

(c) Find the spherical coordinate formula for $*d*d$ by the same method.

Problem 4.3.4. (a) Compute the operator $d*d* - *d*d$ on a 1–form, in terms of cartesian coordinates on R^3.

(b) From part (a) derive the formula for the laplacian of a vector field on R^3: $\nabla^2\theta = $ grad div $\theta - $ curl curl θ.

(c) Show that $d*d* - *d*d$ is $\pm$ the laplacian on forms of all orders on R^3. (Note that d is 0 on 3–forms.)

4.4.   Interior Products

The *interior product by* X is an operator $i(X)$ on p–forms for every vector field X. It maps a p–form into a $(p - 1)$–form; essentially this is done by fixing the first variable of the p–form θ at X, leaving the remaining $p - 1$ variables free to be the variables of $i(X)\theta$ (except for a normalizing factor p). In formulas, for vector fields $X_1, \ldots, X_{p-1}$,

$$[i(X)\theta](X_1, \ldots, X_{p-1}) = p\theta(X, X_1, \ldots, X_{p-1}).$$

For 0–forms we define $i(X)f = 0$.

Example. We compute $i(\partial_1)$ on the basis p-forms $dx^{i_1}\cdots dx^{i_p}$, where $i_1 < \cdots < i_p$.

(a) If $i_1 \neq 1$ we have for $j_1 < \cdots < j_{p-1}$,

$$i(\partial_1)(dx^{i_1}\cdots dx^{i_p})(\partial_{j_1}, \ldots, \partial_{j_{p-1}}) = p\, dx^{i_1}\cdots dx^{i_p}(\partial_1, \partial_{j_1}, \ldots, \partial_{j_{p-1}})$$
$$= 0 \qquad \text{(see Problem 4.2.1)}.$$

Thus $i(\partial_1)(dx^{i_1}\cdots dx^{i_p}) = 0$ since its components are all 0.

(b) If $i_1 = 1$ we have that $p\, dx^1\, dx^{i_2}\cdots dx^{i_p}(\partial_1, \partial_{j_1}, \ldots, \partial_{j_{p-1}})$ is 0 if $(i_2, \ldots, i_p) \neq (j_1, \ldots, j_{p-1})$ and it is $p/p! = 1/(p-1)!$ if $(i_2, \ldots, i_p) = (j_1, \ldots, j_{p-1})$. Since these are the same values which $dx^{i_2}\cdots dx^{i_p}$ has on $(\partial_{j_1}, \ldots, \partial_{j_{p-1}})$ it follows that

$$i(\partial_1)(dx^1\, dx^{i_2}\cdots dx^{i_p}) = dx^{i_2}\cdots dx^{i_p}.$$

The action of $i(\partial_1)$ on all other forms now can be obtained by using the linearity of $i(\partial_1)$, the latter being obvious from the definition.

Proposition 4.4.1. *The operator $i(X)$ is a derivation of forms, that is, for a p-form θ and a q-form τ it satisfies the product rule:*

$$i(X)(\theta \wedge \tau) = i(X)\theta \wedge \tau + (-1)^p \theta \wedge i(X)\tau.$$

[As with d, if we think of $i(X)$ as having degree -1, then in passing over the p-form θ we get a factor of $(-1)^p$.]

Proof. The operator $i(X)$ is purely algebraic, so that the value of $i(X)\theta$ at a point depends only on the values of X and θ at that point. In particular, if X is 0 at a point, then both sides of the product rule formula are 0 at that point. Hence we only need consider further the case where $X \neq 0$, so we might as well choose coordinates such that $X = \partial_1$.

If we write θ and τ in terms of their coordinate expressions and expand both sides of the product formula using the distributive law for wedge products and the linearity of $i(\partial_1)$, it becomes clear that we only need prove the formula for the cases where θ and τ are coordinate forms $dx^{i_1}\cdots dx^{i_p}$ and $dx^{j_1}\cdots dx^{j_q}$, respectively. For this there are four subcases depending on whether i_1 and j_1 equal 1 or not. These subcases can be dealt with using the above Example and the details are left as an exercise. ∎

Multiplication of these interior product operators is skew-symmetric, that is, $i(X)i(Y) = -i(Y)i(X)$:

$$i(X)i(Y)\theta(\cdots) = pi(Y)\theta(X, \ldots)$$
$$= p(p-1)\theta(Y, X, \ldots)$$
$$= -p(p-1)\theta(X, Y, \ldots)$$
$$= -i(Y)i(X)\theta(\cdots).$$

It follows that the operation $i(X)i(Y)$ depends only on $X \wedge Y$, so we define $i(X \wedge Y) = i(X)i(Y)$. Then we extend linearly to obtain $i(A)$ for every skew-symmetric contravariant tensor A of degree 2. The operator $i(A)$ maps p–forms into $(p - 2)$–forms. Similarly we can define $i(B)$, for any skew-symmetric contravariant tensor B of degree r, mapping p–forms into $(p - r)$–forms.

Since $i(X)$ is a derivation it is determined by its action on 0–forms and 1–forms. One need only express an arbitrary p–form in terms of 0–forms and 1–forms and apply the product rule repeatedly.

Since Lie derivatives are brackets in one case, and the exterior derivative operator d is given in terms of brackets and evaluations of forms on vector fields by (c) in Section 4.3, it is not too surprising that there is a relation between the operators L_X, $i(X)$, and d, operating on forms.

Theorem 4.4.1.   *On differential forms, Lie derivatives are given by the operator equation*

$$L_X = i(X)d + di(X).$$

(We also remember this as $L = id + di$.)

Proof. We have seen that L_X is a derivation of degree 0 of skew-symmetric tensors; that is, it preserves degree and satisfies the product rule (see Section 3.6). We shall show that $i(X)d + di(X)$ also is a derivation:

$$
\begin{aligned}
[i(X)\,d + di(X)](\theta \wedge \tau) &= i(X)(d\theta \wedge \tau + (-1)^p \theta \wedge d\tau) \\
&\quad + d(i(X)\theta \wedge \tau + (-1)^p \theta \wedge i(X)\tau) \\
&= i(X)\,d\theta \wedge \tau + (-1)^{p+1}\,d\theta \wedge i(X)\tau \\
&\quad + (-1)^p i(X)\theta \wedge d\tau + (-1)^{2p}\theta \wedge i(X)\,d\tau \\
&\quad + di(X)\theta \wedge \tau + (-1)^{p-1}i(X)\theta \wedge d\tau \\
&\quad + (-1)^p\,d\theta \wedge i(X)\tau + (-1)^{2p}\theta \wedge di(X)\tau \\
&= (i(X)d + di(X))\theta \wedge \tau + \theta \wedge (i(X)\,d + di(X))\tau.
\end{aligned}
$$

Thus if L_X and $i(X)d + di(X)$ agree on 0–forms and 1–forms, then they agree on all p–forms.

On 0–forms we have $L_X f = Xf$, whereas $i(X)\,df + di(X)f = i(X)\,df + d0 = df(X) = Xf$.

On a 1–form df we have $L_X\,df = d(Xf)$, since $L_X\langle Y, df\rangle = X\langle Y, df\rangle = X(Yf)$.

On the other hand,

$$
\begin{aligned}
L_X\langle Y, df\rangle &= \langle L_X Y, df\rangle + \langle Y, L_X\,df\rangle \\
&= \langle [X, Y], df\rangle + \langle Y, L_X\,df\rangle \\
&= XYf - YXf + \langle Y, L_X\,df\rangle,
\end{aligned}
$$

so $\langle Y, L_X\, df \rangle = YXf = \langle Y, d(Xf) \rangle$. But

$$[i(X)d + di(X)]\, df = i(X)\, d^2f + di(X)\, df$$
$$= 0 + d(Xf).$$

We do not need to check values on the more general 1–forms $g\, df$ because of the product rule being satisfied by each operator.

Corollary. *The operators d and L_X commute on forms; that is, for every p–form θ, $dL_X\theta = L_X\, d\theta$.*

Proof. The formula for L_X and the fact that $d^2 = 0$ give $di(X)\, d\theta$ for both sides. ∎

When written in the form

$$(p + 1)\, d\theta(X, \ldots) = [L_X\theta - d(i(X)\theta)](\ldots),$$

the relation gives a means of determining d on p–forms from Lie derivatives and d on $(p - 1)$–forms. This suggests that when we wish to develop some property of d and we have some corresponding property of Lie derivatives, we should try an induction on the degree of the forms involved.

Problem 4.4.1. (a) Using the fact that L_X commutes with contractions show that

(1) If θ is a 1–form: $(L_X\theta)(Y) = X\theta(Y) - \theta[X, Y]$,

(2) If θ is a 2–form: $(L_X\theta)(Y, Z) = X\theta(Y, Z) - \theta([X, Y], Z) - \theta(Y, [X, Z])$.

(b) Use $(p + 1)\, d\theta(X, \ldots) = L_X\theta(\ldots) - d(i(X)\theta)(\ldots)$ to prove the second and third formulas of Section 4.3(c).

4.5.   Converse of the Poincaré Lemma

Consider the following commonly accepted results from vector analysis in E^3.

(a) If grad $f = 0$, then f is constant.

(b) If curl $X = 0$, then there is a function f such that $X = $ grad f.

(c) If div $X = 0$, then there is a vector field Y such that curl $Y = X$.

(d) For every function f there is a vector field X such that div $X = f$.

Each of these statements is defective, although (d) only requires a modest differentiability assumption. Such differentiability assumptions cannot repair (a), (b), and (c), however, since their major defect lies in the failure to specify certain topological assumptions on the domains of definition of f and X. We give counterexamples to (a), (b), and (c) below. In (a) it must be assumed that the domain of f is connected; in (b) that the domain of X is simply connected, that is, every simple closed curve in the domain of X is the boundary curve of a surface of finite extent in the domain of X; in (c) that every compact surface in the domain of X must be the boundary of a bounded region in the domain of X.

The purpose of this section is to unify and generalize the corrected versions of (a), (b), (c), and (d). This is achieved by translating the results to statements about p–forms. The conditions on the domains are replaced by a single stronger condition which implies all the special cases: We assume that the domain is a coordinate cube for some coordinate system x^i.

Before proceeding with the general theorem let us give some examples which show the necessity of some restrictive hypothesis on the domains. These examples will parallel (a), (b), and (c) above, but will be given in terms of forms.

Examples.   (a) If we define f on $R^3 - \{(0, b, c) \mid b, c \in R\}$ by

$$f(x, y, z) = \begin{cases} 1 & \text{if } x > 0, \\ -1 & \text{if } x < 0, \end{cases}$$

then f is C^∞ and $df = 0$, but f is not constant. This is possible because the domain of f is not connected.

  (b) On $R^3 - \{(0, 0, c) \mid c \in R\}$ we define a 1–form

$$\tau = \frac{-y\,dx + x\,dy}{x^2 + y^2}.$$

We have $d\tau = 0$, but locally $\tau = d\theta$, where θ is any single-valued determination of the cylindrical angle variable. Since this angle cannot be defined continuously throughout the domain of τ, it is impossible to find a function f such that $df = \tau$. (A more convincing argument is that $\int_\gamma df = fp_1 - fp_0$, where p_0 and p_1 are the initial and final points of the curve γ, so if γ is closed, then $\int_\gamma df = 0$. However, if γ is the counterclockwise oriented unit circle in the xy plane, then $\int_\gamma \tau = 2\pi$.) The domain of τ is not simply connected since a curve around the z axis cannot be filled with a surface in the domain of τ.

  (c) Let $r = (x^2 + y^2 + z^2)^{1/2}$ and define the 2–form τ on $R^3 - \{0\}$ by

$$\tau = \frac{1}{r^3}(x\,dy\,dz + y\,dz\,dx + z\,dx\,dy).$$

Then it is easily checked that $d\tau = 0$. However, there is no 1–form α defined on $R^3 - \{0\}$ such that $d\alpha = \tau$, for by Stokes' theorem we have $\int_{S^2} d\alpha = 0$, for every 1–form α, where S^2 is the central positively-oriented unit sphere ($r = 1$). However, the restriction of τ to S^2 is the area element of S^2, so we have $\int_{S^2} \tau = 4\pi = $ area of S^2. Note that S^2 is a compact surface which is not the boundary of a bounded region in the domain of τ. However, the domain of τ is simply connected.

[Our definition and notation for line and surface integrals in terms of forms, as well as Stokes' theorem, will be given later, in Sections 4.8 and 4.9. The translation to the usual vector formulation is in (**b**):

$$\int_\gamma \tau = \int_\gamma (x^2 + y^2)^{-1} \cdot (-y\mathbf{i} + x\mathbf{j}) \cdot d\mathbf{r}$$

and in (**c**):

$$\int_{S^2} \tau = \iint_{S^2} r^{-3}\mathbf{r} \cdot d\mathbf{\sigma}.]$$

A p–form τ is *closed* if $d\tau = 0$; we say, for $p > 0$, that τ is *exact* if there is a $(p - 1)$–form θ such that $d\theta = \tau$; a 0–form is *exact* if it is constant. If for every m in the domain of τ there is a neighborhood U of m such that $\tau|_U$, the restriction of τ to U, is exact, then we say that τ is *locally exact*. It is obvious that exactness implies local exactness. Axiom (**3**) for d, the Poincaré lemma, shows that local exactness implies closedness. Indeed, if $\tau|_U = d\theta_U$, then $d\tau|_U = d(\tau|_U) = d^2\theta_U = 0$, and since this holds for some U about every point, $d\tau = 0$. Our aim is to prove a local converse of the Poincaré lemma:

Theorem 4.5.1.   *If τ is a closed p–form, $p = 1, \ldots, d$, then for every cubical coordinate neighborhood $U = \{m \mid a^i < x^im < b^i\}$ contained in the domain of τ, there is a $(p - 1)$–form θ defined on U such that $d\theta = \tau|_U$. A closed 0–form defined on U is constant on U. In particular, every closed p–form is locally exact.*

Proof. For a 0–form τ, $d\tau = 0$ means that in U, $\partial_i\tau = 0$, $i = 1, \ldots, d$. It then follows that τ is constant along any C^∞ curve in U, and since U is connected, τ is constant in U.

Without loss of generality we may assume that the origin $0 \in R^d$ corresponds to some point in U under $(x^1, \ldots, x^d)$; that is, $a^i < 0 < b^i$, $i = 1, \ldots, d$.

To complete the proof we will construct what is known as an *algebraic homotopy H* of d on forms defined on U. This means that H is a linear transformation of p–forms into $(p - 1)$–forms, $p = 0, 1, \ldots, d$, such that for every p–form τ we have

$$Hd\tau + dH\tau = \tau;$$

that is, $Hd + dH$ is the identity map on forms. Once we have such a homotopy H it is trivial to solve the problem of finding θ, since $d\tau = 0$ gives $dH\tau = \tau$; thus we let $\theta = H\tau$.

We define H on terms of the type $\alpha = f(x^1, \ldots, x^d)dx^{i_1}\cdots dx^{i_p}$, where $i_1 < \cdots < i_p$, by

$$H\alpha = [\int_0^1 f(0, \ldots, 0, tx^{i_1}, x^{i_1+1}, \ldots, x^d)\, dt]x^{i_1}\, dx^{i_2}\cdots dx^{i_p}.$$

If α is a 0-form we let $H\alpha = 0$. Then H is extended to all forms by linearity.

It requires a rather lengthy computation to verify that $Hd\alpha + dH\alpha = \alpha$. Taking the exterior derivatives involves partial differentiation of an integral with respect to parameters in the integrand. A standard theorem of advanced calculus justifies taking the partial derivative operators inside the integral signs. Other than this one needs to observe that

$$x^i\,\partial_i f(0,\ldots,0,tx^i,x^{i+1},\ldots,x^d) = \frac{d}{dt}f(0,\ldots,0,tx^i,x^{i+1},\ldots,x^d)$$

and that

$$tx^i\,\partial_i f(0,\ldots,0,tx^i,x^{i+1},\ldots,x^d) + f(0,\ldots,0,tx^i,x^{i+1},\ldots,x^d)$$
$$= \frac{d}{dt}[tf(0,\ldots,0,tx^i,x^{i+1},\ldots,x^d)]$$

(i not summed in either case).

The details are left as an exercise. ∎

Examples. We show how H performs to give the results (b), (c), (d) above.

(b) Suppose $\tau = f\,dx + g\,dy + h\,dz$, $d\tau = 0$, and τ is defined on a cubical region of R^3. From $d\tau = 0$ we have $f_y = g_x$, $f_z = h_x$, and $g_z = h_y$, where the subscripts indicate partial derivatives. Then $\theta = H\tau$ is the 0-form given by

$$\theta(x,y,z) = x\int_0^1 f(tx,y,z)\,dt + y\int_0^1 g(0,ty,z)\,dt + z\int_0^1 h(0,0,tz)\,dt.$$

From this we get

$$d\theta = \left(\int_0^1 [f(tx,y,z) + xtf_x(tx,y,z)]\,dt\right)dx$$
$$+ \left(\int_0^1 [xf_y(tx,y,z) + g(0,ty,z) + ytg_y(0,ty,z)]\,dt\right)dy$$
$$+ \left(\int_0^1 [xf_z(tx,y,z) + yg_z(0,ty,z) + h(0,0,tz) + zth_z(0,0,tz)]\,dt\right)dz$$
$$= \left(\int_0^1 \frac{d}{dt}[tf(tx,y,z)]\,dt\right)dx + \left(\int_0^1 [xg_x(tx,y,z) + \frac{d}{dt}(tg(0,ty,z))]\,dt\right)dy$$
$$+ \left(\int_0^1 [xh_x(tx,y,z) + yh_y(0,ty,z) + \frac{d}{dt}(th(0,0,tz))]\,dt\right)dz$$
$$= (f(x,y,z) - 0)\,dx + ([g(x,y,z) - g(0,y,z)] + [g(0,y,z) - 0])\,dy$$
$$+ ([h(x,y,z) - h(0,y,z)] + [h(0,y,z) - h(0,0,z)] + [h(0,0,z) - 0])\,dz$$
$$= f\,dx + g\,dy + h\,dz.$$

(c) If $\tau = f\,dx\,dy + g\,dx\,dz + h\,dy\,dz$, then $d\tau = 0$ gives $f_z - g_y + h_x = 0$. For $\theta = H\tau$ we have

$$\theta = \left(\int_0^1 f(tx,y,z)\,dt\right)x\,dy + \left(\int_0^1 g(tx,y,z)\,dt\right)x\,dz + \left(\int_0^1 h(0,ty,z)\,dt\right)y\,dz.$$

The verification that $d\theta = \tau$ requires the same technique, so it is omitted.

(d) If $\tau = f\,dx\,dy\,dz$ and

$$\theta = \left(\int_0^1 f(tx, y, z)\,dt\right) x\,dy\,dz,$$

then it is obvious that $d\theta = \tau$.

Remarks. (a) There is no reason why a^i, b^i cannot be $-\infty, +\infty$. In particular the coordinate range may be all of R^d.

(b) It should be clear that the solution for θ is not unique, except when τ is a 1-form. In fact, if α is an arbitrary $(p - 2)$-form, then $d(\theta + d\alpha) = d\theta + d^2\alpha = \tau$. Moreover, there is nothing special about the operator H; it even depends on the order in which the coordinates are numbered. A more general construction of such homotopies is given in H. Flanders, *Differential Forms*, Academic Press, New York, 1963.

(c) We have already indicated that the sort of domain on which a closed p-form is always exact depends on p. If $p = 0$, then the domain must be connected; if $p = 1$, simply connected; if $p = 2$, spherical surfaces must be deformable to a point; etc. More generally, de Rham has proved a theorem which equates the number of "independent" closed, nonexact p-forms defined globally on a manifold with the pth Betti number B_p of the manifold. The Betti numbers are the same topological invariants encountered in Morse theory (cf. Section 3.10). By this we mean that there are closed p-forms $\tau_1, \ldots, \tau_B, (B = B_p)$ such that:

(1) A linear combination $\sum a_i \tau_i$ with constant a_i's is exact only if all the $a_i = 0$.

(2) For any closed p-form τ there are constants a_i such that $\tau - \sum a_i \tau_i$ is exact.

For example, the first Betti number of $R^2 - \{0\}$ is $B_1 = 1$, and the closed 1-form $(x\,dy - y\,dx)/r^2 = \tau_1$ is the only independent one. In fact, if $\dot\tau$ is any other closed 1-form on $R^2 - \{0\}$ and $c = (2\pi)^{-1} \int_{S^1} \tau$, then $\tau - c\tau_1$ is exact, where S^1 is the central counterclockwise-oriented unit circle. Indeed, by the choice of c, $\int_{S^1} \tau - c\tau_1 = 0$, and hence by Green's theorem in R^2, $\int_\gamma \tau - c\tau_1 = 0$ for every closed curve γ. Thus the line integral $\int \tau - c\tau_1$ is independent of path, so an indefinite integral makes sense and is a 0-form f such that $df = \tau - c\tau_1$.

We may use de Rham's theorem in either direction. If we know something about the Betti numbers of a manifold (for example, by applying Morse theory), we may assert the existence of so many closed forms. Conversely, if we can display some independent closed forms we know that the Betti numbers are at least that great.

Example. Let $M = $ the torus. The angle variables θ and φ giving the amount of rotation in either direction around the torus are defined only up to multiples of 2π, but for any choice the differentials $\tau_1 = d\theta$ and $\tau_2 = d\varphi$ are the same, and hence globally defined, even though θ and φ cannot be. Since τ_1 and τ_2 are locally exact they are closed. The integrals of τ_1 and τ_2 along curves measure the amount of smooth change of θ and φ along the curve. The integral of any *exact* form around a closed curve is 0, so if $a_1\tau_1 + a_2\tau_2$ were exact, then

$$\int_{\gamma_1} a_1\tau_1 + a_2\tau_2 = 2\pi a_1 = 0,$$
$$\int_{\gamma_2} a_1\tau_1 + a_2\tau_2 = 2\pi a_2 = 0,$$

where γ_1 and γ_2 are the sides, identified in pairs, of a square used to represent M (see Figure 14). Thus τ_1 and τ_2 are independent and the Betti number B_1 of M

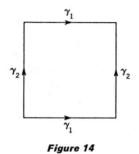

Figure 14

is at least 2. By using the right Morse function (one with only two saddle points) we can prove that $B_1 \leq 2$, which determines B_1 and shows that there are no more independent closed 1–forms.

Problem 4.5.1. Find a vector field X such that curl $X = y\mathbf{i} + z\mathbf{j} + x\mathbf{k}$.

Problem 4.5.2. What are the partial differential equations in terms of co-ordinates for which Theorem 4.5.1 asserts there are local solutions in the case $p = 2$? What are the integrability conditions?

Problem 4.5.3. Generalize Examples (b) and (c) to higher dimensions by finding a radially symmetric $(d-1)$–form on $R^d - \{0\}$ which is closed but not exact.

4.6.   Cubical Chains

The objects over which we integrate p–forms are somewhat more general than p-dimensional oriented submanifolds: We integrate over oriented C^∞ p-cubes and formal sums of them (chains). Of course, the domain of integration in a problem arising in applications is not usually given as a chain, so that in applying this integration theory one must develop the skill of realizing commonly encountered domains as chains, that is, parametrizing the domains. In mathematical applications one rarely parametrizes domains specifically but rather uses the fact that a broad class of domains are parametrizable.

A *rectilinear p-cube* $(p > 0)$ in R^p is a closed cubical neighborhood with respect to cartesian coordinates:

$$U = \{(u^1, \ldots, u^p) \mid b^i \leq u^i \leq b^i + c^i, i = 1, \ldots, p\},$$

where the b^i and c^i are given constants with $c^i > 0, i = 1, \ldots, p$. We do not allow infinite values for the bounds on the u^i, so U is closed and bounded, hence compact.

A C^∞ *p-cube* α in a manifold M is a C^∞ map $\alpha: U \to M$, where U is a rectilinear p-cube. (The meaning of C^∞ on a closed set is that there is some C^∞ extension to an open set $U+$ containing U.)

An *oriented p*-cube is a pair (α, ω), where α is a p-cube and ω is an orientation of R^p. According to the definition in Appendix 3.C, ω is then an atlas of charts on R^p related to each other by positive jacobian determinants. However, for our purposes it is better to express the orientation in terms of p–forms. If coordinates $x^1, \ldots, x^p$ and $y^1, \ldots, y^p$ are related by jacobian determinant $J = \det(\partial x^i / \partial y^j)$, then $dx^1 \cdots dx^p = J \, dy^1 \cdots dy^p$ (cf. Theorem 2.19.2). Thus we can tell whether coordinate systems are consistently oriented by comparing their "coordinate volume elements" $dx^1 \cdots dx^p$ and $dy^1 \cdots dy^p$. Since one of the two global cartesian systems $u^1, \ldots, u^p$ or $-u^1, u^2, \ldots, u^p$ must be consistently oriented with ω, we choose to identify ω with one of the volume elements $du^1 \cdots du^p$ or $d(-u^1) \, du^2 \cdots du^p = -du^1 \cdots du^p$.

If $-\omega$ is the orientation opposite to ω, then we say that $(\alpha, -\omega)$ is the *negative of* (α, ω).

We complete our definitions to include the case $p = 0$ by defining a 0-*cube* in M to be a point $m \in M$, and an *oriented* 0-cube to be a point paired with $+1$ or -1, that is, $(m, +1)$ or $(m, -1)$; these are *negatives* of one another.

If U, as above, is the domain of a p-cube α, we define the $(p-1)$-*faces* of α to be the $(p-1)$-cubes $\alpha_{i\varepsilon}, i = 1, \ldots, p, \varepsilon = 0, 1$, defined by

$$\alpha_{i\varepsilon}(v^1, \ldots, v^{p-1}) = \alpha(v^1, \ldots, v^{i-1}, b^i + \varepsilon c^i, v^i, \ldots, v^{p-1}),$$

where $b^j \leq v^j \leq b^j + c^j$ for $j = 1, \ldots, i-1$ and $b^{j+1} \leq v^j \leq b^{j+1} + c^{j+1}$ for $j = i, \ldots, p-1$. Thus α has $2p$ such $(p-1)$-faces. The *k-faces* of α, $k = 0, \ldots, p-2$, are defined recursively to be the k-faces of the $(k+1)$-faces of α. They are written $\alpha_{i_1 \varepsilon_1 i_2 \varepsilon_2 \cdots i_h \varepsilon_h}$ to avoid the more cumbersome notation $(\cdots (\alpha_{i_1 \varepsilon_1})_{i_2 \varepsilon_2} \cdots)_{i_h \varepsilon_h}$, where $h = p - k$. In particular, the *vertices* or 0-*faces* of α are the points $\alpha(b^1 + \varepsilon_1 c^1, \ldots, b^p + \varepsilon_p c^p)$, so there are 2^p vertices.

To define the $(p-1)$-*faces* of an oriented p-cube (α, ω) we provide $\alpha_{i\varepsilon}$ with an orientation $\omega_{i\varepsilon}$. Since we want $(-\omega)_{i\varepsilon} = -(\omega_{i\varepsilon})$ we take this as part of the definition and restrict our attention to $\omega = du^1 \cdots du^p$. Then we let

$$\omega_{i\varepsilon} = (2\varepsilon - 1)(-1)^{i-1} dv^1 \cdots dv^{p-1}.$$

On the face of U on which $u^i = b^i + \varepsilon c^i$, the coordinate vector of the co-ordinate $(2\varepsilon - 1)u^i$ is directed *outward* from the interior of U. If we follow $(2\varepsilon - 1)u^i$ by $(2\varepsilon - 1)(-1)^{i-1}v^1, v^2, \ldots, v^{p-1}$, we obtain a system consistent with ω, since the sign $(-1)^{i-1}$ compensates for the shift in position of u^i. Thus we have chosen the orientation on the boundary faces of U in accordance with the "outward pointing normal" convention (see Figure 15).

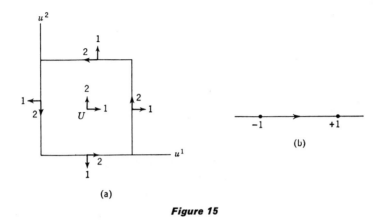

(a)

(b)

Figure 15

The 0-*faces* of an oriented 1-cube (α, du^1) are defined to be $(\alpha(b^1), -1)$ and $(\alpha(b^1 + c^1), +1)$.

It is interesting and important that it is not possible to consistently define oriented $(p - 2)$-faces of an oriented p-cube. In fact, we have

Proposition 4.6.1. *Let α be a p-cube $(p > 1)$, ω an orientation of α, and $1 \le i < j \le p$. Then the $(p - 2)$-cubes $\alpha_{i\delta(j-1)\varepsilon}$ and $\alpha_{j\varepsilon i\delta}$ are the same $(p - 2)$-face of α, and the orientations given it by ω through $\alpha_{i\delta}$ and $\alpha_{j\varepsilon}$ are negatives, that is, $\omega_{i\delta(j-1)\varepsilon} = -\omega_{j\varepsilon i\delta}$.*

Proof. It is evident that $\alpha_{i\delta(j-1)\varepsilon}$ and $\alpha_{j\varepsilon i\delta}$ are both obtained from α by restricting u^i and u^j to be $b^i + \delta c^i$ and $b^j + \varepsilon c^j$, respectively. The signs attached to the first of the remaining coordinates, which determine the orientations $\omega_{i\delta(j-1)\varepsilon}$ and $\omega_{j\varepsilon i\delta}$ in the case $\omega = du^1 \cdots du^p$, are $(2\delta - 1)(-1)^{i-1}(2\varepsilon - 1)(-1)^j$ and $(2\varepsilon - 1)(-1)^{j-1}(2\delta - 1)(-1)^{i-1}$. These are clearly negatives of each other. ∎

We define p-cubes α and β to be *equivalent* if there is a diffeomorphism φ between open sets containing their domains which maps the (geometric)

k-faces of U (= domain of α) onto the k-faces of V (= domain of β), $k = 0, \ldots, p$, and such that the diagram

commutes; that is, $\beta \circ \varphi = \alpha$. It should be clear that equivalence of p-cubes is an equivalence relation. It is also obvious that equivalent p-cubes have the same range. Simple examples of equivalence are obtained by taking $\varphi =$ translations, multiplication of coordinates by nonzero constants, permutations of coordinates, or a combination of these, and letting $\alpha = \beta \circ \varphi$ for some given β.

If α is equivalent to β via φ and ω is an orientation of α, then since φ is a coordinate map on R^p it is consistently oriented either with $(u^1, \ldots, u^p)$ or with $(-u^1, u^2, \ldots, u^p)$. Depending on which is the case we define (α, ω) to be *equivalent* to (β, ω) or $(\beta, -\omega)$. Again this is an equivalence relation on oriented p-cubes.

Proposition 4.6.2.   *If* (α, ω) *is equivalent to* $(\beta, \delta\omega)$, $\delta = \pm 1$, *then the oriented* $(p - 1)$-*faces of* (α, ω) *are equivalent in some order to those of* $(\beta, \delta\omega)$.

(This follows immediately from the definitions.)

A *p-chain* is a finite formal sum $\sum r_i C_i$ of oriented p-cubes C_i with real numbers r_i as coefficients. A p-chain $\sum r_i C_i$ is *equivalent* to a p-chain $\sum s_j D_j$ if for every oriented p-cube (α, ω) we have

$$\sum \{r_i \mid C_i \text{ is equivalent to } (\alpha, \omega)\}$$
$$- \sum \{r_i \mid C_i \text{ is equivalent to } (\alpha, -\omega)\}$$
$$= \sum \{s_j \mid D_j \text{ is equivalent to } (\alpha, \omega)\}$$
$$- \sum \{s_j \mid D_j \text{ is equivalent to } (\alpha, -\omega)\}.$$

We say that a p-chain $\sum t_i E_i$ is *irreducible* if for every i and j ($i \neq j$), E_i is not equivalent to the negative of E_i, E_j, or the negative of E_j. It follows that every p-chain is equivalent to an irreducible one. For each p, we allow the empty sum, called the *null p-chain*, or simply, the *null chain* 0, as a possibility. Chains can be added to each other and multiplied by real numbers in an obvious way, and these operations are compatible with the equivalence relation.

For a 0-chain we combine the orienting signs with the coefficients and simply write it as a sum of numbers times points: $\sum r_i m_i$.

For each p we see that the set of p-chains is a vector space over the reals.

Define x to be a *regular* point of α if $x \in U^0$, the interior of the cubical range U of α, and if α_* is nonsingular at x. Denote by U^r the set of regular points of α.

The geometric meaning of p-chains comes from the possibility of representing by them a region in a p-dimensional oriented submanifold N. We do this only for irreducible p-chains with coefficients $t_i = 1$. We say that an oriented p-cube (α, ω) *parametrizes* a region S in N if α is 1–1 on U^r, S is the range of α $(S = \alpha U)$, and whenever α_* is nonsingular at x and $v_1, \ldots, v_p$ is a basis of R_x^p which is consistent with the orientation ω, then the basis $\alpha_* v_1, \ldots, \alpha_* v_p$ of $N_{\alpha x}$ is consistent with the orientation of N. [A basis of tangent vectors v_i at x is consistent with ω if there is a coordinate system consistently oriented with ω such that $\partial_i(x) = v_i$.] If a p-cube parametrizes a region, so also does any equivalent p-cube.

An irreducible p-chain (α_i, ω_i) *parametrizes* a region S of N if

(a) Each (α_i, ω_i) parametrizes a region S_i of N.

(b) S is the union of the S_i.

(c) For every $i \neq j$, $\alpha_i U_i^r$ and $\alpha_j U_j^r$ are disjoint, where U_i is the domain of α_i.

Note that we have not required the p-cubes to match along the faces in any regular way, although such a matching is reasonable for parametrization of a manifold with boundary, defined below (cf. Theorem 4.6.2).

A reducible p-chain *parametrizes* S if an equivalent irreducible p-chain parametrizes S. Any other equivalent irreducible p-chain will also parametrize S.

Examples. (a) A constant map $\alpha \colon U \to M$ is a p-cube. Since $\alpha(u^1, \ldots, u^p) = \alpha(-u^1, \ldots, u^p)$, $p > 0$, (α, ω) is equivalent to its negative $(\alpha, -\omega)$.

(b) Triangles, tetrahedra, and their higher dimensional analogues, p-simplexes, can be parametrized by a p-cube. For example, the p-simplex with vertices $(0, \ldots, 0)$ and the unit points $(1, 0, \ldots, 0)$, $(0, 1, 0, \ldots, 0)$, $\ldots$, $(0, \ldots, 0, 1)$ is the range of the p-cube in R^p defined on $U \colon 0 \leq u^i \leq 1$, by

$$\alpha(u^1, \ldots, u^p) = (u^1, u^2[1 - u^1], u^3[1 - u^1][1 - u^2], \ldots,$$
$$u^p[1 - u^1][1 - u^2] \cdots [1 - u^{p-1}]).$$

All interior points of the domain of α are regular. The same formula defines an extension of α to an open set (all of R^p, in fact) containing U, so α is C^∞.

This example, and the fact that a p-cube can be decomposed into p-simplexes, shows that nothing essential can be gained or lost by basing a theory on simplexes rather than on cubes.

(c) The polar coordinate map $\alpha(r, \theta) = (r \cos \theta, r \sin \theta)$ parametrizes the closed unit disk by a 2-cube defined on the rectangle $0 \leq r \leq 1, 0 \leq \theta \leq 2\pi$

(see Figure 16). The regular points are the interior points. The four faces are given by

$\alpha_{10}\theta = (0\cos\theta, 0\sin\theta) = (0, 0)$, so α_{10} is a constant 1-cube,

$\alpha_{11}\theta = (\cos\theta, \sin\theta)$, so α_{11} parametrizes a circle,

$\alpha_{20}r = \alpha_{21}r = (r, 0)$, so α_{20} and α_{21} are equivalent and parametrize a unit segment of the x-axis. Note, however, that for either orientation ω of α, the oriented faces $(\alpha_{20}, \omega_{20})$ and $(\alpha_{21}, \omega_{21})$ are negatives of one another.

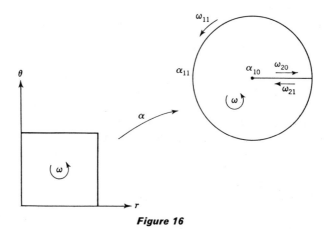

Figure 16

(d) We generalize **(c)** by defining a p-cube which parametrizes the unit p-ball, which has as its topological boundary in R^p the unit sphere S^{p-1}:

$$\alpha(r, \theta_1, \ldots, \theta_{p-1}) = (r\cos\theta_1, r\sin\theta_1\cos\theta_2, r\sin\theta_1\sin\theta_2\cos\theta_3, \ldots,$$
$$r\sin\theta_1\sin\theta_2\cdots\sin\theta_{p-2}\cos\theta_{p-1}, r\sin\theta_1\cdots\sin\theta_{p-1}),$$

where $0 \leq r \leq 1, 0 \leq \theta_i \leq \pi$ for $i = 1, \ldots, p - 2$; and $0 \leq \theta_{p-1} \leq 2\pi$. The face α_{11}, on which $r = 1$, is a parametrization of the $(p - 1)$-dimensional submanifold S^{p-1} of R^p. The face α_{10} is constant. For $2 \leq i \leq p - 1$ the faces $\alpha_{i\varepsilon}$ are constant as functions of $\theta_i, \ldots, \theta_{p-1}$, and so when given an orientation they are equivalent to their negatives. Finally, α_{p0} and α_{p1}, on which θ_{p-1} equals 0 and 2π, respectively, are equal, but for either orientation ω of α, ω_{p0} and ω_{p1} are opposite to each other.

Problem 4.6.1.   In the parametrization of the tetrahedron, Example **(b)** in the case $p = 3$, show that every 2-face is either a parametrization of a triangle or, when given an orientation, equivalent to its negative. Moreover, each of the triangular faces of the tetrahedron is parametrized by just one 2-face of α.

The *boundary* of an oriented p-cube (α, ω) is the $(p - 1)$-chain consisting of the sum of all the (oriented) $(p - 1)$-faces, $\sum_{i=1\cdots, p, \varepsilon=0,1}(\alpha_{i\varepsilon}, \omega_{i\varepsilon})$. It is denoted by $\partial(\alpha, \omega)$. The *boundary* of a p-chain $\sum r_i C_i$ is $\partial \sum r_i C_i = \sum r_i \partial C_i$. Thus ∂ is a linear operator from the vector space of p-chains to the vector space of $(p - 1)$-chains, called the *boundary operator*. It also behaves well with respect to equivalence; that is, if p-chain C is equivalent to p-chain D, then ∂C is equivalent to ∂D.

Proposition 4.6.3.   *For any p-chain $C, p > 1, \partial\partial C = 0$.*

Proof. For an oriented p-cube (α, ω), $\partial\partial(\alpha, \omega) = 0$ follows immediately from Proposition 4.6.1, since the $(p - 2)$-faces cancel in pairs. Then we have

$$\partial\partial C = \partial\partial \sum r_i C_i = \partial \sum r_i \partial C_i = \sum r_i \partial\partial C_i = 0. \quad \blacksquare$$

For a 1-cube (α, ω), the boundary consists of, roughly, the final point minus the initial point. Thus the sum of the coefficients of $\partial(\alpha, \omega)$ is 0. In general, we define the sum of the coefficients of a 0-chain C to be its *Kronecker index*, denoted by $IC = I(\sum r_i m_i) = \sum r_i$. It follows that for any 1-chain D, $I \partial D = 0$. The converse is not true, but the condition for it to be true is topological and gives a hint of the relation between chain algebra and topology:

Proposition 4.6.4.   *A manifold M is connected iff every 0-chain C such that $IC = 0$ is the boundary of some 1-chain D: $\partial D = C$.*

Proof. Suppose M is connected and C is a 0-chain such that $IC = 0$. Then $C = \sum r_i m_i$, where $\sum r_i = 0$. Choose a point $m_0 \in M$. For each m_i we may choose a C^∞ curve α_i from m_0 to m_i since M is connected. We may assume that α_i is parametrized from 0 to 1, so that α_i is a 1-cube defined on $[0, 1]$. Then the 1-chain $D = \sum r_i(\alpha_i, du)$ has boundary

$$\partial D = \sum r_i \, \partial(\alpha_i, du) = \sum r_i(\alpha_i(1) - \alpha_i(0))$$
$$= \sum r_i m_i - \left(\sum r_i\right) m_0 = \sum r_i m_i = C.$$

Conversely, if M is not connected, then for each connected component M_τ of M we define a partial Kronecker index I_τ on 0-chains by $I_\tau \sum r_i m_i =$ the sum of those r_i such that $m_i \in M_\tau$. Since a 1-cube is entirely in M_τ or entirely without, we still have $I_\tau \partial D = 0$ for every 1-chain D. Now let M_0 and M_1 be different components and choose $m_0 \in M_0$ and $m_1 \in M_1$. Then $m_1 - m_0$ is a 0-chain C such that $IC = 0$, but $I_1(m_1 - m_0) = 1$, so $m_1 - m_0$ cannot be a boundary. $\blacksquare$

A large class of regions can be parametrized by chains. We shall state some theorems to that effect without proof, since they involve topological techniques beyond the scope of this book.

Theorem 4.6.1.   *Let M be a compact, oriented manifold of dimension d. Then*

there is a d-chain C in M which parametrizes M itself and for which ∂C is
equivalent to 0.

Another important class of parametrizable regions are the compact,
orientable manifolds with boundary. A subset $N^- = N \cup B$ of a manifold M
is a p-dimensional *submanifold with boundary B* if

(a) N is an open submanifold of a p-dimensional submanifold $N+$ of M.
(b) B is a $(p - 1)$-dimensional submanifold of $N+$.
(c) B is the topological boundary of N with respect to the topology of $N+$.
(d) At each point $b \in B$ there are coordinates $x^i, i = 1, \ldots, p$, on a neighborhood U of b in $N+$ such that $B \cap U = \{n \mid x^1 n = 0\}$ and $N \cap U =$
$\{n \mid x^1 n < 0\}$.

If N is oriented, then B has a corresponding *induced orientation*, the one such
that whenever the coordinates x^i as in (d) are consistent with the orientation of
N, then the coordinates $x^2, \ldots, x^p$, restricted to B, are consistent with the
orientation on B.

Theorem 4.6.2. *If N^- is a compact, oriented manifold with boundary B, then
there is a chain C which parametrizes N^- such that ∂C parametrizes B with the
induced orientation.*

[Theorem 4.6.2 implies Theorem 4.6.1 as the special case where B is empty.
Their proofs follow from the *triangulation* theorem of S. Cairns (*Bull. Am.
Math. Soc.*, 1961). This theorem says that N^- can be decomposed into pieces
diffeomorphic to simplexes and fitting together nicely. Then by using the
mappings of Example (b) we can get the parametrizing chain.]

Volume Elements. If M is an oriented d-dimensional manifold we define a
volume element on M to be a d–form Ω which is defined on all of M, is never 0,
and is consistent with the orientation of M in the following sense: For every
coordinate system x^i on M which is consistently oriented with M, the co-
ordinate expression for Ω is $\Omega = f\,dx^1 \cdots dx^d$, where f is a positive C^∞ function. For any positive C^∞ function g defined on all of M, $g\Omega$ is a volume
element on M if Ω is, and, conversely, any two volume elements are positive
C^∞ multiples of each other. Moreover, any d–form is a C^∞ multiple of Ω.

That a volume element always exists on an oriented manifold can be shown
by using a technical device known as a partition of unity to smooth out the
local coordinate volume elements $dx^1 \cdots dx^d$ into a globally defined one.
Alternatively, a riemannian metric defines an inner product on d–forms, and
since the d–forms at a point form a one-dimensional space, there are only two
of unit length. Only one of these is consistent with the orientation and the field
of these gives a d–form called the *riemannian volume element.* If $\theta_1, \ldots, \theta_d$ is a
local orthonormal basis of 1–forms, then $\Omega = \theta_1 \wedge \cdots \wedge \theta_d$ is a local expression
for Ω. For example, on E^3 it is $dx\,dy\,dz$.

If Ω is a volume element on M and (α, ω) is a d-cube which parametrizes a region of M, then the consistency of the orientation given by α at its regular points gives us the fact that $\alpha^*\Omega = f\omega$, where $f \geq 0$ and $f > 0$ at the regular points.

Examples.   (e) Let $M = R^3$ and let N^- be the closed cylindrical surface: $N^- = \{(x, y, z) \mid x^2 + y^2 = 1, 0 \leq z \leq 1\}$. Then N^- is a two-dimensional submanifold with boundary consisting of two circles. For $N+$ we may take the infinite cylinder $\{(x, y, z) \mid x^2 + y^2 = 1\}$. N^- may be parametrized by a single 2-cube defined by

$$\alpha(u, v) = (\cos u, \sin u, v), \qquad 0 \leq u \leq 2\pi, \quad 0 \leq v \leq 1.$$

(f) In Example (d) the range of α, the closed solid ball N^- in R^p, is a p-dimensional submanifold of R^p with boundary S^{p-1}. We may let $N+ = R^p$. If ω is either orientation of R^p, hence of N, then (α, ω) is a parametrization of N^- such that $\partial(\alpha, \omega)$ parametrizes S^{p-1}. (It should be checked that the orientations do match. If $\omega = du^1 \cdots du^p$ is the notation for the orientation in the range of α, then we might better write $\omega' = dr\, d\theta_1 \cdots d\theta_{p-1}$ for the same orientation of R^p, now viewed as the domain of α.) The $(p-1)$-chain $\partial(\alpha, \omega)$ is reducible since $(\alpha_{10}, \omega_{10})$ and each $(\alpha_{i\varepsilon}, \omega_{i\varepsilon})$, $i = 2, \ldots, p-1$, are constant in one or more variables, so are equivalent to their negatives, hence to 0; moreover, $(\alpha_{p0}, \omega_{p0})$ and $(\alpha_{p1}, \omega_{p1})$ are negatives of each other. Thus $\partial(\alpha, \omega)$ is equivalent to the irreducible chain consisting of one $(p-1)$-cube $(\alpha_{11}, \omega_{11})$, which parametrizes S^{p-1} with the induced orientation. It follows that $\partial(\alpha_{11}, \omega_{11})$ is equivalent to $\partial\partial(\alpha, \omega) = 0$, so $(\alpha_{11}, \omega_{11})$ is a parametrization of S^{p-1} of the type mentioned in Theorem 4.6.1.

We indicate briefly the relation between the algebra of chains and the Betti numbers mentioned in Section 3.10 (Morse theory) and Section 4.5 (de Rham's theorem). A p-cycle is a p-chain Z such that ∂Z is equivalent to the null chain 0. A p-boundary is a p-chain B which is equivalent to ∂C for some $(p+1)$-chain C. The pth (real coefficients) Betti number of M is the integer B_p such that there are B_p p-cycles $Z_1, \ldots, Z_{B_p}$ for which

(a) the only linear combination $\sum r_i Z_i$ which is a p-boundary is the trivial one with all $r_i = 0$, and

(b) for every p-cycle Z there is a linear combination $\sum r_i Z_i$ such that $Z - \sum r_i Z_i$ is a p-boundary. [If M is not compact, there may be no finite number of Z_i satisfying (b), in which case we say $B_p = \infty$.]

By analyzing more carefully the method of proof of Proposition 4.6.4, it can be shown easily that B_0 is the number of connected components of M. Moreover, if M is simply connected, then $B_1 = 0$, but not conversely. If d is the dimension of M, then $B_p = 0$ for $p > d$. If M is compact and orientable, then the Betti numbers are symmetric; that is, $B_p = B_{d-p}$ (Poincaré duality).

4.7.   Integration on Euclidean Spaces

We review here material which can be found in every book on advanced calculus, at least in the two- and three-dimensional cases.

The (standard) *measure* of a rectilinear p-cube

$$U = \{(u^1, \ldots, u^p) \mid a^i \leq u^i \leq b^i\}$$

is $\mu_p U = (b^1 - a^1)(b^2 - a^2)\cdots(b^p - a^p)$. Thus the function μ_p assigns numbers to these cubical subsets of R^p. The *Riemann integral* of a real-valued function f defined on U, if it exists, is

$$\int_U f\, d\mu_p = \lim_{U_j \to 0} \sum_{j=1}^{N} f(x_j)\mu_p U_j,$$

where U has been broken up into N smaller p-cubes U_j and a point x_j has been chosen in each U_j. By the limit existing we mean that it must be possible to make the sum be as close as we please to the supposed limiting value by choosing all the U_j's sufficiently small, no matter what choice of the x_j's is made. The integral can be proved to exist if f is continuous.

This definition is quite natural from the viewpoint of applications, where it is thought of as generalizing the situation for a constant function f. For example, if the density of a substance is constant, the mass is obtained by multiplying the volume (measure) by the density. For variable density f, which is usually assumed to be continuous, it is quite natural to think of the mass as being given approximately by the sum of products $f(x_j)\mu_3 U_j$, where the U_j are small cubes on which f has practically constant value $f(x_j)$, $x_j \in U_j$. Thus the Riemann integral $\int_U f\, d\mu_3$ is a reasonable definition of the mass in the cube U. Most physical applications of integration start with a definition of a quantity by a similar process.

However, such limits of sums are difficult to evaluate (although approximations obtained by computers are being used more and more). For this reason they are related to entirely different objects, *iterated single integrals*. This method of evaluation of Riemann integrals of functions of several variables is used so invariably that frequently the method of evaluation is confused with the definition. For the same reason, superfluous integral signs are used to denote the Riemann integral. The justification of the method of evaluation goes under the name

Fubini's Theorem.   *If f is continuous on U, then the definite integrals*

$$f_p(u^1, \ldots, u^{p-1}) = \int_{a^p}^{b^p} f(u^1, \ldots, u^{p-1}, u^p)\, du^p$$

are continuous functions of the parameters $u^1, \ldots, u^{p-1}$, *and the Riemann integral of f is given by*

$$\int_U f \, d\mu_p = \int_{U_{p-1}} f_p \, d\mu_{p-1},$$

where the $(p-1)$-*cube* $U_{p-1} = \{(u^1, \ldots, u^{p-1}) \mid a^i \le u^i \le b^i, i = 1, \ldots, p-1\}$. *It follows by iteration that*

$$\int_U f \, d\mu_p = \int_{a^1}^{b^1} \left(\cdots \int_{a^{p-1}}^{b^{p-1}} \left(\int_{a^p}^{b^p} f(u^1, \ldots, u^p) \, du^p \right) du^{p-1} \cdots \right) du^1.$$

(Of course, this merely reduces the problem back to one of a similar sort, the evaluation of definite single integrals, which are themselves defined as limits of Riemann sums. For these we have a similar situation: They are almost invariably evaluated by applying the fundamental theorem of calculus, which relates them to the process of finding *antiderivatives*.)

Although convenient for definitive purposes, restricting the domains of functions to be rectilinear cubes is not adequate for most applications. To define the integral of a function f on a more general bounded domain D, we enclose D in a rectilinear cube U and let

$$\int_D f \, d\mu_p = \int_U \Phi_D f \, d\mu_p,$$

where Φ_D is the *characteristic function* of D, defined by

$$\Phi_D x = \begin{cases} 1 & \text{if } x \in D, \\ 0 & \text{if } x \notin D. \end{cases}$$

Again, this definition is not very convenient for evaluative purposes, so it is customary to reduce integrals on D to integrals on a cube by finding a 1–1 C^1 map of a cube onto D and applying the

Change of Variable Theorem for Riemann Integrals.   *If E and D are regions in* R^p, $\varphi: E \to D$ *is a 1–1 C^1 map, and* $I = \int_D f \, d\mu_p$ *exists, then* $\int_E (f \circ \varphi) \vert J_\varphi \vert \, d\mu_p$ *exists and equals I, where J_φ is the jacobian determinant of* φ; *if φ is given by equations* $u^i = F^i(v^1, \ldots, v^p)$, $i = 1, \ldots, p$, *where the u^i are the cartesian coordinates on D, and the v^i are the cartesian coordinates on E, then*

$$J_\varphi(v^1, \ldots, v^p) = \det(\partial_j F^i(v^1, \ldots, v^p)), \ \partial_j = \partial/\partial v^j.$$

Note that if φ is orientation-preserving at points where φ_* is nonsingular, then $J_\varphi \ge 0$ and we may omit the absolute-value signs. Note also that at the singular points of φ_*, $J_\varphi = 0$, so the theorem may be strengthened slightly by only requiring that φ be 1–1 on the regular set of φ, that is, where φ_* is nonsingular.

We illustrate this change of variable theorem in the case $p = 2$ by showing how it can be used to give a common form of Fubini's theorem where the interior limits are functions of the more exterior variables rather than constants. Suppose that

$$D = \{(x, y) \mid a \leq x \leq b, \text{ and for each } x, h(x) \leq y \leq k(x)\},$$

where h and k are given C^1 functions such that $h(x) \leq k(x)$ for $a \leq x \leq b$. We map the rectangle $E = \{(u, v) \mid a \leq u \leq b, 0 \leq v \leq 1\}$ onto D by $\varphi: E \to D$ which has equations $x = u$, $y = v \cdot k(u) + (1 - v)h(u)$ (see Figure 17). Then

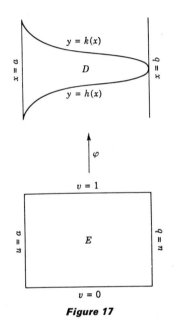

Figure 17

$J_\varphi = 1 \cdot (k(u) - h(u)) - 0 \cdot (\cdots) = k(u) - h(u) \geq 0$. By the change of variable theorem followed by Fubini's theorem,

$$\int_D f(x, y) \, d\mu_2 = \int_E f(u, vk(u) + (1 - v)h(u))(k(u) - h(u)) \, d\mu_2$$

$$= \int_a^b \left(\int_0^1 f(u, vk(u) + (1 - v)h(u))(k(u) - h(u)) \, dv \right) du.$$

Now each of the interior integrals, for each value of u, can be transformed by the change of variable theorem for one variable; keeping u fixed and letting $y = vk(u) + (1 - v)h(u)$ we have $dy = (k(u) - h(u)) \, dv$, $y = h(u)$ when $v = 0$, and $y = k(u)$ when $v = 1$, so the interior integral becomes $\int_{h(u)}^{k(u)} f(u, y) \, dy$.

Now the change of dummy variable, x for u, in the exterior integral yields the usual form of Fubini's theorem for integrals on D,

$$\int_D f(x, y) \, d\mu_2 = \int_a^b \left(\int_{h(x)}^{k(x)} f(x, y) \, dy \right) dx.$$

4.8.   Integration of Forms

When we turn to the integration of forms, the new element of orientation is injected. This arises naturally in applications. For example, the work done in traversing a curve under the influence of a force field depends, in sign, on the direction along which the curve is traveled. It would be natural, from a physical viewpoint, to formulate the definition in terms of limits of sums. However, the usual difficulties encountered in handling such sums are magnified by the need to integrate on curved objects. By now we should anticipate that such integrals would be evaluated by means other than the definition. So instead of formulating such a limit-of-sums definition we give a definition in terms of Riemann integrals, for which the evaluation problem has been resolved already by Fubini's theorem.

Let θ be a p–form defined on a region of a manifold M which contains the range of an oriented p-cube (α, ω), where $\alpha: U \to M$. Then we pull back θ to U, using the map α, to get a p–form $\alpha^*\theta$ defined on U. Recall from Section 3.9 that, in terms of coordinates, finding the expression for $\alpha^*\theta$ amounts to a straightforward substitution of the coordinate formulas for α into the co-ordinate expression for θ. Since we have chosen to consider ω as being $\pm du^1 \cdots du^p$, ω is a basis for p–forms on R^p. Thus we have an expression $\alpha^*\theta = f\omega$, where f is a C^∞ real-valued function on U. If we define an inner product $\langle \ , \ \rangle_p$ on p–forms on R^p by letting ω be unitary, that is, $\langle \omega, \omega \rangle_p = 1$, then $f = \langle \alpha^*\theta, \omega \rangle_p$. The definition of the integral of θ on (α, ω) is then

$$\int_{(\alpha, \, \omega)} \theta = \int_U \langle \alpha^*\theta, \omega \rangle_p \, d\mu_p.$$

The integral of a 0–form θ on a 0-cube $m \in M$ is defined to be the value θm of θ on m. The integral of a p–form θ on a p-chain $\sum r_i C_i$ is defined in the most obvious way in terms of the integrals on p-cubes:

$$\int_{\sum r_i C_i} \theta = \sum r_i \int_{C_i} \theta.$$

Examples.   (a) The circle $S^1 = \{(x, y) \mid x^2 + y^2 = 1\}$ in R^2, with the counter-clockwise orientation, is parametrized by (α, du), where α is defined on $[0, 2\pi]$ by $\alpha(u) = (\cos u, \sin u)$. The coordinate equations for α are thus $x = \cos u$, $y = \sin u$. If $\theta = (x \, dy - y \, dx)/(x^2 + y^2)$ then

$$\alpha^*\theta = [\cos u \, d(\sin u) - \sin u \, d(\cos u)]/[\cos^2 u + \sin^2 u]$$
$$= \cos^2 u \, du + \sin^2 u \, du = du.$$

Now we have $\langle du, du \rangle_1 = 1$, so

$$\int_{(\alpha, du)} \theta = \int_{[0, 2\pi]} 1 \, d\mu_1 = \int_0^{2\pi} du = 2\pi.$$

(b) The sphere S^2 has been parametrized in Example **(d)**, Section 4.6, with the 2-cube α defined on $U = [0, \pi] \times [0, 2\pi]$ by equations

$$(x, y, z) = (\cos u, \sin u \cos v, \sin u \sin v)$$
$$= \alpha(u, v).$$

(The notation there was: $p = 3$, $\alpha_{11} = \alpha$, $\theta_1 = u$, $\theta_2 = v$, $r = 1$.) We define the *positive orientation* of S^2 to be the one for which the coordinates y, z, restricted to S^2, is a consistently oriented system in a neighborhood of $(1, 0, 0)$. This follows the outward-pointing normal convention in that ∂_x points out from the ball bounded by S^2 in R^3 and x, y, z define what we consider to be the positive orientation on R^3. Then $(\alpha, du \, dv)$ is a parametrization of this positively oriented S^2.

In Example **(c)**, Section 4.5, we defined a 2–form τ on $R^3 - \{0\}$ by $\tau = (x \, dy \, dz + y \, dz \, dx + z \, dx \, dy)/r^3$. We compute $\alpha^* \tau$ by substituting the following and employing Grassmann algebra:

$$\alpha^* \, dx = -\sin u \, du,$$
$$\alpha^* \, dy = \cos u \cos v \, du - \sin u \sin v \, dv,$$
$$\alpha^* \, dz = \cos u \sin v \, du + \sin u \cos v \, dv,$$
$$\alpha^* r^3 = 1,$$
$$\alpha^*(dx \, dy) = \alpha^* \, dx \wedge \alpha^* \, dy = \sin^2 u \sin v \, du \, dv,$$
$$\alpha^*(dy \, dz) = \cos u \sin u \cos^2 v \, du \, dv - \sin u \cos u \sin^2 v \, dv \, du$$
$$= \cos u \sin u \, du \, dv,$$
$$\alpha^*(dz \, dx) = \sin^2 u \cos v \, du \, dv,$$
$$\alpha^* \tau = [\cos^2 u \sin u + \sin^3 u \cos^2 v + \sin^3 u \sin^2 v] \, du \, dv$$
$$= \sin u \, du \, dv.$$

The surface integral of τ on $(\alpha, du \, dv)$ "$=$" S^2 (cf. the remark following Theorem 4.8.2) is now easily evaluated:

$$\int_{(\alpha, du \, dv)} \tau = \int_U \sin u \, d\mu_2 = \int_0^\pi \int_0^{2\pi} \sin u \, dv \, du = 4\pi.$$

(c) A 2–chain representing the three faces of a tetrahedron pictured in Figure 18 is $C = C_1 + C_2 + C_3$, where the $C_i = (\alpha_i, du \, dv)$ are given by $\alpha_1(u, v) = (0, u, (1 - u)v)$, $\alpha_2(u, v) = ((1 - u)v, 0, u)$, and $\alpha_3(u, v) = (u, (1 - u)v, 0)$, where $0 \le u \le 1, 0 \le v \le 1$ defines their common domain U. For $\theta = dy \, dz + dx \, dy$ we have

$$\alpha_1^* \theta = du \wedge d[(1 - u)v] = (1 - v) \, du \, dv,$$
$$\alpha_2^* \theta = 0 \qquad \text{(each term has a } d0 = 0\text{)},$$
$$\alpha_3^* \theta = du \wedge d[(1 - u)v] = (1 - u) \, du \, dv.$$

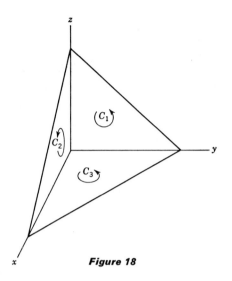

Figure 18

Thus

$$\int_C \theta = \int_U (1 - u)\, d\mu_2 + \int_U 0\, d\mu_2 + \int_U (1 - u)\, d\mu_2$$

$$= 2\int_0^1 \int_0^1 (1 - u)\, du\, dv = 1.$$

Translation to Vector Notation. For oriented integrals on E^2 and E^3 the customary vector notation is not difficult to translate to the notation of forms. In fact, it will be found that the common methods of evaluating vector integrals have the translation to forms concealed in them.

For line integrals the vector notation is $\int_C \mathbf{F} \cdot d\mathbf{r} = \int_C \mathbf{F} \cdot \mathbf{T}\, ds$, where $\mathbf{F}$ is a vector field, $\mathbf{r}$ is the displacement vector, $\mathbf{T}$ is the unit tangent field along C, and s is arc length on C. The notation of forms is in common use and follows immediately by substituting $\mathbf{F} \cdot d\mathbf{r} = F^1 dx + F^2\, dy + F^3\, dz$, where F^1, F^2, and F^3 are the components of $\mathbf{F}$.

For line integrals in the plane another type is encountered, the line integral giving the flux of a vector field across a curve. The vector notation is $\int_C \mathbf{F} \cdot \mathbf{N}\, ds$, where $\mathbf{N}$ is the unit normal for the positive direction across C. It is transformed to the other type by applying the Hodge star operator, which in E^2 is merely a rotation by $\pi/2$. Thus $*\mathbf{N} = \mathbf{T}$, and since $\mathbf{F} \cdot \mathbf{N} = *\mathbf{F} \cdot *\mathbf{N}$, $\int_C \mathbf{F} \cdot \mathbf{N}\, ds = \int_C *\mathbf{F} \cdot d\mathbf{r}$.

For surface integrals in E^3 the vector notation is $\iint_S \mathbf{F} \cdot d\sigma = \iint_S \mathbf{F} \cdot \mathbf{N}\, d\sigma$, where $\mathbf{N}$ is the orienting unit normal and $d\sigma$ is the area element. Again, the $*$ operator can be used to give an oriented "unit tangent" to S, $*\mathbf{N}$, which is equal to $\mathbf{E}_1 \wedge \mathbf{E}_2$ if $\mathbf{E}_1, \mathbf{E}_2$ is an orthonornal basis of the tangent space of S consistent with the orientation of S. To compensate, we apply $*$ to $\mathbf{F}$ also, and obtain for the integral on S: $\iint_S \mathbf{F} \cdot \mathbf{N}\, d\sigma = \int_S F_1\, dy\, dz + F_2\, dz\, dx + F_3\, dx\, dy$, where $F_j =$

$F^i \, \delta_{ij} = F^j$ since $\partial_i \cdot \partial_j = \delta_{ij}$. Thus the form we integrate comes from $*\mathbf{F}$ by using the metric to lower indices.

In volume integrals the orientation is not usually mentioned, since it is invariably taken to be the "positive orientation" $dx \, dy \, dz$. With this convention the customary notation and ours almost coincide:

$$\iiint_V f \, dx \, dy \, dz = \iiint_V f \, dV = \int_V f \, d\mu_3.$$

Note that $*f = f \, dx \, dy \, dz$, so the $*$ operator may have use here, especially in combination with d to give "div."

Independence of Parametrization. To assure that the integrals we have defined have geometric meaning it is necessary to establish two results on independence of parametrization. The first is that the integrals of a p-form θ on equivalent p-cubes are the same. The second is that the integral of θ on a parametrization of an "oriented subset" (and, in particular, of an oriented submanifold with boundary) is independent of parametrization. The first allows us to ignore the distinction, in integration theory, between equivalent p-chains. The second allows us to define the integral of a form on an oriented subset.

Theorem 4.8.1. *If (α, ω) and $(\beta, \delta\omega), \delta = \pm 1$, are equivalent p-cubes, then for any p-form θ defined on the range of α and β,*

$$\int_{(\alpha,\omega)} \theta = \int_{(\beta,\delta\omega)} \theta.$$

Proof. Since (α, ω) and $(\beta, \delta\omega)$ are equivalent, there is a diffeomorphism $\varphi\colon U \to V$ such that $\beta \circ \varphi = \alpha$, where $\alpha\colon U \to M$ and $\beta\colon V \to M$. It follows immediately from the chain rule $(\beta_* \circ \varphi_* = \alpha_*$, cf. Problem 1.8.2) and the alternative definition of φ^* (Proposition 3.9.4) that $\varphi^* \circ \beta^* = \alpha^*$. Thus we have, if $\alpha^*\theta = f\omega$ and $\beta^*\theta = g\delta\omega, f\omega = \varphi^*(\beta^*\theta) = \varphi^*(g\delta\omega) = (g \circ \varphi)\varphi^*(\delta\omega)$. However, φ must carry coordinates on V which are consistently oriented with $\delta\omega$ into coordinates on U which are consistently oriented with ω. This means that the sign of the jacobian of φ is the same as that of δ.

Let the equations for φ be $v^i = F^i(u^1, \ldots, u^p)$, $i = 1, \ldots, p$, where u^i are the cartesian coordinates on U, v^i those on V. Then, if say $\omega = dv^1 \cdots dv^p = du^1 \cdots du^p$ (as forms on R^p),

$$
\begin{aligned}
\varphi^*(\delta\omega) &= \delta\varphi^*(dv^1 \cdots dv^p) \\
&= \delta\varphi^* \, dv^1 \wedge \cdots \wedge \varphi^* \, dv^p \\
&= \delta(\partial_{i_1} F^1 \, du^{i_1}) \wedge \cdots \wedge (\partial_{i_p} F^p \, du^{i_p}) \\
&= \delta \det (\partial_i F^j) \, du^1 \cdots du^p \\
&= \delta J_\varphi \omega.
\end{aligned}
$$

The same equation (between the first and last) obtains if $\omega = -du^1 \cdots du^p$, since this merely inserts $-$ signs in the intermediate quantities. Thus $f\omega = (g \circ \varphi)\varphi^*(\delta\omega) = (g \circ \varphi) \mid J_\varphi \mid \omega$, since the signs of J_φ and δ are the same; that is, $f = (g \circ \varphi) \mid J_\varphi \mid$. Now by the change of variable theorem,

$$\int_{(\beta, \delta\omega)} \theta = \int_V g \, d\mu_p = \int_U (g \circ \varphi) \mid J_\varphi \mid d\mu_p = \int_U f \, d\mu_p = \int_{(\alpha, \omega)} \theta. \quad \blacksquare$$

Corollary. *If C and D are equivalent p-chains, then $\int_C \theta = \int_D \theta$.*

Proof. Besides an obvious equality of sums of integrals over equivalent cubes this requires an additional triviality: $\int_{(\alpha, -\omega)} \theta = -\int_{(\alpha, \omega)} \theta$.

Theorem 4.8.2. *If p-chains C and D both parametrize the same region S of a p-dimensional oriented submanifold N and θ is a p-form defined on S, then*

$$\int_C \theta = \int_D \theta.$$

Outline of Proof. According to the above Corollary we may replace C and D by equivalent chains, so we may assume they are irreducible, say, $C = \sum_{i=1}^h (\alpha_i, \omega_i)$ and $D = \sum_{j=1}^k (\beta_j, \omega_j')$, where α_i is defined on U_i and β_j is defined on V_j. Let U_i^r and V_j^r be the corresponding sets of regular points. It is important for the proof to use the fact that integration over S may be accomplished by integration over the subset consisting of the common part of the ranges of the α_i's and β_j's on their regular sets, that is, on

$$\left(\bigcup_i \alpha_i U_i^r\right) \cap \left(\bigcup_j \beta_j V_j^r\right) = \bigcup_{i,j} (\alpha_i U_i^r \cap \beta_j V_j^r).$$

The reason we may ignore the remainder of S is that the nonregular points correspond to "sets of measure zero" in S, that is, the nonregular points do not contribute to the integral over S; the boundary points of U_i and V_j are lower dimensional and may be enclosed in slabs of arbitrarily small measure, and in a neighborhood of a singular point of α_{i*}, α_i is approximated by the tangent map α_{i*} which maps the tangent space to a lower-dimensional subspace. The same situation prevails for β_j. Making use of these facts reduces the proof to the following computation:

$$\int_C \theta = \sum_i \int_{(\alpha_i, \omega_i)} \theta$$

$$= \sum_i \int_{U_i^r} \langle \alpha_i^*\theta, \omega_i \rangle_p \, d\mu_p$$

$$= \sum_{i,j} \int_{(\alpha_i^{-1}\beta_j V_j^r) \cap U_i^r} \langle \alpha_i^*\theta, \omega_i \rangle_p \, d\mu_p$$

$$= \sum_{i,j} \int_{V_j^r \cap \beta_j^{-1}\alpha_i U_i^r} \langle \beta_j^*\alpha_i^{-1*}\alpha_i^*\theta, \omega_j' \rangle_p \, d\mu_p$$

$$\int_{\partial(\alpha,\omega)} \theta = \sum_{i,\varepsilon} \int_{(\alpha_{i\varepsilon},\ \omega_{i\varepsilon})} \theta$$

$$= \sum_{i,\varepsilon} \int_{U_i} (2\varepsilon - 1)f_i \Big|_{u^i = b^i + \varepsilon c^i} d\mu_{p-1}$$

$$= \sum_i \int_{U_i} f_i \Big|_{u^i = b^i}^{u^i = b^i + c^i} d\mu_{p-1}.$$

On the other hand, $\int_U \langle \alpha^* \, d\theta, \omega \rangle_p \, d\mu_p = \sum_i \int_U \partial_i f_i \, d\mu_p$. We apply the first step of Fubini's theorem to $\int_U \partial_i f_i \, d\mu_p$ with the ith variable as the variable of integration, obtaining

$$\int_U \partial_i f_i d\mu_p = \int_{U_i} \int_{b^i}^{b^i + c^i} \partial_i f_i(u^1, \ldots, u^i, \ldots, u^p) \, du^i \, d\mu_{p-1}$$

$$= \int_{U_i} [f_i(u^1, \ldots, b^i + c^i, \ldots, u^p)$$
$$- f_i(u^1, \ldots, b^i, \ldots, u^p)] \, d\mu_{p-1},$$

where the second step follows from the fundamental theorem of calculus. The terms now match those of $\int_{\partial(\alpha,\omega)} \theta$. ∎

The following corollary is used to obtain Green's formulas in E^2 and E^3.

Corollary 1. (Integration by Parts.)   *Under the same hypothesis, if f is a real-valued function defined on the domain of θ,*

$$\int_C df \wedge \theta = \int_{\partial C} f\theta - \int_C f \, d\theta.$$

More generally, if θ is a p–form, τ a q–form, and C a $(p + q + 1)$-chain, then

$$\int_C d\theta \wedge \tau = \int_{\partial C} \theta \wedge \tau - (-1)^p \int_C \theta \wedge d\tau.$$

(The proof follows immediately from the fact that d is a derivation.)

Corollary 2.   *If θ is a $(d-1)$–form defined on a compact oriented d-dimensional manifold M, then $\int_M d\theta = 0$ and $d\theta = 0$ at some point.*

Proof. By Theorem 4.6.1 there is a d-chain C in M which parametrizes M and for which ∂C is equivalent to the null chain 0. Thus $\int_M d\theta = \int_C d\theta = \int_{\partial C} \theta = \int_0 \theta = 0$.

To prove the last part we may assume that M is connected, for otherwise we would consider the restriction of θ to a connected component of M. Let Ω be a volume element for M. Then $d\theta = f\Omega$, where f is a real-valued C^∞ function on M. For any p-cube (α_i, ω_i) in C, $\langle \alpha_i{}^* d\theta, \omega_i \rangle_d = f \circ \alpha_i \langle \alpha_i{}^*\Omega, \omega_i \rangle_d$ has the same sign at a regular point x of α_i as $f(\alpha_i x)$. Since

$$\sum_i \int_{(\alpha_i, \omega_i)} \langle \alpha_i{}^* d\theta, \omega_i \rangle_d \, d\mu_d = 0,$$

f is either identically 0 or not of the same sign everywhere. In the latter case the general intermediate-value theorem (Proposition 0.2.7.4) tells us that f must be zero at some point. $\blacksquare$

Examples.   Stokes' theorem is often used in conjunction with a riemannian structure. On an oriented riemannian manifold which is not compact it is sometimes possible to find a $(d-1)$–form θ such that $d\theta$ is the riemannian volume element. It follows from Stokes' theorem that the integral of θ on the boundary of a region is the d-dimensional volume of the region. In E^2 we may take $\theta = x\,dy$, $-y\,dx$, or $ax\,dy - by\,dx$, where $a + b = 1$. In E^3 we can use $x\,dy\,dz$, $y\,dz\,dx$, or $z\,dx\,dy$. The integral of $x\,dy$ in the positive direction around a simple closed curve in E^2 gives the area enclosed by the curve. The integral of $x\,dy\,dz$ on the boundary of a region in E^3 gives the volume of the region.

On a riemannian manifold a generalization of the laplacian, $\partial_x^2 + \partial_y^2 + \partial_z^2$, on E^3 is defined in terms of the Hodge star operator by $\nabla^2 f = * d* df$. (∇^2 is called the Laplace-Beltrami operator.) We can use Stokes' theorem to produce uniqueness theorems for the elliptic partial differential equations associated with ∇^2. For example, Poisson's equation $\nabla^2 f = \rho$ has a unique solution up to an additive constant, if any at all, on a compact orientable connected manifold. In particular, the only *harmonic functions*, that is, solutions of $\nabla^2 f = 0$, are constants. The proof uses the fact that for a 1–form θ, $\theta \wedge *\theta$ is a nonnegative multiple of the volume element and is 0 only at points where $\theta = 0$. For any two solutions f_1, f_2 of $\nabla^2 f = \rho$, $g = f_1 - f_2$ is harmonic; that is, $d* dg = 0$. We integrate $0 = g\,d(* dg)$ by parts, and since ∂M is equivalent to the null chain 0, $0 = \int_M g\,d* dg = -\int_M dg \wedge * dg$. But the only way that the integral of a nonnegative multiple of the volume element can vanish is for the integrand to vanish identically. That is, $dg \wedge * dg = 0$, from which it follows that $dg = 0$ so g is constant on connected components of M.

Problem 4.9.1.   Show that one case of Stokes' theorem is the fundamental theorem of calculus.

Problem 4.9.2.   Find an $(n-1)$-form θ on E^n such that $d\theta = du^1 \cdots du^n$ and which is radially symmetric, that is, can be derived from $r^2 = \sum (u^i)^2$.

Problem 4.9.3. Extend the uniqueness theorem for Poisson's equation to the case of a function on a manifold with boundary for which the values on the boundary are specified. The solution is unique without the freedom to add a constant. The value of $* dg$ on the boundary is essentially the normal derivative of g at the boundary, so we also get a uniqueness theorem when the value of the normal derivative on the boundary is specified.

Problem 4.9.4. Suppose that θ is a p-form on M such that for every $(p + 1)$-cube C in M, $\int_{\partial C} \theta = 0$. Show that θ is closed. (*Hint:* See Problem 4.8.1.)

Problem 4.9.5. Recall that for each p the set of p-chains is a vector space $\mathscr{C}_p$ over R. An element of the dual space $\mathscr{C}^p = \mathscr{C}_p{}^*$ of $\mathscr{C}_p$ is called a p-*cochain*. Define the *coboundary operator* $\partial^* : \mathscr{C}^p \to \mathscr{C}^{p+1}$ by

$$(\partial^* f)C_{p+1} = f(\partial C_{p+1}).$$

The $(p + 1)$-cochain $\partial^* f$ is called the coboundary of the p-cochain f. The operator ∂^* is linear and its square is the null operator: $\partial^* \, \partial^* f = 0$.

Problem 4.9.6. For each p–form τ defined globally on a manifold M there is a corresponding mapping $f_\tau : \mathscr{C}_p \to R$ defined by

$$f_\tau C_p = \int_{C_p} \tau.$$

Show that f_τ is a p-cochain and that the function which sends τ into f_τ is linear.

Problem 4.9.7. A cochain f is a *cocycle* if $\partial^* f = 0$. Show that if τ is a closed p–form, then f_τ is a cocycle.

Problem 4.9.8. A cochain f is a *coboundary* if there is a cochain g such that $\partial^* g = f$. Show that if τ is exact, then f_τ is a coboundary.

4.10. Differential Systems

Many systems of partial differential equations have a geometric formulation in terms of differential forms. The principal reason for this is very simple. For example, let x, y, z, p, q be coordinates on R^5 and let S be a two-dimensional submanifold on which x and y can be used as coordinates (that is, "independent variables"). Then the condition that $p = \partial z / \partial x$ and $q = \partial z / \partial y$ on S is that the differential form $dz - p \, dx - q \, dy$ vanish on S.

A first-order partial differential equation is given by an equation $F(x, y, z, p, q) = 0$. This specifies a hypersurface N of R^5. A solution to the partial differential equation, say, $z = f(x, y)$, determines a two-dimensional parametric submanifold S of N, $z = f(x, y)$, $p = (\partial_x f)(x, y)$, $q = (\partial_y f)(x, y)$. This surface S is an integral submanifold of the three-dimensional distribution

D on N given by the equation $dz - p\,dx - q\,dy = 0$. More formally, D is specified by $D(n) = \{t \mid t \in N_n \text{ and } \langle t, dz - p\,dx - q\,dy \rangle = 0\}$. Of course, there are a few difficulties with degeneracy: The points where $F = 0$ and $dF = 0$ simultaneously must be eliminated because N is not usually a manifold in a neighborhood of such a point; if there is a point $n \in N$ where dF is proportional to $dz - p\,dx - q\,dy$, then the subspace $D(n)$ specified above is all of N_n and D is not uniformly three-dimensional; finally, the solution surface must be chosen so that x and y can be taken as coordinates on it. It will not usually occur that the distribution D is completely integrable.

The above formulation and its generalizations have not been used extensively to study partial differential equations. Some results have been obtained using this means for analytic equations, by the great mathematician E. Cartan. However, we consider that the geometric setting gives important insight into what one should expect by way of solutions. In the following we shall abstract from the above example, defining structures dual to distributions (see Sections 3.11 and 3.12). In particular, we shall obtain a dual formulation of Frobenius' theorem.

A *k-dimensional codistribution* Δ on a manifold M is a function which assigns to $m \in U \subset M$ a k-dimensional subspace $\Delta(m)$ of the cotangent space $M_m{}^*$. It is C^∞ if its domain U is open and for each $m \in U$ there is a neighborhood V of m and 1-forms $\omega^1, \ldots, \omega^k$ defined on V such that at each $n \in V$ the subspace $\Delta(n)$ is spanned by $\omega^1(n), \ldots, \omega^k(n)$. To each k-dimensional codistribution Δ there is the *associated* $(d - k)$-*dimensional distribution* D given by $D(m) = \{t \mid t \in M_m, \langle t, \omega \rangle = 0 \text{ for every } \omega \in \Delta(m)\}$, and vice versa, for each $(d - k)$-dimensional distribution D there is the *associated k-dimensional codistribution* Δ, given by $\Delta(m) = \{\omega \mid \omega \in M_m{}^*, \langle t, \omega \rangle = 0 \text{ for every } t \in D(m)\}$. Clearly, if D is associated to Δ, then Δ is associated to D. The D and Δ associated in this way are said to *annihilate* each other. If one is C^∞, so is the other.

A submanifold N of M is an *integral submanifold* of a codistribution Δ if N is an integral submanifold of the associated distribution. A codistribution is *completely integrable* if the associated distribution is completely integrable.

The local version of Frobenius' theorem (Theorem 3.12.1) is that for a completely integrable distribution D there are coordinates x^i such that $D(m)$ is spanned by $\partial_1(m), \ldots, \partial_h(m)$ and the integral submanifolds of dimension $h = d - k$ are the coordinate slices $x^\alpha = c^\alpha$, $\alpha = h + 1, \ldots, d$. It follows that the associated codistribution Δ is spanned by the dx^α in the coordinate neighborhood.†

---

† It should be evident that "$x^\alpha = c^\alpha$" and "$dx^\alpha = 0$" convey practically the same information. For this reason the codistribution formulation predominates historically. Distributions were usually denoted in terms of a local 1-form basis ω^α of the associated codistribution by writing $\omega^\alpha = 0$. Before the formalization in terms of tangent vector spaces and dual spaces the 1-forms seem to have been thought of as infinitesimal displacements in the dual vector directions. Thus $\omega^\alpha = 0$ indicates that displacement is allowed only in the directions of the distribution D.

Any other 1–form belonging to Δ can be expressed as $\omega = f_\alpha \, dx^\alpha$, where α is a summation index running from $h + 1$ to d. (We will also use β as a summation index with this range.) The exterior derivative $d\omega = df_\alpha \wedge dx^\alpha$ is a "linear combination"† of the dx^α with 1–form coefficients df_α. If ω^α is another local basis of Δ, then $dx^\alpha = g_\beta^\alpha \omega^\beta$, where (g_β^α) is a nonsingular matrix of C^∞ functions. Then we have $d\omega = df_\alpha \wedge dx^\alpha = (g_\beta^\alpha \, df_\alpha) \wedge \omega^\beta$, which is a linear combination of the ω^β's. Thus a necessary condition that Δ be completely integrable is that $d\omega$ be a linear combination of a local basis ω^α for every ω belonging to Δ. That this condition is also sufficient is the dual formulation of Frobenius' theorem, which follows.

Theorem 4.10.1. *A C^∞ codistribution Δ is completely integrable iff for every 1–form ω belonging to Δ the 2–form $d\omega$ is locally a linear combination $\tau_\alpha \wedge \omega^\alpha$ of a local 1–form basis ω^α of Δ, where the τ_α are 1–forms.*

Proof. We have already seen that if Δ is completely integrable, then $d\omega$ is such a linear combination.

Suppose, conversely, that $d\omega$ is such a linear combination whenever $\omega \in \Delta$. In particular, for a local basis ω^α of Δ, the 2–forms $d\omega^\alpha = \tau_\beta^\alpha \wedge \omega^\beta$ for some 1–forms τ_β^α. Then for any vector fields $X, Y \in D$, the associated distribution, we have $2 \, d\omega^\alpha(X, Y) = \tau_\beta^\alpha(X)\omega^\beta(Y) - \tau_\beta^\alpha(Y)\omega^\beta(X) = 0$. But by the intrinsic formula for d, Section 4.3(3), we have $2 \, d\omega^\alpha(X, Y) = X\omega^\alpha(Y) - Y\omega^\alpha(X) - \langle [X, Y], \omega^\alpha \rangle = -\langle [X, Y], \omega^\alpha \rangle$, since the derivatives of $\omega^\alpha(Y) = 0$ and $\omega^\alpha(X) = 0$ are 0. Thus $[X, Y]$ is annihilated by a basis of Δ and therefore $[X, Y] \in D$. We have shown that D is involutive, so by the vector version of Frobenius' theorem, D is completely integrable, hence also Δ. ∎

Another way of stating the integrability condition of Theorem 4.10.1 is that $d\omega(X, Y) = 0$ for all $X, Y \in D$; that is, D *annihilates* $d\omega$. More generally, the tangent spaces of an integral submanifold N (of any dimension) annihilate $d\omega$ whenever $\omega \in \Delta$. Indeed, if $I: N \to M$ is the inclusion map, then the fact that N is an integral submanifold means that $I^*\omega = 0$ for all $\omega \in \Delta$. Thus we have $d(I^*\omega) = I^*(d\omega) = 0$; that is, $d\omega(X, Y) = 0$ for all vector fields X, Y tangent to N. This leads us to restrictions on the tangent spaces of an integral submanifold N in order that some given vectors be tangent to N, as in the following.

---

† This type of linear combination does not have unique coefficients as it does in the scalar coefficient case. The degree of nonuniqueness is measured exactly by Cartan's lemma, Problem 2.18.7.

Theorem 4.10.2.   *Let Δ be a C^∞ codistribution and let X be a C^∞ vector field belonging to the associated distribution D. Then for every integral submanifold N to which X is tangent, the forms $i(X)\,d\omega$, where $\omega \in \Delta$, annihilate the tangent spaces of N. In particular, if for some $X \in D$ and $\omega \in \Delta$, $i(X)\,d\omega$ does not belong to Δ, then ω is not completely integrable.*

Proof. For any other vector field Y on N we have from above $d\omega(X, Y) = 0$. But then $\langle Y, i(X)\,d\omega \rangle = 2\,d\omega(X, Y) = 0$. If there is an $X \in D$ and an $\omega \in \Delta$ such that $i(X)\,d\omega \notin \Delta$, then there can be no h-dimensional integral submanifold N of D through points at which $i(X)\,d\omega \notin \Delta$. Indeed, X would be tangent to such a manifold, so by the first result N would be an integral submanifold of the $(k + 1)$-dimensional codistribution spanned by Δ and $i(X)\,d\omega$, making $\dim N \le h - 1$.   ∎

Remark. Once the 2–forms $d\omega$ have been obtained, the restrictions on the tangent space of an integral submanifold are given algebraically and thus may be applied point by point. The above theorem could have been stated for a vector $x \in N_n$ at a single point n or the tangent vectors to a curve in N just as well as the vector field X. What this means in terms of codistributions arising from partial differential equations is that when boundary or initial values are given, the tangent space of a solution surface may be restricted along those values. In fact, the directions which do not give sufficiently many restrictions are exceptional and are considered to be improper as tangents to the boundary value submanifold. They are called the *characteristics* of the system.

First-order Partial Differential Equations. We want to consider a first-order partial differential equation (PDE) for a dependent variable z and n independent variables $x^1, \ldots, x^n$. Letting $p_i = \partial_i z$, such a first-order PDE will be given in terms of a C^∞ function F on an open subset of R^{2n+1} by

$$F(x^1, \ldots, x^n, p_1, \ldots, p_n, z) = 0.$$

It is no more difficult to consider simultaneously the equations $F = $ constant. We use $i, j, \ldots$ as summation indices running from 1 to n. A solution $z = f(x^1, \ldots, x^n)$ will determine an n-dimensional submanifold of R^{2n+1} by the additional equations $p_i = \partial_i f(x^1, \ldots, x^n)$. This submanifold is an integral submanifold of the two-dimensional codistribution Δ spanned by $\omega^0 = dF$ and $\omega^1 = dz - p_i\,dx^i$. Of course, we must restrict to the open submanifold M of R^{2n+1} on which ω^0 and ω^1 are (pointwise) linearly independent. Conversely, any n-dimensional integral submanifold of Δ on which $x^1, \ldots, x^n$ are coordinates yields a solution of the PDE. Let D be the associated distribution.

Since $d\omega^0 = d^2F = 0$, only $d\omega^1 = dx^i\,dp_i$ need be used to give restrictions on the tangent spaces of integral submanifolds as in Theorem 4.10.2. Let us determine those vector fields X such that $i(X)\, d\omega^1 \in \Delta$, that is, the *characteristic vectors* of Δ. Letting ∂_i, P^i, and ∂_z be the coordinate vector fields of the coordinates x^i, p_i, and z, we may write $X = X^i\partial_i + Q_iP^i + X_z\partial_z$. Our assumptions are that $X \in D$, $X \neq 0$, and $i(X)\, d\omega^1 = f\omega^0 + g\omega^1$ for some functions f and g. These give us equations $\omega^0(X) = \omega^1(X) = 0$ and $X^i\, dp_i - Q_i\, dx^i = f(F_i\, dx^i + G^i\, dp_i + F_z\, dz) + g(dz - p_i\, dx^i)$, where $G^i = P^iF$. Thus the components of X and the functions f and g satisfy the following:

$$F_iX^i + G^iQ_i + X_zF_z = 0,$$
$$-p_iX^i + X_z = 0,$$
$$Q_i = -fF_i + gp_i,$$
$$X^i = fG^i,$$
$$0 = fF_z + g.$$

The first of these equations is a consequence of the remaining ones, and we can solve the latter for X, obtaining

$$X = f(G^i\partial_i - [F_i + p_iF_z]P^i + p_iG^i\partial_z).$$

From this we conclude first that a characteristic vector is unique up to a scalar multiple, since we have been able to solve for X on the assumption that it is characteristic. Moreover, we are free to choose any $f \neq 0$, and having done so the solution for X is not the contradictory solution $X = 0$ at any point. For if $G^i\partial_i - [F_i + p_iF_z]P^i + p_iG^i\partial_z = 0$, then $G^i = 0$ and $F_i = -p_iF_z$, from which we obtain $\omega^0 = dF = F_z(-p_i\, dx^i + dz) = F_z\omega^1$, showing that ω^0 and ω^1 would be linearly dependent at points where $X = 0$. Finally, we observe that the 1-forms $i(X)\, d\omega^1 = f(\omega^0 - F_z\omega^1)$ are unique up to a scalar multiple. We incorporate these results into the following theorem.

Theorem 4.10.3. (a) *The two-dimensional codistribution*

$$\Delta = \{dF, \omega^1 = dz - p_i\, dx^i\}$$

of a first-order PDE has a unique one-dimensional distribution Φ of characteristic vectors.

   (b) *The distribution Φ is spanned by the vector field $X = G^i\partial_i - [F_i + p_iF_z]P^i + p_iG^i\partial_z$, where $dF = F_i\, dx^i + G^i\, dp_i + F_z\partial_z$.*

   (c) *If D is the associated distribution of Δ, then the linear map $i(\cdot)\, d\omega^1: D \to T^*M$ is nonsingular and its range intersects Δ in a one-dimensional codistribution which is the image of Φ under $i(\cdot)\, d\omega^1$.*

   (d) *If $E(m)$ is a k-dimensional subspace of $D(m)$ which contains no nonzero vector of $\Phi(m)$, then $\Delta(m)$ and $i(E(m))\, d\omega^1$ span a $(k+2)$-dimensional subspace of M_m^*.*

   (e) *If N is an n-dimensional integral submanifold of Δ, then for every $m \in N$, N contains a characteristic curve (= an integral submanifold of Φ) through m.*

Proof. Parts **(a)**, **(b)**, and **(c)** have been proved above; Part **(d)** follows immediately from **(c)**, since $i(E(m))\,d\omega^1$ is a k-dimensional subspace of M_m^* which intersects $\Delta(m)$ only in 0.

Suppose that N is an n-dimensional integral submanifold of Δ and that $m \in N$ exists such that $X(m) \notin N_m$. We apply **(d)** to $E(m) = N_m$, with $k = n$. By Theorem 4.10.2, N_m is annihilated by $\Delta(m) + i(N_m)\,d\omega^1$. But $\Delta(m) + i(N_m)\,d\omega^1$ has dimension $n + 2$ and so annihilates only a space of dimension $2n + 1 - (n + 2) = n - 1$, which is a contradiction. Hence we must have $X(m) \in N_m$ for every $m \in N$. If $I: N \to M$ is the inclusion map this means that X is I-related to a vector field X_N on N. The integral curves of X_N are mapped by I into integral curves of X, and the range of an integral curve of X is a characteristic curve. This proves **(e)**. ∎

Remark. Part **(e)** of Theorem 4.10.3 is basically a uniqueness theorem, but as is frequently the case with uniqueness theorems, it gives information about existence of solutions. In particular, if an $(n - 1)$-dimensional integral submanifold which is transversal to Φ can be found, then it can be pushed along the characteristic curves to produce an n-dimensional integral submanifold. When $n = 2$ it is easy to realize one-dimensional integral submanifolds transversal to Φ as the integral curves of a vector field $Y \in D$ which is independent of X.

Theorem 4.10.4. *Let P be an $(n - 1)$-dimensional integral submanifold of Δ such that $X(m) \notin P_m$ for every $m \in P$ and let $\{\mu_t\}$ be the flow of X. Then $N = \{\mu_t p \mid p \in P, \mu_t p \text{ is defined}\}$ is an n-dimensional integral submanifold of Δ.*

Proof. First we indicate how to show that N is a submanifold. If $\mu_t p$ is defined, then there is a coordinate neighborhood V of p in P and $\varepsilon > 0$ such that $\mu_s q$ is defined whenever $q \in V$ and $t - \varepsilon < s < t + \varepsilon$. We then take as coordinates of $\mu_s q$ the $n - 1$ coordinates of q and the number s.

Any tangent vector in N_m is a linear combination of $X(m)$ and a vector of the form $\mu_{t*} Y$, where $Y \in P_p$ and $m = \mu_t p$. Since $X \in D$, it suffices to show that $\mu_{t*} Y \in D(m)$ for all such Y, or equivalently, $\omega^0(\mu_{t*} Y) = \omega^1(\mu_{t*} Y) = 0$. Fix Y and let $ft = \omega^0(\mu_{t*} Y)$ and $gt = \omega^1(\mu_{t*} Y)$. Because P is an integral submanifold of Δ, we have $f0 = g0 = 0$. We shall express the derivatives of f and g in terms of the Lie derivatives of ω^0 and ω^1 with respect to X. In fact, we have $L_X \omega^\alpha = d/dt(0)(\mu_t^* \omega^\alpha)$. Thus

$$f's = \frac{d}{dt}(0)f(t + s) = \frac{d}{dt}(0)(\omega^0(\mu_{t*}\mu_{s*} Y))$$
$$= \frac{d}{dt}(0)(\mu_t^* \omega^0(\mu_{s*} Y)) = L_X \omega^0(\mu_{s*} Y),$$

and similarly, $g's = L_X\omega^1(\mu_{s*}Y)$. Now we use the formula $L_X = i(X)d + di(X)$ (Theorem 4.4.1). Since $\omega^0 = dF$, $L_X\omega^0 = i(X)d^2F + di(X)\omega^0 = 0 + d0 = 0$, and it follows that $f' = 0$, f is constant, and hence $f = 0$.

We have already seen that $i(X)d\omega^1 = \omega^0 - F_z\omega^1$. Thus $L_X\omega^1 = i(X)d\omega^1 + di(X)\omega^1 = i(X)d\omega^1 = \omega^0 - F_z\omega^1$. Applying this to $\mu_{s*}Y$ we obtain the fact that g satisfies a linear first-order ordinary differential equation:

$$g's = fs - F_z(\mu_s p)gs = -F_z(\mu_s p)gs.$$

But the initial value is 0, so it can be shown that $g = 0$.   ■

Problem 4.10.1. If Δ and X are as above, show that there are local bases θ^0, θ^1 of Δ such that $i(X)d\theta^0 = 0$ and $i(X)d\theta^1 = 0$.

Problem 4.10.2. Show that there are many candidates at each point m for the tangent space of an n-dimensional integral submanifold of Δ. Specifically, choose a basis $X_0 = X(m)$, $X_k \in D(m)$, $k = 1, \ldots, n-1$, such that for each k the choice of X_k is made from the subspace annihilated by $\omega^0, \omega^1, i(X_1)d\omega^1$, $\ldots, i(X_{k-1})d\omega^1$.

CHAPTER 5

Riemannian and
Semi-riemannian Manifolds

5.1. Introduction

For a given manifold we would like to recover and construct as many geometric notions as possible from our experience. These may include such
notions as distance, angle, parallel lines, straight lines, and one which is trivial
in the euclidean case—parallel translation along a curve. The choice of which
features we will try to generalize to arbitrary manifolds will be determined by
a desire to include the torus, the surface of an egg or pear, and even more
irregular surfaces. On these manifolds, the notion of parallel lines, as well as
some of the properties of straight lines, seem meaningless. However, the
concepts of angle, distance, length, and the shortest curve joining two points
are still meaningful. We shall find that the concept of parallel translation of
tangent vectors along a curve is basic, and that such a notion is associated in a
natural way to a reasonable idea of length of a vector. To see what it might
mean, picture a curve on a surface in E^3 with a tangent to the surface at the
initial point of the curve. In general, this tangent cannot be pushed along
the curve so that it remains parallel in E^3 since we want it to be tangent to the
surface always. We can require, however, that whatever turning it does is only
that necessary to keep it tangent to the surface, so that at any instant the rate
of change of the tangent will be a vector normal to the surface.

Closely related to the concept of parallel translation is the notion of the
absolute or covariant derivative of a vector field in a given direction or along
a curve. This is a measure of the deviation of the vector field from the field
displaced by parallel translation. In E^d, a vector field is parallel if its components are constants when referred to a cartesian basis. A measure of how
much a vector field is turning is given by the derivatives of its components,
which are themselves the components of a vector in the direction of turning.
For a vector field on a surface we want only that part representing a twisting

within the surface itself, so we project the derivative vector orthogonally onto the tangent plane of the surface. This agrees with our previous notion of parallelism since the projection is zero if and only if the turning is in a direction normal to the surface.

In E^d, besides possessing the property of minimizing distances, a straight line has its field of velocity vectors parallel along the line, provided the parameter is proportional to distance. When the parallel translation on an arbitrary manifold is the one associated with some metric, the distance-minimizing property of a curve is a consequence of the parallel velocity field property, and, except for changes in parametrization, the converse is also true. To get a notion corresponding to that of a straight line in E^d even when only parallelism and not distance is given, a geodesic will be defined as a curve with a parallel velocity field.

To keep track of these and other notions the diagram below is useful. An arrow should be read "leads to."

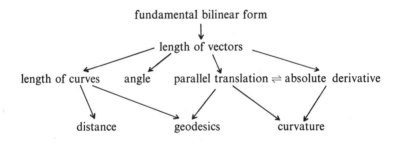

There are some exceptions to this diagram in that lengths of nonzero vectors are not always positive (as in the theory of relativity), since then angle has no meaning and length challenges the imagination. For these exceptions the concept of the "energy" of a vector or curve has been found to be a meaningful and effective substitute for length.

5.2.   Riemannian and Semi-riemannian Metrics

We shall follow Riemann's approach in developing a metric geometry for manifolds since such structures occur naturally in physical models and the various notions introduced in Section 5.1 can be defined in terms of this metric. The resulting geometry is therefore intrinsic; that is, the geometrical properties of the manifold are a part of the manifold itself and do not belong to some surrounding space. It is true that one common method of obtaining riemannian structures is by inheritance from an enveloping manifold, such as E^k, but the mechanism of this inheritance is another study, which we shall not undertake here.

Accordingly, let M be a manifold. A *metric* or *fundamental bilinear form* on M is a C^∞ symmetric tensor field b of type $(0, 2)$ defined on all of M which is nondegenerate at every point.

Problem 5.2.1.   If M is connected, show that the index of b is constant on M.

In the nonconnected case we cannot prove that the index of b is constant, so we assume this is always the case. If the index is 0 or $d\,(=\dim M)$, so that the metric is definite, the metric is called a *riemannian metric*, and the resulting geometry is called *riemannian geometry*. The pair (M, b) is then called a *riemannian manifold*. If the index is neither 0 nor d, the metric is said to be *semi-riemannian*. In case the index is 1 or $d - 1$ it is called a *Lorentz metric*. *Minkowski space L^d* is R^d with the Lorentz metric

$$b = du^1 \otimes du^1 - du^2 \otimes du^2 - \cdots - du^d \otimes du^d.$$

If gravitational effects are ignored, L^4 is a model of a "space-time universe," and a study of its geometry gives insight into such relativistic phenomena as "Lorentz-Fitzgerald contraction" and the meaninglessness of "simultaneity."

We shall find it convenient on occasion to employ the symmetric notation $\langle\ ,\ \rangle = b(\ ,\ )$.

A sufficient condition for the existence of a riemannian metric on a manifold is paracompactness (see Section 0.2.11). All of the usual examples have this property and it is difficult to construct one without it. Examples from physics have a profusion of riemannian metrics (see Chapter 6).

The existence of Lorentz and other semi-riemannian metrics depends upon other topological properties; for example, a manifold possesses a Lorentz metric iff it has a C^∞ one-dimensional distribution, that is, a smooth field of line elements. A necessary and sufficient condition that a compact manifold have a smooth field of line elements is that it have vanishing Euler characteristic. If a manifold has a nonzero C^∞ vector field, then that field spans a smooth field of line elements and the manifold has a Lorentz metric. Since the Euler characteristic of an even dimensional sphere is 2, it does not possess a Lorentz metric. However, the torus and the odd-dimensional spheres do possess Lorentz metrics. The higher-dimensional tori have metrics of any given index between 0 and d, but the study of when this happens in general is very difficult.

5.3.   Length, Angle, Distance, and Energy

The notions of length, angle, and distance make good sense only in the riemannian case, so we shall assume for the moment that b is a positive definite metric on a manifold M. The *length* of a tangent vector $v \in M_m$ is defined to be $\|v\| = \langle v, v \rangle^{1/2}$. The *angle* θ between nonzero vectors v and w in M_m is the number θ between 0 and π such that $\cos \theta = \langle v, w \rangle / (\|v\| \cdot \|w\|)$. This is well defined, since $|\langle v, w \rangle| \leq \|v\| \cdot \|w\|$ (see Problem 2.17.5).

The *length* of a curve $\gamma: [a, b] \to M$, denoted by $|\gamma|$, is the integral of the lengths of its velocity vectors:

$$|\gamma| = \int_a^b \|\gamma_* t\| \, dt.$$

Proposition 5.3.1.   *The length of a curve is independent of its parametrization.*

Proof. We first note for every $a \in R$ and $v \in M_m$ the fact that $\|av\| = |a| \|v\|$. This follows easily from the bilinearity of b.

Let $f: [c, d] \to [a, b]$ be a reparametrizing function for a curve $\gamma: [a, b] \to M$; hence $f' > 0$ and the reparametrization is the curve $\tau = \gamma \circ f: [c, d] \to M$. By the chain rule we have $\tau_* = f' \cdot \gamma_* \circ f$, from which

$$\begin{aligned}
|\tau| &= \int_c^d \|\tau_* s\| \, ds \\
&= \int_c^d \|\gamma_*(fs)\| \|(f's)\| \, ds \\
&= \int_a^b \|\gamma_* t\| \, dt \\
&= |\gamma|. \quad \blacksquare
\end{aligned}$$

We define the *parametrization by reduced arc length* as that parametrization $\tau = \gamma \circ f$ of γ for which $\|\tau_*\|$ is constant and defined on the unit interval $[0, 1]$. This parametrization may be obtained as follows when it exists. Let γ be a C^∞ curve defined on $[a, b]$ such that $|\gamma| = L$. We define the reduced arc length function g of γ by

$$gt = \frac{1}{L} \int_a^t \|\gamma_* u\| \, du.$$

The derivative of g is $\|\gamma_*\|/L$, so that g is nondecreasing. If g is increasing, then it is 1–1 and has an inverse $f: [0, 1] \to [a, b]$. Then $\tau = \gamma \circ f$ is a reparametrization of γ such that the length of the part of τ between $\tau 0$ and τh is hL. In general τ is only continuous, but if γ_* never vanishes, then τ is C^∞.

The *distance* between the points m and n on the riemannian manifold (M, b), denoted by $\rho(m, n)$, is the greatest lower bound of the lengths of all parametrized curves from m to n; that is:

(a) $\rho(m, n) \leq |\gamma|$ for any curve γ from m to n.

(b) There are curves joining m and n which have length arbitrarily close or even equal to $\rho(m, n)$. Thus for every $\varepsilon > 0$ there is a curve γ from m to n such that $|\gamma| - \varepsilon < \rho(m, n)$. [If M is not connected, then for m and n in different components the set of lengths $\{|\gamma|\}$, where γ is a curve from m to n, is empty. For such points we set $\rho(m, n) = +\infty$.]

The distance function ρ has the properties:

(1) Positivity: $\rho(m, n) \geq 0$.

(2) Symmetry: $\rho(m, n) = \rho(n, m)$.

(3) The triangle inequality: $\rho(m, p) \leq \rho(m, n) + \rho(n, p)$.

(4) Nondegeneracy: If $\rho(m, n) = 0$, then $m = n$.

Thus, by Section 0.2.2, (M, ρ) is a (topological) metric space.

Properties **(1)**, **(2)**, and **(3)** are easy to establish. The proof of **(4)** depends upon the continuity of b, the validity of **(4)** in euclidean space, and the Hausdorff property of M. We shall limit ourselves to proving **(4)** in the euclidean case (see Theorem 5.4.1).

A curve γ from m to n such that $|\gamma| = \rho(m, n)$ is said to be *shortest*. A shortest curve need not be unique.

If we turn to the semi-riemannian case again, the above notions lose most of their meaning and we utilize instead the notion of the energy of vectors and curves. The *energy* of $v \in M_m$ is defined to be $\langle v, v \rangle$; the *energy* of a curve γ is $E(\gamma) = \int \langle \gamma_* t, \gamma_* t \rangle \, dt$. The terminology of relativity theory is used to specify the possibilities for the signs of energy. Thus a vector v is called *time-like* if the energy of v is positive, *light-like* or *null* if the energy vanishes, and *space-like* if the energy is negative. A curve γ is called *time-like, light-like,* or *space-like* if all its velocity vectors $\gamma_* t$ are of the specified type. The null vectors at m form a hypercone in M_m (see Section 2.21).

The concept of energy is useful in the riemannian case as well as the semi-reimannian case, and it is important, for unifying purposes, to establish a relation between energy and length in the riemannian case. We derive this relation from the Schwartz inequality for integrals:

$$\left(\int_a^b (ft)(gt) \, dt \right)^2 \leq \int_a^b (ft)^2 \, dt \cdot \int_a^b (gt)^2 \, dt.$$

Here we assume that f and g are continuous real-valued functions defined on $[a, b]$. It is known that equality obtains only if one of f, g is a constant multiple of the other. For a curve $\gamma: [a, b] \rightarrow M$, we apply this to the functions $f = 1$ and $g = \|\gamma_*\|$, and our conclusion is

$$|\gamma|^2 \leq (b - a)E(\gamma). \tag{5.3.1}$$

The condition for equality is that g be constant, that is, γ is parametrized proportionally to arc length. Among all the parametrizations of a curve there is none with minimum energy, since we may take $b - a$ arbitrarily large. However, if we fix the parametrizing interval then energy does attain a minimum among all reparametrizations on that interval, and more significantly we obtain a relation between distance and energy, as follows.

Proposition 5.3.2. (a) *The energy of a curve parametrized by reduced arc length is the square of its length.*

(b) *Among all the reparametrizations of a given curve γ on the interval $[0, 1]$ the parametrization by reduced arc length has the least energy.*

(c) *There is a shortest curve from m to n iff there is a curve from m to n parametrized on $[0, 1]$ which has least energy among all such curves.*

Proof. (a) In (5.3.1) the left side, $|\gamma|^2$, is invariant under changes of parametrization. Thus (a) follows by taking $b = 1, a = 0$ and making the parameter proportional to arc length, that is, by using the reduced arc length parametrization.

(b) If γ is the reduced arc length parametrization of a curve and τ is some other parametrization on $[0, 1]$, then $E(\gamma) = |\gamma|^2 = |\tau|^2 \le E(\tau)$, so γ has the least energy for the $[0, 1]$-parametrizations.

(c) Suppose γ is a curve from m to n which is shortest. We may assume that γ is parametrized by reduced arc length.† Then for any other curve τ from m to n parametrized on $[0, 1]$, we have by (5.3.1) and the minimality of $|\gamma|$, $E(\gamma) = |\gamma|^2 \le |\tau|^2 \le E(\tau)$. Thus γ has the least energy among such curves.

Conversely, let γ have the least energy among $[0, 1]$-parametrized curves from m to n. For any† other such curve τ, let τ' be the reduced arc length reparametrization of τ. Then we have $|\gamma|^2 = E(\gamma) \le E(\tau') = |\tau'|^2 = |\tau|^2$, so that γ is shortest among such curves. ∎

Remarks. (a) If we wish to develop a theory of shortest curves in a riemannian manifold it suffices to consider least-energy curves among those parametrized on $[0, 1]$. Since the latter makes sense on a semi-riemannian manifold as well, we use it rather than the more geometric notion of length.

(b) If the circle S^1 is given a riemannian metric then there will be pairs of "opposite" points for which two shortest curves from one to the other are obtained, the two arcs of equal length into which S^1 is separated by the removal of the points. This nonuniqueness of shortest curves is a common occurrence in global riemannian geometry, but it can be proved that there is a unique shortest curve from a given point to those points which are "sufficiently near."

(c) There may be points m and n in a riemannian manifold for which no shortest curve exists, even if we eliminate the obvious counterexample of a nonconnected manifold with m and n in different components (see Problem 5.4.1).

---

† To make these arguments rigorous it must be shown that there is no loss in discarding curves not having a reduced arc length parametrization, that is, those for which the velocity vanishes at some points.

5.4.   Euclidean Space

In this section it is shown that the sort of structure introduced by means of the distance function does not violate our intuition by giving a particular riemannian structure on R^d and showing that the shortest curve joining two points is what it ought to be, a straight line.

Let u^i be the cartesian coordinates on R^d. At every point $m \in R^d$, $\partial/\partial u^i(m) = \partial_i(m)$ is a basis of R^d_m, so we may define b by specifying its components as real-valued functions on R^d. We set $b_{ij} = \delta_{ij}$. For $v = v^i \partial_i(m)$, $w = w^i \partial_i(m) \in R^d_m$ this means

$$\langle v, w \rangle = \sum_{i=1}^{d} v^i w^i,$$

which is the usual formula for the dot product in R^d_m.

The metric b is called the *standard flat metric* on R^d and $E^d = (R^d, b)$ is called *(ordinary) euclidean d-space.*

A C^∞ curve in R^d is given by d real-valued C^∞ functions $\gamma s = (f^1 s, \ldots, f^d s)$. Since $f^i = u^i \circ \gamma$, the velocity field of γ is $\gamma_* = f^{i'} \cdot \partial_i \circ \gamma$. Thus

$$\|\gamma_* s\| = \left(\sum_{i=1}^{d} (f^{i'} s)^2 \right)^{1/2},$$

and the length of γ is

$$|\gamma| = \int_a^b \left(\sum_{i=1}^{d} (f^{i'} s)^2 \right)^{1/2} ds.$$

This is the classical formula for the length of γ from $m = \gamma a$ to $n = \gamma b$ in E^d.

If γ is a straight-line segment from $m = (m^1, \ldots, m^d)$ to $n = (n^1, \ldots, n^d)$ with the usual parametrization:

$$\begin{aligned} \gamma s &= m + sv \\ &= (m^1 + sv^1, \ldots, m^d + sv^d), \end{aligned}$$

where $0 \le s \le 1$, $v = n - m$, then the $f^{i'} = v^i$ are constant for each $i = 1, \ldots, d$. Consequently, $|\gamma| = (\sum (v^i)^2)^{1/2}$, which is the usual formula for the distance from m to n. We also know by condition (**a**) in the definition of ρ that

$$\rho(m, n) \le |\gamma| = \left(\sum (v^i)^2 \right)^{1/2}.$$

The claim is that $\rho(m, n) = |\gamma|$, in accordance with condition (**b**). We show that there are no shorter curves from m to n, so that $\rho(m, n)$ is the usual distance in E^d.

Theorem 5.4.1. *Let γ be the straight-line segment in E^d from m to n and τ any other curve from m to n. Then $|\tau| \ge |\gamma|$, with equality holding iff τ is a reparametrization of γ. Thus the shortest curve joining two points in E^d is the straight-line segment joining the two points.*

Proof. We decompose τ_* into two orthogonal components, one parallel to γ and the other perpendicular to γ. Then we show that the integral of the length of the parallel component alone is at least as great as $|\gamma|$. Thus τ will be longer than γ if the perpendicular component is not always zero or if the parallel component is not always in the right direction. Now let us make this precise.

The "constant" unit field in the direction of γ is $X = \alpha^i \partial_i$, where $\alpha^i = v^i/|\gamma|$ and $v^i = n^i - m^i$. The parallel component of τ_* is $\langle \tau_*, X \circ \tau \rangle X \circ \tau$, and it has length $g = \langle \tau_*, X \circ \tau \rangle$ since $\|X\| = 1$. If θ is the angle between X and τ_* we have

$$\|\tau_*\| \cos\theta = \|\tau_*\| \|X\| \cos\theta = g \leq \|\tau_*\|,$$

and equality holds only if $\tau_* = gX \circ \tau$ and $g \geq 0$.

Let $\tau s = (f^1 s, \ldots, f^d s)$, $a \leq s \leq b$. Then $\tau_* s = (f^{i\prime} s)\partial_i(\tau s)$ and $gs = \sum \alpha^i f^{i\prime} s = \sum (\alpha^i f^i)' s$. Since $\tau a = m$ and $\tau b = n$ we have $f^i a = m^i$ and $f^i b = n^i$. Thus

$$|\tau| = \int_a^b \|\tau_* s\|\, ds \geq \int_a^b gs\, ds = \sum \alpha^i (f^i b - f^i a)$$

$$= \sum \alpha^i v^i$$

$$= |\gamma|.$$

If equality holds, then $\tau_* = gX \circ \tau$ and $g \geq 0$. Then $f^{i\prime} s = \alpha^i gs$, so $f^i s = m^i + \alpha^i \int_a^s gt\, dt = m^i + \alpha^i hs$ (defining hs). Thus $\tau s = \gamma(hs/|\gamma|)$, which shows that τ is a reparametrization of γ. ∎

Problem 5.4.1. Let M be R^2 with the closed line segment from $(-1, 0)$ to $(1, 0)$ removed and let the metric on M be the restriction of the euclidean metric to M. (Observe that M is a manifold and the domain of the distance function excludes $[(-1, 0), (1, 0)]$.)

(a) What is the distance in M from the point $(0, 1)$ to the point $(-1, -1)$? Is there a curve in M between these two points having this distance as its length?

(b) Which pairs of points in M are at the same distance apart as they are in E^2?

5.5.   Variations and Rectangles

In Section 5.4 it was shown that the shortest curve between two points in E^d is the straight-line segment. Thus if we have a one-parameter family of curves γ_t with the same endpoints and same parameter interval, such that γ_0 is the line segment, then $E(\gamma_t)$ has a minimum when $t = 0$. Hence if $E(\gamma_t)$ is a differentiable function of t, its derivative must vanish when $t = 0$. This property of a straight line is a likely candidate for a definition of a "straight line" or geodesic in a metric manifold, and is in fact the one often used. However, the definition of a geodesic given below will be closer to the notion of not bending. A straight line does not bend; that is, its velocity field is a parallel field (see Section 5.12).

We begin by defining the idea of a smooth one-parameter family of curves which will be called a C^∞ rectangle.

A C^∞ *rectangle* Q is a C^∞ map of a rectangle in R^2 into a manifold M. Thus the domain of Q will be of the form $[a, b] \times [c, d]$. Usually we shall have $c = 0$ (see Figure 19).

The curves γ_t given by fixing t and varying s, $\gamma_t s = Q(s, t)$, are called the *longitudinal curves* of Q. The curve γ_c is called the *base curve* of Q.

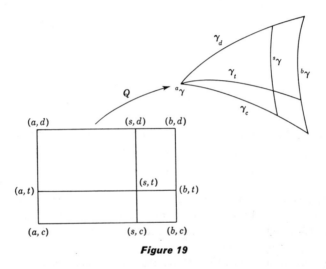

Figure 19

The curves ${}^s\gamma$ given by fixing s and varying t, ${}^s\gamma t = Q(s, t)$, are called the *transverse curves* of Q. The *initial* and *final* transverse curves are ${}^a\gamma$ and ${}^b\gamma$, respectively.

The *vector field associated with* Q is the "vector field," denoted V, along the base of Q with value at each point of the base curve equal to the velocity vector of the transverse curve through that point. It is not a vector field, strictly speaking, but a map $V: [a, b] \rightarrow TM$. In symbols, the definition is $V(s) = {}^s\gamma_* c$.

Proposition 5.5.1.   *Let γ be a C^∞ curve and V a vector field along γ whose components in any coordinate system are C^∞ functions of the parameter of γ. Then there is a C^∞ rectangle Q with γ as its base curve and V as its associated vector field.*

We shall not prove this proposition. However, Q is easily constructed if γ lies in a coordinate system, and, since γ is compact, the general case may be handled by piecing together the parts of Q from each coordinate system in a finite number of systems covering γ. There is no unique choice for Q.

It is occasionally easier to work with *broken C^∞ rectangles*. These are continuous maps of a rectangle in R^2, $[a, b] \times [c, d]$, with $[a, b]$ divided into a finite number of intervals $[a, s_1], [s_1, s_2], \ldots, [s_{k-1}, b]$, such that the map is a C^∞ rectangle when restricted to each subrectangle $[s_{h-1}, s_h] \times [c, d]$, where $s_0 = a, s_k = b$, and $h = 1, \ldots, k$. The associated vector field is then said to be a *broken C^∞ vector field* along the base curve γ_c.

If the initial and final transverse curves of Q are the constant curves, $^a\gamma t = m$ and $^b\gamma t = n$ for every t, then Q is called a *variation among curves from m to n*. If we are interested in comparing the base curve γ_c with other curves from m to n, then Q is called a *variation of γ_c*. In these cases $V(a) = 0$ and $V(b) = 0$. Conversely, if V is a vector field along the curve γ with $V(a) = 0$ and $V(b) = 0$, then there is a variation with V as its associated vector field. We call V an *infinitesimal variation* of γ.

A (broken) C^∞ curve γ in the riemannian manifold M is said to be *length-critical* if $d|\gamma_t|/dt(0) = 0$ for every variation of γ having $\gamma = \gamma_0$. It is *length-minimizing* if $|\gamma_t|$ is a minimum when $t = 0$ for every variation of γ such that $\gamma = \gamma_0$. We define *energy-critical* and *energy-minimizing* similarly, replacing $|\gamma_t|$ by $E(\gamma_t)$.

These notions of length-minimizing and energy-minimizing are not quite the same as the notions of shortest and least-energy in Section 5.3. A shortest (least-energy) curve has the least length (energy) among all possible curves between the endpoints, whereas length (energy)-minimizing involves a comparison with only those curves passing through some neighborhood of the given curve. Thus a curve may be length-minimizing but not shortest, because there may be a shorter curve which follows a different sort of path in a topological sense, say, by going around a different "hole" in the space. On the other hand, it is obvious that shortest (least-energy) curves are always length (energy)-minimizing, since the derivative of a function of t at an absolute minimum always vanishes.

Since $|\gamma_t|$ is independent of the parametrization of γ_t, the notions of length-critical and length-minimizing are not properties of a parametrized curve but rather of a curve thought of as a collection of points. On the other hand, when we replace length by energy the parametrization becomes significant, so that the parametrization of an energy-critical curve has a special meaning. In the case of non-light-like curves, in particular for all curves in the riemannian case, this special parametrization is proportional to arc length, but even for light-like energy-critical curves there are special parametrizations to fit the metric.

A length (energy)-minimizing curve is also length (energy)-critical but not conversely. For an example to illustrate the latter fact one can take an arc of a great circle which goes more than halfway around an ordinary euclidean sphere in E^3, parametrized by reduced arc length. There are shorter nearby curves between the endpoints of such an arc so it is not length-minimizing, but it is length-critical. In Section 5.13 we shall see that in the riemannian case an energy-critical curve is also length-critical, so nothing is lost in emphasizing energy, whereas we gain generality by including the semi-riemannian case and also allowing curves which have vanishing velocity at some points.

5.6. Flat Spaces

A coordinate system on a semi-riemannian manifold (M, b) is called *affine* if the components of b are constant. A semi-riemannian manifold is said to be *flat* if there is an affine coordinate system at every point. We shall show later that this is equivalent to the vanishing of a certain tensor called the *curvature tensor*. The euclidean and Minkowski spaces are flat.

Theorem 5.6.1. *In a flat space the energy-critical curves are those corresponding to straight lines in any affine coordinate system.*

Proof. We develop an analytic condition for a curve γ to be energy-critical. Since the range of γ is compact, we cover it by a finite number of affine coordinate systems. Thus we may assume that the domain $[a, b]$ is subdivided by points $a_0 = a, a_1, \ldots, a_n = b$, such that γ maps each interval $[a_{\alpha-1}, a_\alpha]$, $\alpha = 1, \ldots, n$, into an affine coordinate neighborhood.

Let us fix our attention on one such interval $[a_{\alpha-1}, a_\alpha]$. A variation Q of γ has as its expression in terms of affine coordinates x^i a set of d functions $f^i = x^i \circ Q$ of two variables s and t, where $a_{\alpha-1} \le s \le a_\alpha$, and t runs through a neighborhood of 0. By the assumption that the x^i are affine, the metric b is given in terms of them by $b = b_{ij}\, dx^i\, dx^j$, where the b_{ij} are constant and form a nonsingular symmetric matrix. The velocity field of the curve γ_t is then $\gamma_{t*} = f^i_s \cdot \partial_i \circ \gamma_t$, where $f^i_s = \partial f^i / \partial s$, so that the energy of γ_t is a sum of n integrals of the form

$$E_\alpha(t) = \int_{a_{\alpha-1}}^{a_\alpha} b_{ij} f^i_s f^j_s(s, t)\, ds.$$

The condition that $\gamma = \gamma_0$ be energy-critical is that $\sum_\alpha E'_\alpha(0) = 0$. In computing $E'_\alpha(0)$ it is permissible to differentiate under the integral sign with respect to t:

$$E'_\alpha(0) = 2 \int_{a_{\alpha-1}}^{a_\alpha} b_{ij} f^i_s f^j_{st}(s, 0)\, ds, \qquad (5.6.1)$$

where the subscripts s, t on f^j again indicate partial derivatives. This integral can be integrated by parts, letting $u = f^i_s$ and $dv = f^j_{ts}\, ds$, to obtain

$$E'_\alpha(0) = 2b_{ij}[f^i_s f^j_s(a_\alpha, 0) - f^i_s f^j_s(a_{\alpha-1}, 0)]$$
$$- 2\int_{a_{\alpha-1}}^{a_\alpha} b_{ij} f^i_{ss} f^j_t(s, 0)\, ds$$
$$= 2[\langle \gamma_*(a_\alpha), V(a_\alpha)\rangle - \langle \gamma_*(a_{\alpha-1}), V(a_{\alpha-1})\rangle]$$
$$- 2\int_{a_{\alpha-1}}^{a_\alpha} \langle A_\alpha(s), V(s)\rangle\, ds, \qquad (5.6.2)$$

where V is the vector field associated with Q and $A_\alpha = f^i_{ss}(\cdot, 0)\partial_i \circ \gamma$ is the "acceleration" of γ. [Acceleration, in the sense of second derivatives of coordinate components, is not ordinarily an invariant notion, but in a flat space when attention is restricted to affine coordinates this acceleration is an (affine) invariant. We shall not need this fact in the following but a direct proof is possible; see Problems 5.6.1 and 5.6.2.] When these expressions for $E'_\alpha(0)$ are added, the initial terms telescope, leaving only a piece of the first and last, $2[\langle \gamma_*(b), V(b)\rangle - \langle \gamma_*(a), V(a)\rangle]$, which vanishes due to the fact that $V(b) = 0$ and $V(a) = 0$.

Thus γ is energy-critical iff the sum of the integrals

$$E'(0) = -2\sum_\alpha \int_{a_{\alpha-1}}^{a_\alpha} \langle A_\alpha(s), V(s)\rangle\, ds$$

vanishes for every choice of vector field V along γ such that $V(a) = 0$ and $V(b) = 0$. The advantage of this form of the energy derivative is that the variation enters infinitesimally, as a vector field. From it we now conclude that the accelerations A_α of γ must vanish identically. For if $A_\alpha(c) \neq 0$ for some α and c, then we can first choose a vector v at $\gamma(c)$ such that $\langle A_\alpha(c), v\rangle \neq 0$, since b is nondegenerate. Then we extend v to a C^∞ vector field V_1 along γ and by continuity determine a subinterval $[c_1, c_2]$ of $[a_{\alpha-1}, a_\alpha]$ containing c on which $\langle A_\alpha(s), V_1(s)\rangle \neq 0$. If we multiply V_1 by a C^∞ hump function h which vanishes outside $[c_1, c_2]$ [see Example (b) in 1.5] we obtain a suitable infinitesimal variation $V = hV_1$ for which only one of the integrals is nonzero, and its integrand $\langle A_\alpha, V\rangle = h\langle A_\alpha, V_1\rangle$ does not change sign. But then the integral cannot vanish, which is a contradiction.

Thus for γ to be energy-critical it is necessary that the accelerations A_α all vanish identically. But the affine coordinate components of A_α are the second derivatives $f^i_{ss}(\cdot, 0)$ of the components $f^i(\cdot, 0) = x^i \circ \gamma$ of γ. Hence an energy-critical curve must have linear affine components; that is, $(x^i \circ \gamma)s = u^i s + v^i$ for some constants u^i and v^i. This holds for every affine coordinate system at any point of γ since such a coordinate system can be included among the n chosen ones.

Conversely, if $x^i \circ \gamma$ is a linear function of s for every affine coordinate system in a covering of γ, then the accelerations A_α all vanish identically, and consequently the energy first variation $E'(0)$ is zero. $\blacksquare$

Corollary 1. *In a flat space the velocity field of an energy-critical curve γ has constant pointwise energy; that is, $\langle \gamma_*, \gamma_* \rangle$ is constant. In particular, an energy-critical curve which is space-like, light-like, or time-like at one point remains so at all points. Finally, in the riemannian case energy-critical curves are parametrized proportionally to arc length.*

The proofs are trivial.

Corollary 2. *In a flat riemannian space an energy-critical curve is length-critical.*

Proof. Let γ be an energy-critical curve. By Corollary 1, γ is parametrized proportionally to arc length on an interval $[a, b]$. If Q is any rectangle with γ as the base curve, we may reparametrize the longitudinal curves proportionally to arc length without altering γ, obtaining a new rectangle Q_1. Let the lengths of the longitudinal curves be $L(t)$ and their energies be $E(t)$ (the parameter t refers to the longitudinal curves γ_t in Q_1). Then since the condition for equality in (5.3.1) is satisfied, $L(t)^2 = (b - a)E(t)$. But $E'(0) = 0$ since γ is energy-critical, so $2L(0)L'(0) = 0$. Thus either $L'(0) = 0$ for all such Q and γ is length-critical, or $L(0) = 0$ and γ is a constant curve, which is also length-critical. $\blacksquare$

Corollary 3. *If a curve in a flat space is energy-critical, then any segment of the curve is energy-critical. Conversely, if a (nonbroken) C^∞ curve can be subdivided into segments which are all energy-critical, then the whole curve is energy-critical.*

This follows from the fact that the vanishing of acceleration with respect to some coordinate system is a local condition.

Remark. Although flat spaces are themselves very special, Theorem 5.6.1, its proof, and the corollaries generalize without essential change to all semi-riemannian spaces. What is lacking at this point is a notion of differentiation of vector fields to generalize the differentiation performed to get (5.6.1) and (5.6.2). In particular we need a notion of the intrinsic acceleration of a curve. Such a notion of differentiation is discussed abstractly without being related to a metric in the following three sections. Then in Section 5.11 we discuss how a metric leads naturally to a notion of differentiation with properties adequate to carry out the generalization of the proof of Theorem 5.6.1. Thus we will reach the conclusion that a curve is energy-critical iff it is a geodesic (Theorem 5.13.1).

Problem 5.6.1. If x^i and y^i are overlapping affine coordinate systems for the metric b, show that $\partial^2 x^i/\partial y^j\,\partial y^k = 0$ for all i, j, k, so that $a^i_j = \partial x^i/\partial y^j$ are constants in every connected component of the intersection of the coordinate domains. Hence $x^i = a^i_j y^j + b^i$ in such connected components, where a^i_j and b^i are constants. That is, the coordinate changes are affine also.

Outline. Define $Y_i = \partial/\partial y^i$ and $Y_{ij} = [\partial^2 x^k/\partial y^j\,\partial y^i]\,\partial/\partial x^k$. Show that $Y_k\langle Y_i, Y_j\rangle = \langle Y_{ik}, Y_j\rangle + \langle Y_i, Y_{jk}\rangle = 0$ and hence that the quantities $T_{ijk} = \langle Y_i, Y_{jk}\rangle$ are skew-symmetric in i and j, symmetric in j and k. From Problem 2.17.1, $T_{ijk} = 0$. Since the Y_i are a basis $Y_{jk} = 0$.

Problem 5.6.2. Prove that the acceleration field A_α of γ is independent of the affine coordinates used (so the subscript α may be dropped).

Problem 5.6.3. Prove the converse of Corollary 2—that a length-critical curve is energy-critical and hence linear in terms of affine coordinates. (To obtain differentiability of the length function on longitudinal curves, assume that the velocity field never vanishes. Reparametrize with respect to arc length and then follow the pattern of proof as in Theorem 5.6.1.)

An infinitesimal variation V along a non-light-like energy-critical curve γ may be split into two components TV and $\perp V$, where TV is tangent to γ and $\perp V$ is perpendicular to γ. The tangential part TV indicates a tendency to reparametrize γ, but not to change its range. The change in energy due to such a tangential variation TV is indicated by $E''(0)$, where $E(t)$ is the energy of the longitudinal curve γ_t of a rectangle attached to TV. This part of the second energy variation will be found to have the same sign as $\langle\gamma_*, \gamma_*\rangle$. The second derivative of energy for the rectangles attached to the other part $\perp V$ is more informative about the geometry neighboring γ, so it pays to study the *second variation of normal variations*, as such second derivatives are called. In Lorentz manifolds (hence in relativity theory) it can be shown that the time-like geodesics, the so-called *world lines*, have negative second normal variations, and hence these curves *maximize* energy with respect to normal variations. We shall not carry this topic further except to give a special case as the following problem.

Problem 5.6.4. Show that the time-like straight line segments in Minkowski space are energy-maximizing for normal variations.

5.7. Affine Connexions

It is possible to introduce an invariant type of differentiation on a manifold called covariant differentiation, and when this is done the manifold is said to have an *affine connexion* or to be *affinely connected*. An affine connexion can be obtained quite naturally from a semi-riemannian structure (see Section 5.11), or from other special structures such as a parallelization (see Problem 5.7.3) or an atlas of affinely related coordinates (see Problem 5.7.5). Sometimes it is convenient to choose an affine connexion to use as a tool. However, there is no unique affine connexion on a manifold.

Affine connexions arose historically as an abstraction of the structure of a riemannian space. The name may be due to the idea that nearby tangent spaces are connected together by linear transformations, so that differences between vectors in different spaces may be formed and the limit of difference quotients taken to give derivatives. Originally the operation of covariant differentiation was conceived of as a modification of partial differentiation by adding in corrective terms to make the result invariant under change of coordinates. We prefer to introduce affine connexions axiomatically in a somewhat broader context than is done classically. This additional generality is required to make covariant derivatives of vector fields along curves sensible. A preliminary discussion of vector fields over maps follows. A vector field along a curve (see Section 5.5) is the special case in which the map is a curve.

Suppose $\mu: N \to M$ is a C^∞ map of a manifold N into a manifold M. A *vector field X over μ* is a C^∞ map $X: N \to TM$ such that for every $n \in N$, $X(n) \in M_{\mu n}$. An ordinary C^∞ vector field on an open subset E of M is then a vector field over the inclusion map $i: E \to M$. In most of that which follows, the classical notions can be obtained by specializing μ to the identity map $i: M \to M$.

We single out two special cases of vector fields over a map $\mu: N \to M$.

(a) The *restriction of a vector field X on M to μ* is the composition of the maps $\mu: N \to M$ and $X: M \to TM$ and is thus denoted $X \circ \mu$.

(b) The *image of a vector field Y on N under μ* is denoted $\mu_* Y$ and defined by $(\mu_* Y)(n) = \mu_*(Y(n))$.

It follows that vector fields Y on N and X on M are μ-related iff $\mu_* Y = X \circ \mu$.

The vector fields over μ can be added to each other and multiplied by C^∞ functions on N in the usual pointwise fashion. That is, if X and Y are vector fields over μ and $f: N \to R$, then $(X + Y)(n) = X(n) + Y(n)$ and $(fX)(n) = f(n)X(n)$.

If $X_1, \ldots, X_d$ is a local basis of vector fields in a neighborhood $U \subset M$ (for example, the X_i could be coordinate vector fields ∂_i), then for every vector field Y over μ and every n such that $\mu n \in U$ we may write $Y(n) = f^i(n)X_i(\mu n)$. This defines d real-valued functions f^i on $\mu^{-1}U$. It is easily seen that the f^i are C^∞, for if $\{\omega^i\}$ is the dual basis of 1–forms on U, then $f^i = \omega^i \circ Y$, which is a composition of the C^∞ maps $Y: N \to TM$ and $\omega^i: TU \to R$. Hence an arbitrary vector field over μ has the local form $Y = f^i X_i \circ \mu$. Thus the restrictions to μ of a local basis of vector fields on M gives a local basis of vector fields over μ, where the components are C^∞ real-valued functions on N. Henceforth we shall handle local questions in terms of a local basis $\{X_i\}$ of vector fields over μ without assuming that these X_i are obtained by restriction from a local basis on M as above, but the method of restriction remains the principal way of obtaining such local bases over μ.

An *affine connexion* D on $\mu: N \to M$ is an object which assigns to each $t \in N_n$ an operator D_t which maps vector fields over μ into $M_{\mu n}$ and satisfies the following axioms.† We require them to be valid for all $t, v \in N_n$, X, Y vector fields over μ, C^∞ functions $f: N \to R$, $a, b \in R$, and C^∞ vector fields Z on N.

(1) Linearity in t: $D_{at+bv}X = aD_t X + bD_v X$.
(2) Linearity over R of D_t: $D_t(aX + bY) = aD_t X + bD_t Y$.
(3) D_t is a derivation: $D_t(fX) = (tf)X(n) + (fn)D_t X$.
(4) Smoothness: The vector field $D_Z X$ over μ defined by $(D_Z X)(n) = D_{Z(n)}X$ is C^∞.

The value $D_t X$ is called the *covariant derivative* of X with respect to t.

An *affine connexion on* M is an affine connexion on the identity map $i: M \to M$.

Since we shall deal only with affine connexions, we shall refer to them simply as "connexions."

The covariant derivative operator D_t is local in the following sense. If X and Y are vector fields over μ such that $X = Y$ on some neighborhood U of $n = \pi t$, then $D_t X = D_t Y$. Indeed, let f be a C^∞ function which is 0 on a smaller neighborhood V of n and 1 outside of U. Then $f \cdot (X - Y) = X - Y$ since $X - Y = 0$ on U, so that

$$
\begin{aligned}
D_t X - D_t Y &= D_t(X - Y) \\
&= D_t f \cdot (X - Y) \\
&= (tf) \cdot (X - Y)(n) + (fn)D_t(X - Y) \\
&= 0.
\end{aligned}
$$

As a consequence we may define the restriction of a connexion D to an open submanifold U of N: If X is a vector field over $\mu|_U$ and $t \in U_n$, then we take a smaller neighborhood V of n such that $X|_V$ has a C^∞ extension X' to N and define $D_t X = D_t X'$. This is independent of the choice of extension X' by the fact we have just proved. We do not distinguish notationally between D and its restriction to U.

---

† In more sophisticated modern notation D is a connexion on the vector bundle over N induced by TM and μ. Moreover, vector fields over μ are cross sections of that vector bundle.

If $X_1, \ldots, X_d$ is a local basis of the vector fields over μ and $Z_1, \ldots, Z_e$ is a local basis of vector fields on N, all defined on $U \subset N$, then we define the *coefficients of the connexion D* with respect to the local bases $\{X_i, Z_\alpha\}$ to be the d^2e functions $\Gamma^i_{j\alpha}$ defined on U by

$$D_{Z_\alpha} X_j = \Gamma^i_{j\alpha} X_i.$$

If Y is a vector field over μ and $Y = f^j X_j$ is its local expression, then for $t = a^\alpha Z_\alpha(n) \in N_n$, $n \in U$ we have

$$
\begin{aligned}
D_t Y &= D_t(f^j X_j) \\
&= (tf^i) X_i(n) + (f^j n) D_t X_j \\
&= [a^\alpha Z_\alpha(n) f^i] X_i(n) + (f^j n) a^\alpha D_{Z_\alpha(n)} X_j \\
&= a^\alpha [Z_\alpha(n) f^i + (f^j n)(\Gamma^i_{j\alpha} n)] X_i(n).
\end{aligned}
$$

Here we have used i and j as summation indices running through $1, \ldots, d$ and α as a summation index running through $1, \ldots, e$. Thus D is determined locally by its C^∞ coefficients $\Gamma^i_{j\alpha}$.

For $t \in N_n$ we also define the *coefficients of D_t* to be the numbers Γ^i_{jt} defined by $D_t X_j = \Gamma^i_{jt} X_i(n)$. With the notation above we have $\Gamma^i_{jt} = a^\alpha \Gamma^i_{j\alpha} n$ and $D_t(f^j X_j) = [tf^i + (f^j n)\Gamma^i_{jt}] X_i(n)$. The map $\omega^i_j: t \to \Gamma^i_{jt}$ is a linear map $\omega^i_j: N_n \to R$ for each $n \in N$, and is thus a 1–form on N. The 1–forms ω^i_j are called the *connexion forms with respect to the basis* $\{X_i\}$. The matrix $(\omega^i_j(t))$ measures the "rate of change" of the basis $\{X_i\}$ with respect to the vector t. Their use is the device favored by the geometer E. Cartan. We can rewrite the formula for covariant differentiation in terms of them as

$$D_Z X = [Z\omega^i(X) + \omega^j(X)\omega^i_j(Z)] X_i$$

since $\omega^i(X)$ are the components of X.

Problem 5.7.1.   Find the law of change for the coefficients of D. That is, if $Y_i = g^j_i X_j$ and $W_\alpha = h^\beta_\alpha Z_\beta$ are new local bases over μ and on N, respectively, determine how the coefficients of D with respect to $\{X_i, Z_\alpha\}$ are related to the coefficients of D with respect to $\{Y_i, W_\alpha\}$. In particular, in the case of a connexion on M, $Z_i = X_i$, and $W_i = Y_i$, show that the coefficients Γ^i_{jk} are not the components of a tensor.

Problem 5.7.2.   Let N be covered by open sets U having local bases $\{X_i, Z_\alpha\}$ and C^∞ functions $\Gamma^i_{j\alpha}$ which satisfy the law of change required for Problem 5.7.1. Prove that there is a connexion having these functions as its coefficients.

Let D be a connexion on M and $\mu: N \to M$ be a C^∞ map. If X is a C^∞ vector field on M and $t \in N_n$ we define

$$(\mu^* D)_t(X \circ \mu) = D_{\mu_* t} X. \tag{5.7.1}$$

This does not define a connexion μ^*D on μ completely, since not every vector field over μ is a restriction $X \circ \mu$. However, the restriction vector fields are basic in that a connexion is determined by (5.7.1), the *connexion μ^*D on μ induced by D.*

Theorem 5.7.1. *If D is a connexion on M and $\mu: N \to M$ is a C^∞ map, then there is a unique connexion μ^*D on μ such that for every vector field X on M and every $t \in N_n$ we have $(\mu^*D)_t(X \circ \mu) = D_{\mu_*t}X$.*

Proof. Uniqueness. Let Y be any vector field over μ and $\{X_1, \ldots, X_d\}$ a local basis of vector fields on $U \subset M$. Then $Y = f^i X_i \circ \mu$, and for $t \in N_n$ it follows that we must have $(\mu^*D)_t Y = (tf^i)X_i(\mu n) + (f^i n)D_{\mu_*t}X_i$. This shows that μ^*D is uniquely determined and gives us a local formula for it.

Existence. We must show that the formula for $(\mu^*D)_t Y$ is consistent. If $\{Y_i\}$ were another local basis, then we would have $X_i = g_i^j Y_j$ and $Y = f^i(g_i^j \circ \mu)Y_j \circ \mu$. The other determination of $(\mu^*D)_t Y$ would then be

$$t[f^i(g_i^j \circ \mu)]Y_j(\mu n) + (f^i n)(g_i^j \mu n)D_{\mu_*t}Y_j$$
$$= (tf^i)(g_i^j \mu n)Y_j(\mu n) + (f^i n)t(g_i^j \circ \mu)Y_j(\mu n) + (f^i n)(g_i^j \mu n)D_{\mu_*t}Y_j$$
$$= (tf^i)X_i(\mu n) + f^i n[(\mu_*t)g_i^j \cdot Y_j(\mu n) + (g_i^j \mu n)D_{\mu_*t}Y_j]$$
$$= (tf^i)X_i(\mu n) + (f^i n)D_{\mu_*t}(g_i^j Y_j)$$
$$= (tf^i)X_i(\mu n) + (f^i n)D_{\mu_*t}X_i.$$

Thus the two determinations of $(\mu^*D)_t Y$ coincide. ∎

Remark. More generally, we can induce a connexion μ^*D on $\varphi \circ \mu: P \to M$ from a connexion D on $\varphi: N \to M$ and a C^∞ map $\mu: P \to N$. The procedure is essentially the same as above, which is the case where $N = M$ and φ is the identity map. The defining property (5.7.1) plus the axioms for a connexion determine μ^*D.

If D is a connexion on $\mu: N \to M$ and $\gamma: (a, b) \to N$ is a curve in N, then we define the *acceleration* of the curve $\tau = \mu \circ \gamma$ to be $A_\gamma = (\gamma^*D)_{d/du}\tau_*$, the covariant derivative of the velocity τ_*.

Example. Let M be a parallelizable manifold and $\{X_1, \ldots, X_d\}$ a parallelization of M (see Appendix 3.B). We define the *connexion of the parallelization* $\{X_i\}$ to be the connexion D on M such that

$$D_t(f^i X_i) = (tf^i)X_i(m),$$

where $t \in M_m$. Thus the coefficients of D with respect to $\{X_i\}$ are all identically zero.

More generally, vector fields $X_1, \ldots, X_d$ over $\mu: N \to M$ are said to be a parallelization of μ if $\{X_i(n)\}$ is a basis of $M_{\mu n}$ for every $n \in N$. If such X_i exist,

μ is said to be *parallelizable*. The *connexion D of a parallelization* $\{X_i\}$ of μ is defined by $D_t(f^iX_i) = (tf^i)X_i(n)$, where $t \in N_n$.

Problem 5.7.3. If $\{X_i\}$ is a parallelization of M and $\mu: N \to M$ is a C^∞ map, show that $\{X_i \circ \mu\}$ is a parallelization of μ. Moreover, if D is the connexion of $\{X_i\}$, then μ^*D is the connexion of $\{X_i \circ \mu\}$.

Problem 5.7.4. If $\{Y_i = g_i^j X_j\}$ is another parallelization of $\mu: N \to M$, show that the connexions of the two parallelizations $\{X_i\}$ and $\{Y_i\}$ are the same iff the g_i^j are constant on connected components of N.

An *affine structure* on a manifold M is an atlas such that every chart in the atlas is affinely related (that is, has constant jacobian matrix) with every other one in the atlas which it overlaps. A manifold having a distinguished affine structure is called an *affine manifold* and the charts which are affinely related to those of the affine structure are called *affine charts*. In each affine coordinate domain the coordinate vector fields form a parallelization of that domain, so there is an associated connexion on each domain.

Problem 5.7.5. Show that the locally defined connexions of the affine co-ordinate vector field parallelizations on an affine manifold are the same on overlapping parts, so there is a unique connexion associated with an affine structure.

Problem 5.7.6. If D is a connexion on $\mu: N \to M$ and we have C^∞ maps $\varphi: P \to N$ and $\tau: Q \to P$, show that $\tau^*(\varphi^*D) = (\varphi \circ \tau)^*D$.

Problem 5.7.7. If the ω_j^i are the connexion forms of D on $\mu: N \to M$ with respect to the local basis $\{X_i\}$ and $\varphi: P \to N$ is a C^∞ map, show that the connexion forms of φ^*D with respect to $\{X_i \circ \varphi\}$ are $\varphi^*\omega_j^i$.

5.8. Parallel Translation

Let D be a connexion on $\mu: N \to M$. A vector field E over μ is said to be *parallel at* $n \in N$ if for every $t \in N_n$,

$$D_t E = 0. \tag{5.8.1}$$

Since D_t is linear in t, it would suffice to require (5.8.1) for only those t running through a basis of N_n. A vector field E is *parallel* if E is parallel at every $n \in N$.

Now let $\{X_i\}$ be a local basis over μ, $\{Z_\alpha\}$ a local basis on N, and $\Gamma_{j\alpha}^i$ the co-efficients of D with respect to these local bases. The local expression for E is then of the form $E = g^i X_i$, where the g^i are C^∞ functions on N. Substituting Z_α for t in (5.8.1) we obtain the local condition for E to be parallel.

$$D_{Z_\alpha} E = (Z_\alpha g^i)X_i + g^i D_{Z_\alpha} X_i,$$
$$= (Z_\alpha g^i + g^j \Gamma^i_{j\alpha})X_i,$$

so E is parallel iff

$$Z_\alpha g^i + g^j \Gamma^i_{j\alpha} = 0, \qquad i = 1, \ldots, d, \quad \alpha = 1, \ldots, e. \qquad (5.8.2)$$

In general there are no solutions to this system of partial differential equations, and hence usually no parallel fields. The integrability condition, that is, the condition under which there will be local bases of parallel fields, is that the "curvature tensor" of D vanish (see Theorem 5.10.3). This condition is satisfied, in particular, for the natural connexion of an affine manifold, because in this case we may choose the affine coordinate vector fields $\{\partial_i\}$ as the local basis over $i: M \to M$. This makes the $\Gamma^i_{j\alpha} = 0$ and hence any constants are solutions for the g^i in (5.8.2).

Example. Consider the circle $S^1 \subset R^2$, as a one-dimensional manifold. It has a standard parallelization X, the counterclockwise unit vector field, which is locally expressible in terms of any determination of the angular coordinate θ as $X = d/d\theta$. Define a connexion D on S^1 by specifying that $D_X X = X$. (If X is regarded as the basis, this is the same as setting $\Gamma^1_{11} = 1$.) If E were a parallel field on S^1, then we would have $E = fX$ for some function f on S^1. The equation $D_X E = 0$ gives $Xf + f = 0$, which locally has solutions $f = Ae^{-\theta}$, where A is constant. This solution does not have period 2π in θ, so there can be no global parallel field for the connexion D. If we omit any point of S^1, there is a parallel field on the remaining open submanifold. (See Problem 5.8.4 for a complete analysis of the connexions on S^1.)

Proposition 5.8.1. *Consider a connexion on $\mu: N \to M$.*

   (a) *If N is connected, a parallel field is determined by its value at a single point.*

   (b) *The set of parallel fields P forms a finite-dimensional vector space of dimension $p \leq d$.*

   (c) *The set of values $\{E(n) \mid E \text{ is a parallel field}\}$ at any $n \in N$ is a p-dimensional subspace $P(n)$ of $M_{\mu n}$.*

   Proof. (a) Fix $n_0 \in N$ and suppose that E is a parallel field. Then for any other point $n \in N$ there is a curve γ from n_0 to n. Let $\gamma_* = f^c Z_\alpha \circ \gamma$ be the local expression for γ_* and $E = g^i X_i$ that of E. If we restrict (5.8.2) to points of γ (by composing it with γ), multiply by f^α, and sum on α, we obtain

$$0 = f^\alpha [Z_\alpha g^i + g^j \Gamma^i_{j\alpha}] \circ \gamma$$
$$= \frac{d}{du}(g^i \circ \gamma) + f^\alpha(\Gamma^i_{j\alpha} \circ \gamma)g^j \circ \gamma.$$

These equations are a system of linear first-order ordinary differential equations for the functions $g^i \circ \gamma$, having C^∞ coefficients $f^\alpha \Gamma^i_{j\alpha} \circ \gamma$. As such, they have a unique solution corresponding to a given set of initial values $g^i(n_0)$, and hence to a given value $E(n_0)$. Thus the values of E along γ, and, in particular, the value at n, are determined by the value at n_0 and the fact that E is parallel.

(b) and **(c)** It is clear that the sum of two solutions E_1, E_2 to (5.8.1) is again a solution, and also that a constant scalar multiple of a solution is again a solution. Thus the set of parallel fields forms a vector space. But for any $n \in N$ the evaluation map $\varepsilon(n): E \to E(n)$, taking P onto $P(n) \subset M_{\mu n}$, is clearly linear and by **(a)** it is 1–1. ∎

In the proof above we have seen that the evaluation map $\varepsilon(n): P \to P(n)$ is an isomorphism from the vector space of parallel fields to the space of their values at n. For two points $n_1, n_2 \in N$ the composition

$$\pi(n_1, n_2) = \varepsilon(n_2) \circ \varepsilon(n_1)^{-1}: P(n_1) \to P(n_2)$$

is called *parallel translation from n_1 to n_2*. Several properties of parallel translation are immediate from the definition:

(a) $\pi(n, n)$ is the identity on $P(n)$.
(b) $\pi(n_1, n_2)$ is a vector space isomorphism of $P(n_1)$ onto $P(n_2)$.
(c) $\pi(n_2, n_3) \circ \pi(n_1, n_2) = \pi(n_1, n_3)$.
(d) $\pi(n_1, n_2)^{-1} = \pi(n_2, n_1)$.

In the case of the ordinary affine structure on R^d, given by the atlas consisting of only the cartesian coordinate system, the parallel translation of the associated connexion is the familiar parallel translation of vectors in R^d. That is, $P(n) = R^d_n$ for every n and in terms of the cartesian coordinate vector fields ∂_i parallel translation leaves components constant:

$$\pi(n_1, n_2)(a^i \partial_i(n_1)) = a^i \partial_i(n_2).$$

We now examine the effect on parallel translation of passing to an induced connexion. Briefly what happens is that parallel translation can be applied to more vectors at fewer points.

Proposition 5.8.2. *Let S be the space of parallel fields of a connexion D on $\mu: N \to M$, let $\varphi: P \to N$ be a C^∞ map, and let Q be the space of parallel fields of the induced connexion $\varphi^* D$.*

(a) *If $E \in S$, then $E \circ \varphi \in Q$.*
(b) *For every $p \in P$, $S(\varphi p)$ is a subspace of $Q(p)$.*
(c) *If φ is a diffeomorphism, then $S(\varphi p) = Q(p)$ for every $p \in P$.*

Proof. Suppose $E \in S$. By (5.7.1), if $t \in P_p$, then $(\varphi^* D)_t(E \circ \varphi) = D_{\varphi_* t}E = 0$, since E is parallel. Hence $E \circ \varphi$ is parallel, which proves (a). Now (b) is trivial. If we apply Problem 5.7.6 to the case where $\tau = \varphi^{-1}$, then (c) follows immediately from (b). ∎

The extreme case is that of a connexion for which parallel translation applies to all tangents. We call such a connexion *parallelizable*, since a basis $\{E_i\}$ of P will be a parallelization of μ. This is somewhat stronger than the satisfaction of the integrability condition (vanishing curvature), except when N is simply connected.

Many properties of connexions can be studied by restricting attention to curves, so the following proposition, which shows that a connexion on a curve is particularly simple, has many uses when applied to induced connexions on curves.

Proposition 5.8.3.   *A connexion on a curve is parallelizable.*

Proof. If D is a connexion on $\gamma: (a, b) \to M$, then the equations for a parallel field $E = g^i X_i$ in terms of a local basis $\{X_i\}$ over γ, the basis d/du on (a, b), and the coefficients $\Gamma^i_j = \Gamma^i_{j1}$ of D are a system of linear ordinary differential equations:

$$\frac{dg^i}{du} + g^j \Gamma^i_j = 0.$$

Hence, choosing an initial point $c \in (a, b)$, there is a solution on an interval about c for any specification of initial values $g^i(c)$, that is, for any given value of $E(c)$. These local solutions may then be extended to all of (a, b) by the usual patching-together method. ∎

When we apply parallel translation with respect to an induced connexion $\gamma^* D$ for a curve $\gamma: (a, b) \to N$ and a connexion D on $\mu: N \to M$ we say that we have *parallel translated vectors along γ with respect to D*. Thus parallel translation along γ gives a linear isomorphism $\pi(\gamma; c, d): M_{\mu\gamma c} \to M_{\mu\gamma d}$ for every $c, d \in (a, b)$. A parallelization $\{E_i\}$ for $\gamma^* D$ is called a *parallel basis field along γ* (for D). It should be clear that unless D is parallelizable the parallel translation from γc to γd depends on the curve γ as well as the endpoints in question. However, Proposition 5.8.2(c) shows that the parametrization of the curve is irrelevant. Specifically, if $\tau = \gamma \circ f$ is a reparametrization of γ, then $\pi(\tau; c', d') = \pi(\gamma; fc', fd')$.

Problem 5.8.1.   For any connexion D on $\mu: N \to M$ show that for a given point $n \in N$ there is a local basis $\{X_i\}$ of vector fields over μ such that each X_i

is parallel at n. Hence for $t \in N_n$ and any vector field $X = f^i X_i$ over μ,
$D_t X = (t f^i) X_i(n)$.

Problem 5.8.2. For any connexion D on $\mu: N \to M$ and vector $t \in N_n$, if we choose a curve γ in N such that $\gamma_*(0) = t$, then we may describe the operator D_t as follows. Let $\{E_i\}$ be a parallel basis field along γ. Then for any vector field X over μ we may express X along γ in terms of the E_i, that is, $X \circ \gamma = f^i E_i$, where the f^i are real-valued functions of the parameter of γ. Show that $D_t X = f^{i\prime}(0) E_i(0)$.

Problem 5.8.3. Let X be the unit field on S^1, as in the example above. For any given constant c define a connexion $^c D$ on S^1 by specifying that $^c D_X X = cX$. (The connexion in the example is $^1 D$; the connexion of the parallelization X is $^0 D$.)

(a) Show that there is no global parallel field for $^c D$ unless $c = 0$.

(b) Show that $^c D$ is the connexion associated with an affine structure on S^1.

(c) If D is any connexion on S^1, then there is a constant c and a diffeomorphism $\mu: S^1 \to S^1$ such that $D = \mu^{*c} D$. (*Hint:* Determine c by the amount that a vector "grows" when it is parallel translated once around S^1. Then define μ by matching corresponding points on integral curves of certain parallel fields.)

Thus we can give a classification, up to equivalence under a diffeomorphism, of all the connexions on a circle. If the diffeomorphism is allowed to be orientation-reversing, then we may take $c \geq 0$.

Problem 5.8.4. Show that a connexion on R is equivalent, up to diffeomorphism, to one of three specific connexions, according to whether a parallel field is (1) complete, (2) has an integral curve extending to ∞ in only one direction, or (3) has an integral curve which cannot be extended to ∞ in either direction.

5.9. Covariant Differentiation of Tensor Fields

If $\mu: N \to M$ is a C^∞ map, we may define tensor fields over μ, in analogy with vector fields over μ, as functions which assign to a point $n \in N$ a tensor over the vector space $M_{\mu n}$. If $\{X_i\}$ is a local basis of vector fields over μ, then the dual basis $\{\omega^i\}$ consists of the 1-forms over μ dual to the X_i at each $n \in$ the domain U of the X_i; that is, for each $n \in U$ the value of ω^i is a cotangent $\omega^i(n) \in M_{\mu n}^*$ and $\{\omega^i(n)\}$ is the basis of $M_{\mu n}^*$ dual to the basis $\{X_i(n)\}$ of $M_{\mu n}$. Then the various tensor products of the X_i and the ω^i form local bases for tensors over μ. Thus a tensor field of type $(1, 1)$ over μ can be written locally as $f_j^i X_i \otimes \omega^j$, where the components f_j^i are C^∞ functions on U.

Now if D is a connexion on μ and $\{X_i\}$ is parallel at n with respect to D, then we *define* the dual basis $\{\omega^i\}$ to be parallel at n also. Covariant differentiation of tensor fields over μ can then be defined as an extension of the result of Problem 5.8.1: If $t \in N_n$, then D_t operates on tensor fields by letting t operate on the components with respect to the basis $\{X_i\}$ which is parallel at n. Thus $D_t(f_j^i X_i \otimes \omega^j) = (tf_j^i)X_i(n) \otimes \omega^j(n)$. Of course, it must be verified that if a different parallel basis at n is used, then the resulting operator D_t is still the same.

Alternatively, covariant differentiation of tensor fields over μ can be defined by generalizing the technique of restricting to a curve and using a parallelization along the curve, as in Problem 5.8.2. Thus if $\{E_i\}$ is a parallel basis along a curve γ and $\{\varepsilon^i\}$ is the dual basis, for any tensor field S over μ we can express the restriction $S \circ \gamma$ in terms of tensor products of the E_i's and ε^i's. If $t = \gamma_*(0)$, then $D_t(f_j^i E_i \otimes \varepsilon^j) = f_j^{i\prime}(0)E_i(0) \otimes \varepsilon^j(0)$. Again, this can be shown to be independent of the choice of curve and parallel basis and furthermore coincides with the definition in terms of a basis which is parallel at just one point. We shall leave the necessary justifications which show that D_t is well defined on tensor fields as exercises.

Problem 5.9.1.   Show that the identity transformation, whose components as a tensor of type $(1, 1)$ are δ_j^i, is parallel with respect to every connexion.

The following proposition lists some automatic consequences of the definition of covariant differentiation of tensor fields.

Proposition 5.9.1.   (a) *If S and T are tensor fields over μ of the same type, then* $D_t(S + T) = D_t S + D_t T$.

(b) *For a real-valued function f on N,* $D_t(fS) = (tf)S(n) + (fn)D_t S$.

(c) *For tensor fields S and T over μ, not necessarily of the same type,* $D_t(S \otimes T) = D_t S \otimes T(n) + S(n) \otimes D_t T$.

(d) *Covariant differentiation commutes with contractions. That is, if C is the operation of contracting a tensor S, then* $C(D_t S) = D_t(CS)$.

(e) *D_t is linear in t:* $D_{at+bv} = aD_t + bD_v$.

If Z is a vector field on N and S is a tensor field over μ, then $D_Z S$ is the tensor field over μ, of the same type as S, defined by $(D_Z S)(n) = D_{Z(n)}S$.

The formulas for covariant differentiation in terms of a local basis are developed next. Since we already have a notation for $D_t X_j$, $\Gamma_{jt}^i X_i = \omega_j^i(t)X_i$, it suffices to obtain the formula for $D_t\omega^i$ and apply the rules of Proposition 5.9.1. However, by Problem 5.9.1, $X_i \otimes \omega^i$ is parallel, so by (c),

$$
\begin{aligned}
0 &= D_t(X_i \otimes \omega^i) \\
&= \Gamma^i_{jt}X_i(n) \otimes \omega^j(n) + X_i(n) \otimes D_t\omega^i \\
&= X_i(n) \otimes [\Gamma^i_{jt}\omega^j(n) + D_{t'}\ ].
\end{aligned}
$$

Since the $X_i(n)$ are linearly independent,

$$
\begin{aligned}
D_t\omega^i &= -\Gamma^i_{jt}\omega^j(n) \\
&= -\omega^i_j(t)\omega^j(n).
\end{aligned}
$$

The expression in terms of a local basis $\{Z_\alpha\}$ on N is an immediate specialization:

$$
D_{Z_\alpha}\omega^i = -\Gamma^i_{j\alpha}\omega^j.
$$

We illustrate the local formula for covariant differentiation of a tensor field of type $(1, 3)$. Let

$$
S = S^i_{jhk}X_i \otimes \omega^j \otimes \omega^h \otimes \omega^k.
$$

Then by repeated applications of the rules of Proposition 5.9.1,

$$
\begin{aligned}
D_{Z_\alpha}S &= (Z_\alpha S^i_{jhk})X_i \otimes \omega^j \otimes \omega^h \otimes \omega^k + S^i_{jhk}\Gamma^p_{i\alpha}X_p \otimes \omega^j \otimes \omega^h \otimes \omega^k \\
&\quad + S^i_{jhk}X_i \otimes (-\Gamma^j_{p\alpha}\omega^p) \otimes \omega^h \otimes \omega^k + S^i_{jhk}X_i \otimes \omega^j \otimes (-\Gamma^h_{p\alpha}\omega^p) \otimes \omega^k \\
&\quad + S^i_{jhk}X_i \otimes \omega^j \otimes \omega^h \otimes (-\Gamma^k_{p\alpha}\omega^p) \\
&= (Z_\alpha S^i_{jhk} + S^p_{jhk}\Gamma^i_{p\alpha} - S^i_{phk}\Gamma^p_{j\alpha} \\
&\quad - S^i_{jhp}\Gamma^p_{k\alpha})X_i \otimes \omega^j \otimes \omega^h \otimes \omega^k.
\end{aligned}
$$

The components

$$
\begin{aligned}
S^i_{jhk|\alpha} &= Z_\alpha S^i_{jhk} + S^p_{jhk}\Gamma^i_{p\alpha} - S^i_{phk}\Gamma^p_{j\alpha} \\
&\quad - S^i_{jpk}\Gamma^p_{h\alpha} - S^i_{jhp}\Gamma^p_{k\alpha}
\end{aligned}
$$

define a "tensor-valued 1–form" on N. If $\{\zeta^\alpha\}$ is the dual basis to $\{Z_\alpha\}$, we may write it

$$
DS = D_{Z_\alpha}S \otimes \zeta^\alpha.
$$

For each $n \in N$, DS is a linear function on N_n with values in $T^1_3M_{\mu n}$: $t \in N_n \to D_tS \in T^1_3M_{\mu n}$. We call DS the *covariant differential* of S.

In the case where D is a connexion on M, so $N = M$, $\mu =$ the identity, and we may take $Z_i = X_i$, $\zeta^i = \omega^i$, the covariant differential becomes a tensor having covariant degree greater by 1. Thus if S is of type $(1, 3)$, then DS is of type $(1, 4)$. As a multilinear function we can give the following intrinsic formula for DS:

$$
DS(\tau, w, x, y, z) = D_zS(\tau, w, x, y),
$$

where $\tau \in M_m{}^*$ and $w, x, y, z \in M_m$.

5.10.   Curvature and Torsion Tensors

Let D be a connexion on $\mu: N \to M$. For every pair of tangent vectors $x, y \in N_n$, a tangent vector $T(x, y) \in M_{\mu n}$ may be assigned, called the *torsion translation* for the pair x, y. The definition is as follows. Let X, Y be extensions of x, y to vector fields on N. Then

$$T(x, y) = D_x(\mu_* Y) - D_y(\mu_* X) - \mu_*[X, Y](n).$$

To show that we have really defined something, we must show that this depends only on x and y and not on the choice of X and Y. In doing this a local expression for T is also obtained. Let $\{X_i\}$ be a local basis over μ at n, let $\{\omega_i\}$ be the dual basis, and ω_j^i the connexion forms for this local basis, so for any $t \in N_n$, $\omega_j^i(t) = \Gamma_{jt}^i$; that is, $D_t X_j = \omega_j^i(t) X_i(n)$. Then we have for any vector field Z over μ, $Z = \omega^i(Z) X_i$. Applying this to $\mu_* X$, $\mu_* Y$, and $\mu_*[X, Y]$ we get

$$\begin{aligned}
D_x(\mu_* Y) &= D_x(\omega^i(\mu_* Y) X_i) \\
&= (x\omega^i(\mu_* Y)) X_i(n) + \omega^i(\mu_* y) D_x X_i \\
&= [x\omega^i(\mu_* Y) + \omega^j(\mu_* y)\omega_j^i(x)] X_i(n), \\
D_y(\mu_* X) &= [y\omega^i(\mu_* X) + \omega^j(\mu_* x)\omega_j^i(y)] X_i(n), \\
\mu_*[X, Y] &= \omega^i(\mu_*[X, Y]) X_i.
\end{aligned}$$

Combining these three we obtain

$$\begin{aligned}
T(x, y) = \{ & x\omega^i(\mu_* Y) - y\omega^i(\mu_* X) - \omega^i(\mu_*[X, Y](n)) \\
& + \omega_j^i(x)\omega^j(\mu_* y) - \omega_j^i(y)\omega^j(\mu_* x)\} X_i(n). \quad (5.10.1)
\end{aligned}$$

Now we note that we can define† 1–forms $\mu^* \omega^i$ on N by the formula $(\mu^* \omega^i)(z) = \omega^i(\mu_* z)$ for any $z \in TN$. The first three terms in the braces of (5.10.1) then become, by (c), Section 4.3, 2 $d\mu^* \omega^i(x, y)$. It is thus independent of the choice of X and Y. The remaining two terms are $2\omega_j^i \wedge \mu^* \omega^j(x, y)$, so we have reduced the formula for T to

$$T(x, y) = 2(d\mu^* \omega^i + \omega_j^i \wedge \mu^* \omega^j)(x, y) X_i(n).$$

The 2–forms on N,

$$\Omega^i = 2(d\mu^* \omega^i + \omega_j^i \wedge \mu^* \omega^j), \quad (5.10.2)$$

are called the *torsion forms of the connexion D*, and the equations (5.10.2) which we have used to define them are called the *first structural equations* (of E. Cartan). The torsion T itself is thus a vector-valued 2–form which may be denoted by

$$T = X_i \otimes \Omega^i. \quad (5.10.3)$$

---

† Since the ω^i are not forms on M, the previous definition for pulling back forms via μ does not have meaning here.

In the case where $X_i = \partial_i \circ \mu$, where the ∂_i are coordinate vector fields on M, $\omega^i = dx^i \circ \mu$ and $\mu^*\omega^i = \mu^* \, dx^i$ in the sense we have previously defined for μ^*. (See Proposition 3.9.4. The "$\circ \mu$" attached to dx^i is merely a means of restricting the domain of dx^i to μN, and this restriction is already included in the operator μ^*.) Hence $d\mu^*\omega^i = \mu^* \, d^2x^i = 0$, so the formula for Ω^i becomes

$$\Omega^i = 2\omega^i_j \wedge \mu^* \, dx^j. \tag{5.10.4}$$

Finally, if μ is the identity on M, then T becomes a tensor of type (1, 2) on M which is skew-symmetric in the covariant variables. The connexion forms are $\omega^i_j = \Gamma^i_{jk} \, dx^k$ and the local expressions for the components of T are classically given as

$$T^i_{jk} = \Gamma^i_{kj} - \Gamma^i_{jk}. \tag{5.10.5}$$

These follow immediately from (5.10.3), which can be expanded using (5.10.5) to give

$$\begin{aligned} T &= X_i \otimes 2(\Gamma^i_{jk} \, dx^k \wedge dx^j) \\ &= X_i \otimes (\Gamma^i_{kj} - \Gamma^i_{jk}) \, dx^j \wedge dx^k. \end{aligned}$$

A connexion for which $T = 0$ is said to be *symmetric*. The name is suggested by the fact that Γ^i_{jk} is symmetric in j and k, or, more generally, from the fact that when X and Y are vector fields over μ such that $[X, Y] = 0$, their covariant derivatives have the symmetry property

$$D_X\mu_* Y = D_Y\mu_* X.$$

Problem 5.10.1.   If D is a connexion on M, then the *conjugate connexion* D^* is defined by

$$D^*_X Y = D_X Y + T(X, Y), \tag{5.10.6}$$

where T is the torsion of D. Show that D^* is actually a connexion on M and that the torsion of D^* is $-T$.

Problem 5.10.2.   If D and E are connexions on $\mu: N \to M$ and f is a function on N into R, show that $fD + (1 - f)E$ is a connexion on μ. [If $t \in N_n$, then $(fD + [1 - f]E)_t = f(n)D_t + (1 - f(n))E_t$.] We call $fD + (1 - f)E$ the *weighted mean of D and E with weights f and* $1 - f$.

Problem 5.10.3.   If D is a connexion on M, show that $^sD = \frac{1}{2}(D + D^*)$ is symmetric and find its coefficients with respect to a coordinate basis in terms of the coefficients of D. The connexion ^sD is called the *symmetrization* of D.

Again turning to a connexion D on $\mu: N \to M$, for every pair of tangents $x, y \in N_n$ a linear transformation $R(x, y): M_{\mu n} \to M_{\mu n}$ may be defined, called the *curvature transformation* of D for the pair x, y. The curvature transformation will give a measure of the amount by which covariant differentiation fails to be commutative. With extensions X and Y of x and y, as above, the definition is [*]

$$R(x, y) = D_{[X,Y](n)} - D_x D_Y + D_y D_x.$$

That is, if $w \in M_{\mu n}$ and W is any vector field over μ such that $W(n) = w$, then

$$R(x, y)w = D_{[X,Y](n)}W - D_x D_Y W + D_y D_x W. \tag{5.10.7}$$

As with torsion, we show that this is independent of the choice of extensions X, Y, and W and simultaneously develop an expression for R in terms of the connexion forms. This time we shall not carry along the evaluation at n, but the tensor character will still become evident.

Taking the terms in order we have

$$
\begin{aligned}
D_{[X,Y]}W &= D_{[X,Y]}\{\omega^i(W)X_i\} \\
&= \{[X,\, Y]\omega^i(W) + \omega^j(W)\omega_j^i[X,\, Y]\}X_i, \\
D_X D_Y W &= D_X(\{Y\omega^i(W) + \omega^j(W)\omega_j^i(Y)\}X_i) \\
&= (X\{Y\omega^i(W) + \omega^j(W)\omega_j^i(Y)\})X_i \\
&\quad + \{Y\omega^k(W) + \omega^j(W)\omega_j^k(Y)\}\omega_k^i(X)X_i \\
&= \{XY\omega^i(W) + \omega_k^i(Y)X\omega^k(W) + \omega^j(W)X\omega_j^i(Y) \\
&\quad + \omega_k^i(X)Y\omega^k(W) + \omega^j(W)\omega_k^i(X)\omega_j^k(Y)\}X_i,
\end{aligned}
$$

and $D_Y D_X W$ is the same except for a reversal of X and Y. In the combination for $R(X, Y)W$, the terms in which the $\omega^i(W)$'s are differentiated by X, Y, or both, cancel. Thus $R(X, Y)W$ is linear in W and depends only on the point-wise values. The remaining terms are

$$
\begin{aligned}
R(X, Y)W &= \{\omega^i_j[X,\, Y] - X\omega^i_j(Y) + Y\omega^i_j(X) \\
&\quad - \omega_k^i(X)\omega_j^k(Y) + \omega_k^i(Y)\omega_j^k(X)\}\omega^j(W)X_i \\
&= 2\{-d\omega_j^i(X,\, Y) - \omega_k^i \wedge \omega_j^k(X,\, Y)\}\omega^j(W)X_i.
\end{aligned}
$$

The 2–forms on N,

$$\Omega_j^i = 2(d\omega_j^i + \omega_k^i \wedge \omega_j^k) \tag{5.10.8}$$

are called the *curvature forms* of D, and the equations (5.10.8) are the *second structural equations* (of E. Cartan). The curvature itself is thus a tensor-valued 2–form of type (1, 1) which we may write

$$R = -X_i \otimes \omega^j \otimes \Omega_j^i. \tag{5.10.9}$$

It can be shown that $R(x, y)$ gives a measure of the amount a tangent vector w, after parallel translation around a small closed curve γ in a two-dimensional surface tangent to x and y, deviates from w. In fact, the result of parallel translation of w around γ gives the vector $w + \alpha R(x, y)w$ as a first approximation, where α is the ratio of the area enclosed by γ to the area of the parallelogram of sides x and y.

The reduction to the classical coordinate formula in the case μ = the identity and $X_i = \partial_i$ follows by letting $\omega_j^i = \Gamma_{jk}^i \, dx^k$ in (5.10.8), computing, and substituting in (5.10.9):

$$
\begin{aligned}
d\omega_j^i &= (\partial_h \Gamma_{jk}^i) \, dx^h \wedge dx^k \\
&= \tfrac{1}{2}(\partial_h \Gamma_{jk}^i - \partial_k \Gamma_{jh}^i) \, dx^h \otimes dx^k, \\
\omega_p^i \wedge \omega_j^p &= \tfrac{1}{2}(\Gamma_{ph}^i \Gamma_{jk}^p - \Gamma_{pk}^i \Gamma_{jh}^p) \, dx^h \otimes dx^k.
\end{aligned}
$$

Thus the components of R, as a tensor of type $(1, 3)$ on M, are

$$
R_{jhk}^i = \partial_k \Gamma_{jh}^i - \partial_h \Gamma_{jk}^i + \Gamma_{pk}^i \Gamma_{jh}^p - \Gamma_{ph}^i \Gamma_{jk}^p. \tag{5.10.10}
$$

Problem 5.10.4. For a connexion on M, derive (5.10.10) directly from (5.10.7) as the components of $R(\partial_h, \partial_k)\partial_j$.

For a connexion on M the curvature tensor (5.10.9) is, as always, skew-symmetric in the last two variables. It makes sense to ask if it is skew-symmetric in the last three variables, but this guess fails completely. In fact, if the connexion is symmetric $(T = 0)$, then the skew-symmetric part of R vanishes, a fact which may be called the *cyclic sum identity of the curvature tensor*. In explicit form this identity may be written

$$
R(X, Y)Z + R(Y, Z)X + R(Z, X)Y = 0,
$$

or

$$
R_{jhk}^i + R_{hkj}^i + R_{kjh}^i = 0. \tag{5.10.11}
$$

Problem 5.10.5. Show that (5.10.11) is equivalent to the vanishing of the skew-symmetric part of R.

The first Bianchi identity generalizes to the case of a symmetric connexion on a map $\mu \colon N \to M$ as follows. By pulling back the covariant part of R to N we obtain a vector-valued tensor $\mu^* R$ of type $(0, 3)$ on N. Specifically, we have

$$
\mu^* R(X, Y, Z) = R(X, Y)\mu_* Z,
$$

where X, Y, Z are vector fields on N. Equivalently, in terms of a local basis

$$
\mu^* R = -X_i \otimes \mu^* \omega^j \otimes \Omega_j^i.
$$

If we apply the alternating operator $\mathscr{A}$, we get the skew-symmetric part, and since Ω_j^i is already a 2–form, it is

$$
\mathscr{A}\mu^* R = -X_i \otimes \mu^* \omega^j \wedge \Omega_j^i.
$$

Theorem 5.10.1 (The cyclic sum identity). *If torsion vanishes, then* $\mu^* \omega^j \wedge \Omega_j^i = 0$; *hence* $\mathscr{A}\mu^* R = 0$.

Proof. If $\Omega^i = 0$ the first structural equation says $d\mu^*\omega^i = -\omega^i_j \wedge \mu^*\omega^j$. Taking the exterior derivative and substituting the second structural equation, $d\omega^i_j = -\omega^i_k \wedge \omega^k_j + \frac{1}{2}\Omega^i_j$ yields

$$d^2\mu^*\omega^i = 0$$
$$= -d\omega^i_j \wedge \mu^*\omega^j + \omega^i_j \wedge d\mu^*\omega^j$$
$$= \omega^i_k \wedge \omega^k_j \wedge \mu^*\omega^j - \frac{1}{2}\Omega^i_j \wedge \mu^*\omega^j - \omega^i_j \wedge \omega^j_k \wedge \mu^*\omega^k$$
$$= -\frac{1}{2}\Omega^i_j \wedge \mu^*\omega^j$$
$$= -\frac{1}{2}\mu^*\omega^j \wedge \Omega^i_j. \quad \blacksquare$$

Problem 5.10.6. The *Ricci tensor* $R_{ij}\,dx^i \otimes dx^j$ of a connexion D on a manifold is the tensor of type $(0, 2)$ obtained by contracting the curvature as follows: $R_{ij} = R^h_{ihj}$. If D is symmetric use (5.10.11) to show that $R_{ij} - R_{ji} = R^h_{hij}$, so there is only one independent contraction of R.

Problem 5.10.7. (a) Let D be a symmetric connexion on a manifold M. Use coordinates x^i such that the ∂_i are parallel at m [hence $\Gamma^i_{jk}(m) = 0$ and covariant derivatives at m coincide with derivatives of components] and (5.10.10) to prove the *Bianchi identity:*

$$R^i_{jhk|p} + R^i_{jkp|h} + R^i_{jph|k} = 0.$$

(b) Interpret DR, the covariant differential of the curvature tensor R, as a tensor of type $(0, 3)$ whose values are tensors of type $(1, 1)$, that is, $DR(x, y, z) = (D_z R)(x, y): M_m \to M_m$ for $x, y, z \in M_m$. Show that the Bianchi identity is equivalent to the fact that the skew-symmetric part of DR vanishes.

Problem 5.10.8. Show that all possible contractions of DR can be obtained from $D(R_{ij}dx^i \otimes dx^j)$, owing to the following consequence of the Bianchi identity:

$$R^h_{ijk|h} = R_{ik|j} - R_{ij|k}.$$

The fact that torsion and curvature behave well under the process of inducing one connexion from another is often used but rarely proved. If $\varphi: P \to N$ is a C^∞ map and D is a connexion on $\mu: N \to M$ with torsion T and curvature R, then we define the pullbacks of T and R to P by

$$\varphi^*T(X, Y) = T(\varphi_*X, \varphi_*Y),$$
$$\varphi^*R(X, Y) = R(\varphi_*X, \varphi_*Y).$$

Thus φ^*T and φ^*R are tensor-valued 2–forms on P with values in TM and $T^1_1 M$, respectively.

Theorem 5.10.2.   *The induced connexion* $\varphi^* D$ *has torsion and curvature* $\varphi^* T$ *and* $\varphi^* R$.

Proof. These are easy consequences of the structural equations and the fact that the connexion forms of the induced connexion are the pullbacks by φ of the connexion forms of D (see Problem 5.7.7).

Corollary.   *An induced connexion of a symmetric connexion is symmetric.*

A connexion is called *flat* if $R = 0$. If the connexion is locally parallelizable, then it is flat. For if there is a local basis of parallel fields $\{E_i\}$, then

$$R(X, Y)E_i = D_{[X,Y]}E_i - D_X D_Y E_i + D_Y D_X E_i$$
$$= 0$$

for any vector fields X, Y. Since the $\{E_i\}$ are a basis $R(X, Y) = 0$. The converse is true (see below), so $R = 0$ is the integrability condition for the equations for parallel fields.

Theorem 5.10.3.   *If* $R = 0$, *then the connexion is locally parallelizable. If in addition* N *is simply connected, then the connexion is parallelizable.*

Proof. At $n_0 \in N$ choose coordinates z^α on $U \subset N$ with n_0 as the origin and choose a basis $\{e_i\}$ of $M_{\mu n_0}$. Parallel translate $\{e_i\}$ along "rays" from n_0 with respect to the coordinates z^α, generating a local basis $\{E_i\}$ on U. By a ray we mean a curve ρ such that $z^\alpha \rho = a^\alpha \cdot s$ for some constants a^α. If $Z_\alpha = \partial/\partial z^\alpha$ the velocity of such a ray is $a^\alpha Z_\alpha \circ \rho$, so the condition on the E_i is

$$a^\alpha D_{Z_\alpha} E_i(\varphi^{-1}(a^1 s, \ldots, a^e s)) = 0,$$

where $\varphi = (z^1, \ldots, z^e)$ is the coordinate map. The E_i are C^∞ because they can be represented as solutions of ordinary differential equations dependent on C^∞ parameters $a^1, \ldots, a^e$. (Note that the E_i are parallel at n_0 and that the procedure works without the assumption $R = 0$.)

Now we use the assumption $R = 0$ to show that the E_i are parallel in all directions, not just the radial directions. If $t \in N_n, n \in U, z^\alpha n = a^\alpha$, and $t = b^\alpha Z_\alpha(n)$, then we define a rectangle τ having as its longitudinal curves rays from n_0, its base curve the ray from n_0 to n, and the final transverse tangent equal to t:

$$\tau(u, v) = \varphi^{-1}((a + bv)u),$$

where $a = (a^1, \ldots, a^e)$ and $b = (b^1, \ldots, b^e)$. Then $\{E_i \circ \tau\}$ is a local basis for vector fields over $\mu \circ \tau$. Let ω_i^j be the connexion forms for the induced connexion $\tau^* D$ with respect to $\{E_i \circ \tau\}$. The domain of τ is an open set in R^2 and thus has local basis $X = \partial/\partial u$ and $Y = \partial/\partial v$. The fields $E_i \circ \tau$ are parallel along the integral curves of X since they correspond to the longitudinal curves of τ which are rays from n_0 in N. Thus

$$\tau^* D_X(E_i \circ \tau) = 0$$
$$= \omega_i^j(X)E_j \circ \tau,$$

and since $\{E_i \circ \tau\}$ is a basis,

$$\omega_j^i(X) = 0.$$

The curve $\tau(0, v) = n_0$ is constant, so $\tau_* Y(0, v) = 0$. Hence

$$\tau^* D_{Y(0, v)}E_i \circ \tau = 0$$

[see Problem 5.7.1(b)]; that is, $\omega_j^i(Y(0, v)) = 0$. The curvature of $\tau^* D$ vanishes (Theorem 5.10.2) so the second structural equations reduce to

$$d\omega_j^i = -\omega_k^i \wedge \omega_j^k.$$

Applying this to (X, Y),

$$X\omega_j^i(Y) - Y\omega_j^i(X) - \omega_j^i[X, Y] = -2\omega_k^i \wedge \omega_j^k(X, Y)$$
$$= 0 \qquad\qquad (5.10.12)$$

since $\omega_k^i(X) = \omega_j^k(X) = 0$. Moreover, $[X, Y] = 0$, so (5.10.12) becomes

$$X\omega_j^i(Y) = 0.$$

Thus the functions $f_j^i(u) = \omega_j^i(Y(u, 0))$ on $[0, 1]$ satisfy $f_j^i(0) = 0$ and $f_j^{i\prime} = 0$, which shows that $f_j^i(1) = 0$. But

$$D_t E_i = D_{t, Y(1, 0)}E_i$$
$$= \tau^* D_{Y(1, 0)}E_i \circ \tau$$
$$= \omega_j^i(Y(1, 0))E_j(n)$$
$$= 0.$$

Hence the E_i are parallel at n.

If N is simply connected, then for any $n \in N$ choose a curve γ such that $\gamma(0) = n_0$, $\gamma(1) = n$ and define $E_i(n) = \pi(\gamma; n_0, n)e_i$. (If N is not connected a base point n_0 must be chosen in each component.) If we can show that $E_i(n)$ is independent of the choice of γ, it will coincide locally with a field of the type defined above and hence be C^∞ and parallel. Thus to show that $\{E_i\}$ is a well-defined parallelization for D, it suffices to show that if σ is another such curve from n_0 to n, then $\pi(\sigma; 0, 1)e_i = \pi(\gamma; 0, 1)e_i$. However, N is simply connected, so γ can be deformed into σ by a rectangle τ such that $\tau(u, 0) = \gamma(u)$, $\tau(u, 1) = \sigma(u)$, and $\tau(0, \cdot)$ and $\tau(1, \cdot)$ are constant curves. The proof now proceeds as above, but with X and Y interchanged. The details are left as an exercise. ∎

Problem 5.10.9. If the Möbius strip M is viewed as a rectangle in R^2 with two opposite edges identified with a twist, show that there is a unique affine · structure for which the restriction of the cartesian coordinates on R^2 is one of the affine charts. Show further that the connexion of this affine structure is flat but not parallelizable.

We close this section with a result indicating the desirability of symmetry of a connexion.

Theorem 5.10.4. *Let D be a symmetric connexion on M and let P be a distribution on M which is spanned locally by parallel fields. Then P is completely integrable and admits coordinate vector fields as a parallel basis.*

Proof. Let $\{E_\alpha\}$, $\alpha = 1, \ldots, p$ be a local parallel basis for P. Then

$$
\begin{aligned}
T(E_\alpha, E_\beta) &= D_{E_\alpha} E_\beta - D_{E_\beta} E_\alpha - [E_\alpha, E_\beta] \\
&= -[E_\alpha, E_\beta] \\
&= 0.
\end{aligned}
$$

Thus P is completely integrable and the E_α themselves are locally coordinate vector fields. ∎

Corollary. *A connexion D on a manifold M is the connexion associated with an affine structure iff T and R both vanish.*

5.11.   Connexion of a Semi-riemannian Structure

The generalization of semi-riemannian structures on manifolds to semi-riemannian structures on maps is straightforward and will not be defined explicitly. Our principal interest will be in such a structure b on a manifold M but if $\mu: N \to M$, then an important tool is the *induced semi-riemannian structure on μ*, which is defined as $b \circ \mu$, viewing b as a tensor field of type $(0, 2)$ on M.

A connexion D on a map $\mu: N \to M$ is said to be *compatible* with the metric $\langle \ , \ \rangle$ on μ if parallel translation along curves in N preserves inner products. Specifically, if γ is any curve in N and $x, y \in M_{\mu\gamma a}$, then

$$
\langle \pi(\gamma; a, b)x, \pi(\gamma; a, b)y \rangle = \langle x, y \rangle
$$

for all a, b in the domain of γ. In particular, $\pi(\gamma; a, b)$ maps an orthonormal basis of $M_{\mu\gamma a}$ into an orthonormal basis of $M_{\mu\gamma b}$, so there are parallel basis fields along γ which are orthonormal at every point. A number of equivalent conditions are given below.

Proposition 5.11.1. *The following are all equivalent.*

(a) *Connexion D is compatible with $\langle \ , \ \rangle$.*
(b) *The metric tensor field $\langle \ , \ \rangle$ is parallel with respect to D.*
(c) *For all vector fields X, Y over μ and all $t \in N_n$,*

$$
t\langle X, Y \rangle = \langle D_t X, Y \rangle + \langle X, D_t Y \rangle. \tag{5.11.1}
$$

(d) *There is an orthonormal parallel basis field along every curve γ in N.*

(e) *For every C^∞ map $\varphi: P \to N$ the induced connexion $\varphi^* D$ is compatible with the induced metric $\langle \ , \ \rangle \circ \varphi$.*

Proof. We have already noted that **(a)** implies **(d)** above, and the reverse implication **(d)** → **(a)** is a simple exercise in linearity.

(d) → (b). We must show that $D_t \langle \ , \ \rangle = 0$ for every t, assuming there is an orthonormal parallel basis along any curve. Choose a curve γ such that $\gamma_*(0) = t$ and let $\{E_i\}$ be such a basis, $\{\varepsilon^i\}$ the dual basis. Then $\langle \ , \ \rangle \circ \gamma = b_{ij}\varepsilon^i \otimes \varepsilon^j$, where the b_{ij} are all 1, -1, or 0. The derivatives of the b_{ij} are all 0, so by the second version of the definition of the covariant derivative of a tensor field, $D_t \langle \ , \ \rangle = 0$.

(b) → (e). This follows from a general rule for covariant derivatives of restriction tensor fields with respect to an induced connexion: $(\varphi^* D)_t S \circ \varphi = D_{\varphi_* t} S$. The general rule follows easily from the special case where S is a vector field by choosing a basis. Then take $S = \langle \ , \ \rangle$.

(a) ↔ (e). The definition of compatibility is little more than the case of (e) where φ is a curve γ, so (a) is a special case of (e). On the other hand, a curve τ in P is pushed into a curve $\gamma = \varphi \circ \tau$ in N by φ. If we have (a), then inner products are preserved under parallel translation along γ, and (because we have essentially the same set of vectors, parallel translation, and metric for those vectors) parallel translation preserves inner products along τ also.

(b) ↔ (c). We may view the evaluation $(X, Y) \to \langle X, Y \rangle$ as a contraction C. By Proposition 5.9.1, C commutes with D_t, from which we derive the value of $D_t \langle \ , \ \rangle$ on (X, Y):

$$(D_t \langle \ , \ \rangle)(X, Y) = t\langle X, Y \rangle - \langle D_t X, Y \rangle - \langle X, D_t Y \rangle.$$

The equivalence of (b) and (c) is now immediate. ∎

Problem 5.11.1. Let $\{F_i\}$ be a local orthonormal basis for a metric $\langle \ , \ \rangle$ over $\mu: N \to M$ and let $\langle F_i, F_i \rangle = a_i$ (no sum), so $a_i = \pm 1$. Show that a connexion D on μ is compatible iff the connexion forms of D with respect to such orthonormal bases satisfy the skew-adjointness property: $\omega_i^j = -a_i a_j \omega_j^i$ (no sum). In particular, for the riemannian case the matrix (ω_j^i) is skew-symmetric.

We mention without proof that there are always compatible connexions with a metric on a map. Some further restriction is needed to force a unique choice of compatible connexion. In the case of a metric on a manifold a restriction which produces uniqueness is given by making the torsion tensor vanish. Besides the analytic simplicity which symmetry gives to a connexion there is a geometric reason why the vanishing of torsion is desirable, which may be roughly explained as follows. Let γ be a curve in M and refer the acceleration A_γ to a parallel basis field along γ: $A_\gamma = f^i E_i$. Then we can find a curve τ in R^d such that the acceleration of τ in the euclidean sense is $\tau'' = (f^1, \ldots, f^d)$. If τ is a *closed* curve in R^d, then a surface S "fitting" γ can be found such that the integral of the torsion (a 2-form!) on S approximates the displacement from the initial to the final point of γ. Thus if torsion vanishes the behavior of short curves can be compared more easily with euclidean curves.

If we attempt to apply the torsion zero condition in the general case of a metric on a map $\mu: N \to M$ we find that it imposes linear algebraic conditions, point by point, on the connexion forms (or coefficients). The solvability of the system is determined by the rank of μ_* at the point. That is, if $q = \dim(\mu_* N_n) = d = \dim M$, then the solution is unique, and if $q = e = \dim N$, then the solution exists (at the point n). Thus to have both existence and uniqueness μ_* must be 1–1, which means that μ is a local diffeomorphism. It is only slightly more restrictive to confine our attention to the case where μ is the identity, that is, to a metric on a manifold.

Theorem 5.11.1. *A semi-riemannian manifold has a unique symmetric connexion compatible with its metric.*

Proof. Uniqueness is demonstrated by developing a formula for $\langle D_X Y, Z \rangle$, using compatibility in the form

$$\langle D_X Y, Z \rangle = -\langle Y, D_X Z \rangle + X \langle Y, Z \rangle, \tag{5.11.1}$$

and torsion zero in the form

$$D_X Y = D_Y X + [X, Y]. \tag{5.11.2}$$

The procedure is to apply (5.11.1) and (5.11.2) alternately, to cyclic permutations of X, Y, Z. At the final step a second copy of $\langle D_X Y, Z \rangle$ appears, giving the formula

$$2\langle D_X Y, Z \rangle = X \langle Y, Z \rangle + Y \langle X, Z \rangle - Z \langle X, Y \rangle$$
$$- \langle X, [Y, Z] \rangle + \langle Y, [Z, X] \rangle + \langle Z, [X, Y] \rangle. \tag{5.11.3}$$

Call the right side of this formula $D(X, Y, Z)$.

To prove existence we observe:

(a) For fixed X and Y, the expression $D(X, Y, Z)$ is a 1–form in Z. When we substitute fZ for Z the terms in which f is differentiated cancel, leaving $D(X, Y, fZ) = fD(X, Y, Z)$. Additivity in Z is obvious. Thus (5.11.3) does not overdetermine $D_X Y$; that is, there is a vector field W such that $2\langle W, Z \rangle = D(X, Y, Z)$ for each X, Y.

(b) Axiom (1) for a connexion, that D_t is linear in t, is similar to (a) and follows from the fact that $D(X, Y, Z)$ is a 1–form in X for fixed Y and Z.

(c) Axiom (2), that D_t is R-linear, follows from the obvious additivity in Y and axiom (3), which is done next.

(d) Axiom (3) is proved by substituting fY for Y and computing to obtain the desired result in the form

$$D(X, fY, Z) = fD(X, Y, Z) + 2(Xf)\langle Y, Z \rangle.$$

(e) It is clear that the formula yields C^∞ results from C^∞ data, so axiom (4) is satisfied.

(f) Compatibility is proved by checking that

$$D(X, Y, Z) + D(X, Z, Y) = 2X\langle Y, Z \rangle.$$

(g) Torsion zero is verified by showing

$$D(X, Y, Z) - D(Y, X, Z) = 2\langle [X, Y], Z \rangle. \quad \blacksquare$$

We call this compatible symmetric connexion the *semi-riemannian connexion*, or, in honor of its discoverer, the *Levi-Civita connexion*.

Corollary.   *A semi-riemannian structure* $\langle \ , \ \rangle \circ \mu$ *induced on* $\mu: N \to M$ *by a metric* $\langle \ , \ \rangle$ *on* M *has a compatible symmetric connexion. It is unique in a neighborhood of any point at which* μ_* *is onto.*

Proof. The existence is shown by μ^*D, where D is the semi-riemannian connexion on M. If μ_* is onto at n, then we can find a local basis over μ consisting of image vector fields μ_*Z_i, where the Z_i are vector fields on N. It follows that every vector field over μ is an image vector field in the basis neighborhood. But the development of (5.11.3) can be carried out as before if attention is restricted to image fields, giving a formula for $2\langle D_X\mu_* Y, \mu_*Z \rangle$, where X, Y, and Z are vector fields on the neighborhood of n. $\blacksquare$

Equations (5.11.3) can be specialized to obtain expressions for the co-efficients of D. In the case of a coordinate basis $\{\partial_i\}$ the results are classical. The functions

$$\langle D_{\partial_i}\partial_j, \partial_k \rangle = [ij, k]$$
$$= \tfrac{1}{2}(\partial_i b_{jk} + \partial_j b_{ik} - \partial_k b_{ij}), \qquad (5.11.4)$$

where the b_{ij} are the components of $\langle \ , \ \rangle$, are called the *Christoffel symbols of the first kind*. The *Christoffel symbols of the second kind* are the coefficients of D as previously defined, the Γ^i_{jk}, and are also denoted by $\{^i_{jk}\}$. They are obtained by raising the index k of $[ij, k]$, since $D_{\partial_i}\partial_j = \Gamma^k_{ji}\partial_k$ gives $[ij, k] = \Gamma^h_{ji}b_{hk}$, and thus

$$\Gamma^h_{ji} = b^{hk}[ij, k]$$
$$= \tfrac{1}{2}b^{hk}(\partial_i b_{jk} + \partial_j b_{ik} - \partial_k b_{ij}), \qquad (5.11.5)$$

where (b^{hk}) is the inverse of the matrix (b_{hk}).

Another natural choice of local basis is an orthonormal basis or *frame* $\{F_i\}$. Its use might be more advantageous, for example, in the case of a riemannian structure on a parallelizable manifold, because in that case the basis can be made global. For frame members as X, Y, and Z the first three terms of

(5.11.3) drop out since the $\langle F_i, F_j \rangle = a_i \delta_{ij}$ (no sum) are constant. The co-efficients of D are given in terms of the *structural functions* c^i_{jk} *for the frame*, that is, the components of their brackets:

$$[F_j, F_k] = c^i_{jk} F_i. \tag{5.11.6}$$

Note that $c^i_{jk} = -c^i_{kj}$. Inserting (5.11.6) in (5.11.3) we obtain

$$\begin{aligned} 2 \langle D_{F_i} F_j, F_k \rangle &= -a_i c^i_{jk} + a_j c^j_{ki} + a_k c^k_{ij} \\ &= 2 a_k \Gamma^k_{ji} \qquad \text{(no sums).} \end{aligned}$$

In this case to lower the indices, solving for the Γ^i_{jk}, is trivial since $(b_{ij}) = (a_i \delta_{ij})$ is its own inverse:

$$\Gamma^i_{jk} = \tfrac{1}{2} a_i (a_i c^i_{kj} + a_j c^j_{ik} + a_k c^k_{ij}) \qquad \text{(no sums).} \tag{5.11.7}$$

Theorem 5.11.2. *The curvature tensor R of a semi-riemannian manifold has the following symmetry properties:*

(a) $\langle R(X, Y)Z, W \rangle = -\langle R(Y, X)Z, W \rangle$.
(b) $\langle R(X, Y)Z, W \rangle = -\langle R(X, Y)W, Z \rangle$.
(c) $\langle R(X, Y)Z, W \rangle = \langle R(Z, W)X, Y \rangle$.
and the first Bianchi identity
(d) $R(X, Y)Z + R(Y, Z)X + R(Z, X)Y = 0$.

(See Problem 2.17.4 for the component version of this theorem, where we use as notation $A_{ijhk} = R_{ijhk} = b_{ip} R^p_{jhk}$. Some further algebraic properties of the curvature tensor are also noted in Problem 2.17.4.)

Proof. Properties (a) and (d) have already been proved in Section 5.10.

Property (b) says that $R(X, Y)$ is a *skew-adjoint linear transformation* with respect to $\langle \ , \ \rangle$. The corresponding property of the matrix (A^i_j) of a linear transformation with respect to a frame $\{F_i\}$ is $A^j_i = -a_i a_j A^i_j$ (no sum). That this property is enjoyed by the connexion form matrix $(\Gamma^i_{jk} dx^k) = (\omega^i_j)$ is immediate from (5.11.7). But the matrix of $R(\cdot, \cdot)$ is the negative of the curvature form matrix $(-\Omega^i_j)$, for which skew-adjointness follows from that of (ω^i_j) and the second structural equations $\Omega^i_j = 2(d\omega^i_j + \omega^i_k \wedge \omega^k_j)$.

The relation (c) follows from (a), (b), and (d), as has been asked for in Problem 2.17.4. Indeed, if we substitute in the relation

$$\langle R(X, Y)Z, W \rangle + \langle R(Y, Z)X, W \rangle + \langle R(Z, X)Y, W \rangle = 0$$

the permutations $(X, W, Y, Z), (Z, W, X, Y)$, and (Y, W, Z, X) of (X, Y, Z, W), then we obtain three similar relations. The sum of the first two minus the sum of the last two gives the desired conclusion. ∎

Problem 5.11.2. Find the components of the curvature of the semi-riemannian connexion

(a) with respect to a coordinate basis $\{\partial_i\}$ in terms of the metric components $b_{ij} = \langle \partial_i, \partial_j \rangle$, and

(b) with respect to a frame $\{F_i\}$ in terms of the structural functions c^i_{jk}, where $[F_j, F_k] = c^i_{jk} F_i$.

Problem 5.11.3. Let g be a symmetric bilinear form on μ which is parallel with respect to a connexion D on μ. [We have (5.11.1) with g in place of $\langle \ , \ \rangle$.] Show that the curvature transformations $R(X, Y)$ of D are skew-adjoint with respect to g.

Problem 5.11.4. A map $\mu : M \to M$ is an *isometry* of a metric $\langle \ , \ \rangle$ if $\mu^* \langle \ , \ \rangle = \langle \ , \ \rangle$. A vector field X is a *Killing field* of $\langle \ , \ \rangle$, or, an *infinitesimal isometry*, if each transformation μ_s of the one-parameter group of X is an isometry of the open subsets of M on which it is defined. Show that X is a Killing field iff $L_X \langle \ , \ \rangle = 0$, where L_X is Lie derivative with respect to X.

Problem 5.11.5. Let D be the semi-riemannian connexion of the metric $\langle \ , \ \rangle$ on M and define $A_X Y = -D_Y X$. For X fixed, A_X is a tensor field of type $(1, 1)$, viewed as a field of linear transformations of tangent spaces. Extend A_X to be a derivation of the whole tensor algebra (see Problem 3.6.8). Show that $A_X = L_X - D_X$. (*Hint:* Two derivations of the tensor algebra, for example, A_X and $L_X - D_X$, will coincide if they coincide when applied to functions and vector fields.)

Problem 5.11.6. Show that X is a Killing field iff A_X is skew-adjoint with respect to $\langle \ , \ \rangle$.

Problem 5.11.7. Show that $A_X = L_X$ at the points where X vanishes.

Problem 5.11.8. Let $\{F_i\}$ be a parallelization on M and define a semi-riemannian metric $\langle \ , \ \rangle$ on M by choosing the a_i and making the F_i ortho-normal. The connexion D of the parallelization $\{F_i\}$ is compatible but it will not generally be the semi-riemannian connexion since its torsion is essentially given by the c^i_{jk}. Under what conditions on the c^i_{jk} will the symmetrized connexion ^sD be the semi-riemannian connexion?

Problem 5.11.9. Use symmetry (b) of Theorem 5.11.2 to show that $R^h_{hji} = 0$ and consequently the Ricci tensor of a semi-riemannian connexion is symmetric; that is, $R_{ij} = R_{ji}$ (see Problem 5.10.6).

Problem 5.11.10.   The *scalar curvature* S of a semi-riemannian connexion is the scalar function $b^{ij}R_{ij} = R^i_i$. Show that $R^j_{i|j}\,dx^i = \frac{1}{2}\,dS$ (see Problem 5.10.8).

5.12.   Geodesics

In E^d a straight line is a curve γ which does not bend; that is, its velocity field is parallel along γ when a particular parametrization is chosen—the linear parametrization. We employ this characterization of a straight line in E^d as motivation for the definition of a geodesic in a manifold M with a connexion D.

A *geodesic* in M is a parametrized curve γ such that γ_* is parallel along γ. Equivalently, the acceleration $A_\gamma = (\gamma^*D)_{d/du}\gamma_* = 0$. If the parametrization is changed it will not remain a geodesic unless the change is affine: $\tau(s) = \gamma(as + b)$, a and b constant, since any other reparametrization will give some acceleration in the direction of γ_*.

The equation for a geodesic is a second-order differential equation $(\gamma^*D)_{d/du}\gamma_* = 0$. The initial conditions for a second-order differential equation are given by specifying a starting point m, where the parameter is 0, and an initial velocity $\gamma_*(0) = v \in M_m$. The conditions on the defining functions of the differential equations will be enough to assert that there is a unique solution for every pair of initial conditions. In fact, a solution will be a C^∞ function of the parameter u, the starting point m, and the initial velocity v.

Another viewpoint is to consider the velocity curves γ_* in the tangent bundle TM. The velocity curves of geodesics are found to be the integral curves of a single vector field G on TM, so the properties mentioned above follow from previous results on integral curves. The following theorem characterizes G intrinsically.

Theorem 5.12.1.   *Let D be a connexion on M, $\pi\colon TM \to M$ the projection taking a vector to its base point, $I\colon TM \to TM$ the identity map on TM viewed as a vector field over π, and π^*D the induced connexion on π. Then there is a unique vector field G on TM such that*

   (a) $\pi_*G = I$ *(note that we have $G\colon TM \to TTM$ and $\pi_*\colon TTM \to TM$).*
   (b) $(\pi^*D)_G I = 0$.

Furthermore, if τ is an integral curve of G, then $\gamma = \pi \circ \tau$ is a geodesic in M and $\tau = \gamma_*$. There is no other vector field on TM whose integral curves are the velocity fields of all geodesics.

Proof. Equations **(a)** and **(b)** are linear equations for $G(t)$ for each $t \in TM$, so they can be considered pointwise without loss of generality. At a given $t \in TM$, with $\pi t = m$, we may combine them into one linear function equation by taking the direct sum as follows: $\pi_* : (TM)_t \to M_m$ and $A_I = (\pi^* D)_{(.)} I : (TM)_t \to M_m$ give

$$\pi_* + A_I : (TM)_t \to M_m + M_m \qquad \text{(direct sum)}.$$

(The operator A_I is the negative of that in Problem 5.11.5.) The dimensions of $(TM)_t$ and $M_m + M_m$ are the same, namely $2d$, so to show that there *is* a solution we need only show that it is unique. Now we turn to the coordinate formulation to show that we can solve for G uniquely.

Let x^i be coordinates on M. Then as coordinates on TM we may use $y^i = x^i \circ \pi$ and $y^{i+d} = dx^i$. Let the corresponding coordinate vector fields be $\partial / \partial x^i = X_i$ on M and $\partial / \partial y^i = Y_i$, $\partial / \partial y^{i+d} = Y_{i+d}$ on TM. The vector fields X_i and Y_i are π-related, so a convenient basis for vector fields over π is $X_i \circ \pi = \pi_* Y_i$. For any $t \in M_m$ we have $t = (tx^i) X_i(m) = y^{i+d}(t) X_i(\pi t)$, so the coordinate expression for I is

$$I = y^{i+d} X_i \circ \pi.$$

If the coefficients of D are Γ^i_{jk}, then the coefficients of $\pi^* D$ with respect to $\{X_i \circ \pi, Y_\alpha\}$ are $H^i_{jk} = \Gamma^i_{jk} \circ \pi$ and $H^i_{jk+d} = 0$, since $\pi_* Y_{i+d} = 0$.

Now suppose $G = G^i Y_i + G^{i+d} Y_{i+d}$. Then from (a),

$$\begin{aligned}
\pi_* G &= G^i \pi_* Y_i + G^{i+d} \pi_* Y_{i+d} \\
&= G^i X_i \circ \pi \\
&= I \\
&= y^{i+d} X_i \circ \pi,
\end{aligned}$$

and hence $G^i = y^{i+d}$. From (b) we conclude

$$\begin{aligned}
(\pi^* D)_G I &= (\pi^* D)_G y^{i+d} X_i \circ \pi \\
&= (G y^{i+d}) X_i \circ \pi + y^{i+d} G^\alpha (\pi^* D)_{Y_\alpha} X_j \circ \pi \\
&= (G^{i+d} + y^{j+d} y^{k+d} \Gamma^i_{jk} \circ \pi) X_i \circ \pi \\
&= 0,
\end{aligned}$$

which allows us to solve for G^{i+d}, giving a unique solution

$$G = y^{i+d} Y_i - y^{j+d} y^{k+d} \Gamma^i_{jk} \circ \pi Y_{i+d}. \tag{5.12.1}$$

Now suppose that τ is an integral curve of G and let $\gamma = \pi \circ \tau$. Then $\gamma_* = \pi_* \tau_* = \pi_* (G \circ \tau) = I \circ \tau = \tau$ by (a). Then for the induced connexion on $\gamma = \pi \circ \tau$ we have

$$\begin{aligned}
(\gamma^* D)_{d/du} \gamma_* &= (\gamma^* D)_{d/du} I \circ \tau \\
&= \tau^* (\pi^* D)_{d/du} I \circ \tau \\
&= (\pi^* D)_{\tau_*} I \\
&= (\pi^* D)_G I \circ \tau \\
&= 0
\end{aligned}$$

by (b). Thus γ is a geodesic.

The last statement is now trivial because any vector field is uniquely determined by its totality of integral curves. ∎

A small part of the above proof is the proof of the following lemma which we state for later use.

Lemma 5.12.1. *Let X be a vector field on TM. Then*

(**a'**) *the integral curves of X are the velocity fields of their projections into M iff*
(**a**) $\pi_* X = I$.

Remarks. (**a**) A vector field X on TM such that $\pi_* X = I$ is called a *second-order differential equation* over M. The coordinate expression for such an X has the form

$$X = y^{i+d} Y_i + F^i Y_{i \mid d},$$

and if γ is the projection of an integral curve γ_* of X, then the components $f^i = x^i \circ \gamma$ of γ satisfy a system of second-order differential equations:

$$f^{i''} = {}^c F^i(f^1, \ldots, f^d, f^{1'}, \ldots, f^{d'}),$$

where ${}^c F^i$ is the function on R^{2d} corresponding to F^i under the coordinate map (y^α).

(**b**) The vector field G on TM which gives the geodesics of a connexion D is called the *geodesic spray* of D. For G the components $F^i = G^{i+d}$ are quadratic homogeneous functions of the y^{i+d}. A second-order differential equation X over M such that the F^i are homogeneous quadratic functions of the y^{j+d} is called a *spray* over M. Consequently, for a spray we must have

$$F^i = -y^{j+d} y^{k+d} \Gamma^i_{jk} \circ \pi,$$

for some functions $\Gamma^i_{jk} = \Gamma^i_{kj}$ on M. By a theorem of W. Ambrose, I. Singer, and R. Palais,† the functions Γ^i_{jk} are the coefficients of a connexion D on M; that is, every spray over M is a geodesic spray of some connexion.

(**c**) If γ is a geodesic and $f^i = x^i \circ \gamma$ are its coordinate components, then the coordinate components of γ_* are $y^i \circ \gamma_* = f^i$ and $y^{i+d} \circ \gamma_* = f^{i'}$. By (5.12.1) we get the equations satisfied by the f^i and $f^{i'}$: $f^{i'} = f^{i'}$ for the first d equations, and

$$f^{i''} = -f^{j'} f^{k'} \Gamma^i_{jk} \circ \pi. \tag{5.12.2}$$

The second-order equations (5.12.2) are standard and can be derived easily from the definition of a geodesic.

---

† Sprays, *Ann. Acad. Brazil. Ci.*, **32**, 163–178 (1960).

Theorem 5.12.2. *A connexion is completely determined by its torsion and the totality of all its parametrized geodesics.*

Proof. The torsion determines $\Gamma^i_{jk} - \Gamma^i_{kj}$ and the spray G determines $\Gamma^i_{jk} + \Gamma^i_{kj}$. ∎

Problem 5.12.1.   Given a connexion D and a tensor field S of type $(1, 2)$, skew-symmetric in its covariant indices, there is a unique connexion with the same geodesics as D and whose torsion is equal to S.

Problem 5.12.2.   Show that a connexion D, its conjugate connexion D^*, and the symmetrization ^sD all have the same geodesics.

Problem 5.12.3.   Let x^1 and x^2 be the cartesian coordinates on R^2. Define a connexion D on R^2 by $\Gamma^1_{12} = \Gamma^1_{21} = 1$ and $\Gamma^i_{jk} = 0$ otherwise. Then D is symmetric.

(a) Set up and solve the differential equations for the geodesics in R^2.
(b) Find the geodesic γ with $\gamma(0) = (2, 1)$ and $\gamma_*(0) = \partial_1(2, 1) + \partial_2(2, 1)$.
(c) Do the geodesics starting at $(0, 0)$ pass through all points of R^2?

Problem 5.12.4.   Same as Problem 5.12.3 except that $\Gamma^1_{12} = 1$, $\Gamma^i_{jk} = 0$ otherwise.

Problem 5.12.5.   If every geodesic can be extended to infinitely large values of its parameter, the connexion is said to be *complete*. That is, if the spray G on TM is complete, the connexion is said to be complete. Are the connexions in Problems 5.12.3 and 5.12.4 complete?

Problem 5.12.6.   Show that the geodesics of the connexion of a parallelization $\{X_i\}$ on M are the integral curves of constant linear combinations $a^i X_i$.

5.13.   Minimizing Properties of Geodesics

In this section it is shown that in a riemannian manifold the shortest curve between two points is a geodesic (with respect to the riemannian connexion) provided it exists. More generally, it will be seen that the energy-critical curves in a metric manifold are geodesics.

We first consider the local situation, showing that there is a geodesic segment between a point and all points in some neighborhood. This may be done for any connexion D on a manifold M. For $m \in M$ we define the *exponential map* $\exp_m: M_m \to M$ as follows. If $t \in M_m$ there is a unique geodesic γ such that $\gamma_*(0) = t$. We define $\exp_m t = \gamma(1)$ (see Figure 20). In riemannian terms, we

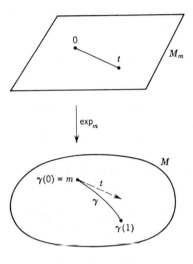

Figure 20

move along the geodesic in the direction t a distance equal to the length of t. To prove that $\exp_m$ is C^∞ we can pass to the tangent bundle TM and use the flow $\{\mu_t\}$ of the spray G. It should be clear that $\exp_m = \pi \circ \mu_1|_{M_m}$, where $\pi: TM \to M$ is the projection, so we have factored $\exp_m$ into the C^∞ maps π and $\mu_1|_{M_m}$.

Since the geodesic with initial velocity αt is the curve τ given by $\tau(s) = \gamma(\alpha s)$, it follows that the rays in M_m starting at 0 are mapped by $\exp_m$ into the geodesics starting at m $\exp_m \alpha t = \tau(1) = \gamma(\alpha)$.

If we choose a basis $b = \{e_i\}$ of M_m we obtain a diffeomorphism $b: R^d \to M_m$ given by $b(x) = u^i(x)\cdot e_i$. The composition with $\exp_m$, $\varphi = \exp_m \circ b$, maps the coordinate axes, which are particular rays, into the geodesics with initial velocities e_i. Thus $\varphi_*(\partial/\partial u^i(0)) = e_i$, which shows that φ_* is nonsingular at 0 and hence φ is a diffeomorphism on some neighborhood of 0. The inverse φ^{-1} is called a *normal coordinate map at m*, and the associated *normal coordinates* are characterized by the fact that the geodesics starting at m correspond to linearly parametrized coordinate rays through 0 in R^d. Since a coordinate map at m must fill all the points of a neighborhood of m we have proved

Proposition 5.13.1. *If M has a connexion D and $m \in M$, then there is a neighborhood U of m such that for every $n \in U$ there is a geodesic segment in U starting at m and ending at n.*

Problem 5.13.1. Show that if x^i are normal coordinates at m for a symmetric connexion D, then the coordinate basis $\{\partial_i\}$ is parallel at m and the coefficients of D with respect to $\{\partial_i\}$ all vanish at m: $\Gamma^i_{jk}(m) = 0$.

Problem 5.13.2. Let $M = C^*$, the complex plane with 0 removed, so that M may be identified with $R^2 - \{0\}$ and has cartesian coordinates x, y and co-ordinate vector fields ∂_1, ∂_2. Let $X = x\partial_1 + y\partial_2$ and $Y = -y\partial_1 + x\partial_2$, so $\{X, Y\}$ is a parallelization of M. Let D be the connexion of this parallelization. Show that the exponential map of D at $m = (1, 0)$ coincides with the complex exponential function if we identify M_m with $C: \alpha\partial_1(m) + \beta\partial_2(m) \leftrightarrow \alpha + i\beta$. That is,

$$\exp_m(\alpha\partial_1(m) + \beta\partial_2(m)) = e^{\alpha + i\beta}.$$

Theorem 5.13.1. *Let γ be a curve in a semi-riemannian manifold M with metric $\langle\ ,\ \rangle$. Then γ is a geodesic iff it is energy-critical.*

Proof. The proof is patterned after that of Theorem 5.6.1, where the acceleration with respect to the semi-riemannian connexion replaces the affine acceleration used there.

Let γ be the base curve of a variation Q defined on $[a, b] \times [c, d]$ with longitudinal and transverse fields $X = Q_*\partial_1$ and $Y = Q_*\partial_2$. The energy function on the longitudinal curves is then

$$E(v) = \int_a^b \langle X(u, v), X(u, v)\rangle\, du.$$

To evaluate $E'(c)$ we differentiate under the integral with respect to v and apply torsion zero:

$$\partial_2\langle X, X\rangle = 2\langle Q^*D_{\partial_2} X, X\rangle.$$
$$= 2\langle Q^*D_{\partial_1} Y, X\rangle.$$

However,

$$\partial_1\langle Y, X\rangle = \langle Q^*D_{\partial_1} Y, X\rangle + \langle Y, Q^*D_{\partial_1} X\rangle,$$

so

$$\partial_2\langle X, X\rangle = 2\partial_1\langle Y, X\rangle - 2\langle Y, Q^*D_{\partial_1} X\rangle.$$

The term $2\partial_1\langle Y, X\rangle$ may be integrated by the fundamental theorem of calculus, giving

$$2\langle Y, X\rangle\Big|_{(a,c)}^{(b,c)} = 0,$$

since $Y(a, c) = 0$ and $Y(b, c) = 0$ follow from the fact that Q is a variation. Thus we are left with the term involving the acceleration

$$A_y(u) = (Q^* D_{\partial_1} X)(u, c)$$

and the infinitesimal variation $Y(\cdot, c)$:

$$E'(c) = -2 \int_a^b \langle Y(u, c), A_y(u) \rangle \, du.$$

Now the proof proceeds exactly as in the affine case, showing that if there were any point at which A_y did not vanish, then a $Y(\cdot, c)$ could be chosen so as to produce a nonzero $E'(c)$. Conversely, if $A_y = 0$, that is, if γ were a geodesic, then $E'(c) = 0$ for all such $Y(\cdot, c)$; that is, γ would be energy-critical. ∎

It is now trivial that a shortest curve between two points in a riemannian manifold, if one exists, is a geodesic. We shall leave as an exercise the proof of the fact that in a small enough neighborhood the geodesic segments from the origin in a normal coordinate system are shortest curves.

An important property, which is a reasonable geometric hypothesis usually assumed in further research and should be checked in specific models, is that of geodesic completeness (see Problem 5.12.5). In the case of riemannian manifolds (but not semi-riemannian) completeness has the consequence that for every pair of points m, n in the same component there is a shortest curve from m to n. A famous theorem of Hopf and Rinow says that geodesic completeness of a riemannian manifold is equivalent to *metric completeness for the distance function*; that is, every Cauchy sequence converges. In particular, a compact riemannian manifold is complete.

5.14.   Sectional Curvature

A *plane section P* at a point m of a manifold M is a two-dimensional subspace of the tangent space M_m. In a semi-riemannian manifold the geodesics radiating from m tangent to P form a surface $S(P)$ which inherits a semi-riemannian structure from that of M (unless P is tangent to or included in the light cone at m, in which case the inherited structure is degenerate. We exclude these types from some of our discussion). By studying the geometry of these surfaces we gain insight into the structure of M. The main invariant of surface geometry is the *gaussian curvature*. A surface in E^3 with positive gaussian curvature is locally cap-shaped. An inhabitant of a cap-shaped surface can detect this property by measuring the length of circles about a point, since a circle of radius r will be shorter than $2\pi r$. A surface in E^3 with negative curvature is saddle-shaped and a circle of radius r is longer than $2\pi r$. For example, on a sphere of radius c in E^3 the circles of radius r have length (see Figure 21)

$$L(r) = 2\pi c \sin \frac{r}{c}$$

$$= 2\pi r - \frac{2\pi r^3}{6c^2} + \cdots.$$

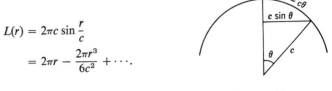

Figure 21

The defect from the euclidean length is about $\pi r^3/3c^2$, which is the gaussian curvature $K = 1/c^2$ multiplied by $\pi r^3/3$. In many cases computing $L(r)$ is an effective method of finding the gaussian curvature of $S(P)$, which we define to be the *sectional curvature $K(P)$ of P*:

$$K(P) = \lim_{r \to 0} \frac{3[2\pi r - L(r)]}{\pi r^3}. \tag{5.14.1}$$

In the semi-riemannian case we cannot define $K(P)$ in this way unless the restriction of $\langle \ , \ \rangle$ is riemannian or the opposite, negative definite. Thus sectional curvature is defined only for space-like or time-like plane sections.

Another possible description of sectional curvature uses the areas of circular disks instead of the lengths of circles. The approximate formula for area is obtained by integrating the length:

$$A(r) = \pi r^2 - K\frac{\pi r^4}{12} + \cdots.$$

We now compute the formula for sectional curvature in terms of the curvature tensor. We shall employ normal coordinates at the center point to gain insight into the nature of normal coordinates.

Let x^i be normal coordinates at m and let $b_{ij} = \langle \partial_i, \partial_j \rangle$ be expanded in a finite Taylor expansion of at least the second order,

$$b_{ij} = a_{ij} + b_{ijk}x^k + b_{ijhk}x^h x^k + \cdots,$$

where $b_{ijhk} = b_{ijkh}$ and the $a_{ij} = b_{ij}(m)$. Since a change from one normal coordinate system at m to another is linear, the tensor whose components with respect to $\{\partial_i(m)\}$ are b_{ijhk} is an invariant of the metric structure. Hence it is the value of a tensor field on M. This tensor field can be expressed in terms of the curvature tensor and its covariant differentials, but we shall not do so here.

The fact that the x^i are normal coordinates implies that along the radial lines $x^i = a^i s$ the velocity field $a^i \partial_i$ is parallel, and in particular, has constant energy $a^i a^j a_{ij}$. For,

$$\langle a^i \partial_i, a^j \partial_j \rangle \mid_{z^i = a^i s} = a^i a^j b_{ij} \mid_{z^i = a^i s}$$
$$= a^i a^j a_{ij} + b_{ijk} a^i a^j a^k s + b_{ijhk} a^i a^j a^h a^k s^2 + \cdots$$
$$= a^i a^j a_{ij},$$

the latter being the value when $s = 0$. The coefficients of each power of s must vanish,

$$b_{ijk} a^i a^j a^k = 0,$$
$$b_{ijhk} a^i a^j a^h a^k = 0, \quad \text{etc.}$$

which says that the symmetric parts of the tensors with components b_{ijk}, b_{ijhk}, etc., vanish. Thus, making use of the symmetry already present, we get

$$b_{ijk} + b_{jki} + b_{kij} = 0, \tag{5.14.2}$$
$$b_{ijhk} + b_{ihjk} + b_{ikjh} + b_{jhik} + b_{hkij} + b_{jkih} = 0. \tag{5.14.3}$$

We shall need the following consequence of (5.14.3). Let v^i and w^i be the components of two vectors at m. Then by some switching of indices we have

$$b_{ijhk} v^i w^j v^h w^k = b_{ikjh} v^i w^j v^h w^k$$
$$= b_{hkij} v^i w^j v^h w^k$$
$$= b_{jhik} v^i w^j v^h w^k.$$

Adding the four quantities in these relations and using (5.14.3) yields

$$A = (b_{ihjk} + b_{jkih}) v^i w^j v^h w^k$$
$$= -2(b_{ijhk} + b_{hkij}) v^i w^j v^h w^k \tag{5.14.4}$$
$$= -2B,$$

where

$$B = (b_{ijhk} + b_{hkij}) v^i w^j v^h w^k$$
$$= (b_{ikjh} + b_{jhik}) v^i w^j v^h w^k. \tag{5.14.5}$$

By Problem 5.13.1 the coefficients of the semi-riemannian connexion are zero at m. Thus by (5.11.4), evaluated at m,

$$a_{hk} \Gamma^h_{ji}(m) = \tfrac{1}{2}(\partial_i b_{jk} + \partial_j b_{ik} - \partial_k b_{ij})(m)$$
$$= \tfrac{1}{2}(b_{jki} + b_{ikj} - b_{ijk}) \tag{5.14.6}$$
$$= 0.$$

Combining (5.14.5), (5.14.2), and the symmetry of b_{ij}, we find

$$b_{ijk} = 0.$$

Now differentiate (5.11.4) with respect to ∂_p:

$$(\partial_p b_{hk}) \Gamma^h_{ji} + b_{hk} \partial_p \Gamma^h_{ji} = \tfrac{1}{2}\partial_p(\partial_i b_{jk} + \partial_j b_{ik} - \partial_k b_{ij})$$

and evaluate at m, using the fact that $(\partial_h \partial_k b_{ij})(m) = 2b_{ijhk}$:

$$a_{hk}(\partial_p \Gamma^h_{ji})(m) = b_{jkip} + b_{ikjp} - b_{ijkp}. \tag{5.14.7}$$

Using (5.14.7) and the coordinate formula for curvature (5.10.10), we find that we can express the curvature tensor, with the contravariant index lowered, at m in terms of the b_{ijhk}:

$$R^i_{jhk}(m) = (\partial_k \Gamma^i_{jh} - \partial_h \Gamma^i_{jk})(m),$$

so

$$
\begin{aligned}
R_{ijhk}(m) &= a_{ip} R^p_{jhk}(m) \\
&= (a_{ip}\partial_k \Gamma^p_{jh} - a_{ip}\partial_h \Gamma^p_{jk})(m) \\
&= b_{jihk} + b_{hijk} - b_{hjik} - b_{jikh} - b_{kijh} + b_{kjih} \quad (5.14.8)\\
&= b_{ihjk} + b_{jkih} - b_{jhik} - b_{ikjh}.
\end{aligned}
$$

Now suppose that P is a plane section at m on which $\langle \ , \ \rangle$ is definite and let $v = v^i \partial_i(m)$ and $w = w^i \partial_i(m)$ be a basis of P which is orthonormal, so $\langle v, v \rangle = \langle w, w \rangle = \delta = \pm 1$. Then the unit vectors in P are all of the form

$$v \cos t + w \sin t,$$

and the circle γ_r or radius r in $S(P)$ is parametrized by t, $0 \le t \le 2\pi$, having coordinates $x^i(\gamma_r t) = (v^i \cos t + w^i \sin t)r$. The velocity field of γ_r is thus

$$\gamma_{r*}t = (-v^i \sin t + w^i \cos t)r\partial_i(\gamma_r t).$$

Continuing, we have

$$
\begin{aligned}
\langle \gamma_{r*}t, \gamma_{r*}t \rangle &= b_{ij}(\gamma_r t)(-v^i \sin t + w^i \cos t)(-v^j \sin t + w^j \cos t)r^2 \\
&= r^2(v^i v^j \sin^2 t - 2v^i w^j \sin t \cos t + w^i w^j \cos^2 t)\cdot \\
&\quad (a_{ij} + b_{ijhk}[v^h v^k \cos^2 t + 2v^h w^k \sin t \cos t + w^h w^k \sin^2 t]r^2 + \cdots) \\
&= r^2\delta + r^4 b_{ijhk}[v^h v^k w^i w^k T(4, 0) \\
&\quad + 2(w^i w^j v^h w^k - v^i w^j v^h v^k)T(3, 1) \\
&\quad + (v^i v^j v^h v^k + w^i w^j w^h w^k - 4v^i w^j v^h w^k)T(2, 2) \\
&\quad + 2(v^i v^j v^h w^k - v^i w^j w^h w^k)T(1, 3) \\
&\quad + v^i v^j w^h w^k T(0, 4)] + \cdots,
\end{aligned}
$$

where we have set $T(p, q) = \cos^p t \sin^q t$. The length of $\gamma_* t$ is the square root of $\delta \langle \gamma_{r*}t, \gamma_{r*}t \rangle = r^2(1 + f(t)r^2 + \cdots)$ and has Taylor series $|\gamma_{r*}t| = r(1 + \frac{1}{2}f(t)r^2 + \cdots)$. When we integrate to find the length of γ_r, we note that the integrals of $T(p, q)$ and $T(q, p)$ are identical, the integrals of $T(3, 1)$ and $T(1, 3)$ are zero, and part of the coefficient of $T(2, 2)$ vanishes by (5.14.3). We have

$$\int_0^{2\pi} T(4, 0) \, dt = \frac{3\pi}{4}$$

$$\int_0^{2\pi} T(2, 2) \, dt = \frac{\pi}{4},$$

so the length of γ_r reduces to

$$|\gamma_r| = 2\pi r + \tfrac{1}{2} \int_0^{2\pi} f(t)dt\, r^3 + \cdots$$
$$= 2\pi r + \tfrac{1}{8}\delta\pi r^3 b_{ijhk}(3w^iw^jv^hv^k + 3v^iv^jw^hw^k - 4v^iw^jv^hw^k) + \cdots.$$

Thus from the definition of sectional curvature (5.14.1) we have

$$K(P) = -\tfrac{3}{8}\delta b_{ijhk}(3w^iw^jv^hv^k + 3v^iv^jw^hw^k - 4v^iw^jv^hw^k)$$
$$= -\tfrac{3}{8}\delta(3b_{ihjk} + 3b_{jkih} - 2b_{ijhk} - 2b_{hkij})v^iw^jv^hw^k$$
$$= -\tfrac{3}{8}\delta(3A - 2B) = 3\delta B,$$

where A and B are as in (5.14.4) and (5.14.5). From (5.14.4), $A = -2B$, we have $B = -(A - B)/3$, so

$$K(P) = -\delta(A - B)$$
$$= -\delta R_{ijhk}v^iw^jv^hw^k, \qquad \text{by (5.14.8)}.$$

However,

$$\langle R(v, w)v, w \rangle = \langle R_{ihk}^p v^hw^kv^i\partial_p(m), w^j\partial_j(m) \rangle$$
$$= -R_{ijhk}v^iw^jv^hw^k,$$

so

$$K(P) = \delta\langle R(v, w)v, w\rangle. \qquad (5.14.9)$$

If we change to a nonorthonormal basis of P, again called $\{v, w\}$, then we must divide by a normalizing factor:

$$K(P) = \frac{\delta\langle R(v, w)v, w\rangle}{\langle v, v\rangle\langle w, w\rangle - \langle v, w\rangle^2}, \qquad (5.14.10)$$

where $\delta = 1$ if $\langle \ , \ \rangle$ is positive definite on P and $\delta = -1$ if $\langle \ , \ \rangle$ is negative definite on P. Frequently, (5.14.10) is used as the definition of sectional curvature.

It follows from Problem 2.17.5 and (5.14.10) that the $K(P)$ for all P at one point and the metric at the point determine the curvature tensor at that point. Thus none of the information carried by the curvature tensor is lost by considering only sectional curvatures.

CHAPTER 6

Physical Applications

6.1. Introduction

The advent of tensor analysis in dynamics goes back to Lagrange, who originated the general treatment of a dynamical system, and to Riemann, who was the first to think of geometry in an arbitrary number of dimensions. Since the work of Riemann in 1854 was so obscurely expressed we find Beltrami in 1869 and Lipschitz in 1872 employing geometrical language with extreme caution. In fact, the development was so slow that the notion of parallelism due to Levi-Civita did not appear until 1917.

Riemannian geometry gradually evolved before the end of the nineteenth century, so that we find Darboux in 1889 and Hertz in 1899 treating a dynamical system as a point moving in a d-dimensional space. This point of view was employed by Painlevé in 1894, but with a euclidean metric for the most part. However, an adequate notation for riemannian geometry was still lacking.

The development of the tensor calculus by Ricci and Levi-Civita culminated in 1900 with the development of tensor methods in dynamics. Their work was not received with enthusiasm, however, until 1916, when the general theory of relativity made its impact.

The main purpose in applying tensor methods to dynamics is not to solve dynamical problems, as might be expected, but rather to admit the ideas of riemannian or even more general geometries. The results are startling. The geometrical spirit which Lagrange and Hamilton tried to destroy in their dynamics is revived; indeed, we see the system moving not as a complicated set of particles in E^3 but rather as a single particle in a riemannian d-dimensional space. The manifold of configurations (configuration space), in which a point corresponds to a configuration of the dynamical system and the manifold of events (configurations and times), in which a point corresponds to a configuration at a given time, will be considered below.

In this chapter the concept of a hamiltonian manifold, that is, a manifold carrying a distinguished closed 2–form of maximal rank everywhere, is introduced (see Section 2.23). An example is given by the tensor bundle $T^*M = T^0_1M$ of any d-dimensional manifold M (see Appendix 3A). In particular, we may take M to be the configuration space (see Section 6.5) of e particles in E^3 and in this case T^*M is known as phase space. The $(6e + 1)$-dimensional manifold obtained by taking the cartesian product of T^*M with R, called state space, is defined and motivates the notion of a contact manifold.

The Hamilton-Jacobi equations of motion

$$\dot{q}^i = \frac{\partial H}{\partial p_i}, \qquad \dot{p}_i = -\frac{\partial H}{\partial q^i},$$

where the p_i and q^i are generalized coordinates and momenta, H the hamiltonian function, and the dot differentiation with respect to the time, are shown to be invariant under a homogeneous contact transformation, that is, a coordinate change preserving the appearance of the 2–form

$$dp_i \wedge dq^i, \qquad i = 1, \ldots, 3e.$$

A contact manifold of dimension d is a manifold carrying a 1–form ω, called a *contact form*, such that $\omega \wedge (d\omega)^r \neq 0$, where $d = 2r + 1$. The 1–form

$$\omega = p_i dq^i - dt, \qquad i = 1, \ldots, 3e,$$

where t is the coordinate on R in $T^*M \times R$, is evidently a contact form.

6.2. Hamiltonian Manifolds

A d-dimensional manifold M, where $d = 2r$, is said to have a *hamiltonian* (or *symplectic*) *structure*, and M is then called a *hamiltonian* (or *symplectic*) *manifold*, if there is a distinguished closed 2–form Ω of maximal rank d defined everywhere on M. The form Ω is called the *fundamental form* of the hamiltonian manifold which is now denoted (M, Ω).

As in riemannian geometry, the nondegenerate bilinear form Ω may be viewed as a linear isomorphism $\Omega_m: M_m \to M_m{}^*$ for each $m \in M$, and hence as a bundle isomorphism† $\Omega: TM \to T^*M$. That is, we may raise and lower indices with respect to Ω, although because of the skew-symmetry of Ω there is now a difference in sign for the two uses of Ω. An explicit version of this isomorphism, which we shall employ in our computations, is given by the

---

† A *bundle isomorphism* is a diffeomorphism from one bundle to another which maps fibers into fibers and is an isomorphism of the fibers. In this case the fibers are the vector spaces M_m and M_m^*.

interior product operators $i(X)$ of Section 4.4. For a vector field X the corresponding 1–form ΩX is given by

$$\Omega X = i(X)\Omega. \qquad (6.2.1)$$

We shall denote the inverse map by $V = \Omega^{-1}: T^*M \to TM$, so that if τ is a 1–form on M, then $V\tau$ is a vector field on M and $\Omega V\tau = \tau$. In the notation ΩX and $V\tau$ it would be more correct to write $\Omega \circ X$ and $V \circ \tau$, indicating their structure as compositions of maps $X: M \to TM$ and $\Omega: TM \to T^*M$, and similarly for $V\tau$.

We define the *Poisson bracket* of the 1–forms τ and θ in terms of the corresponding fields $V\tau$ and $V\theta$ by

$$[\tau, \theta] = i([V\tau, V\theta])\Omega = \Omega[V\tau, V\theta].$$

Clearly the bracket operation on 1–forms is skew-symmetric.

Proposition 6.2.1. *The Poisson bracket of two closed forms is exact.*

Proof. Using the fact that $i(Y)\Omega$ is a contraction of $Y \otimes \Omega$ and that L_X commutes with contractions (Problem 3.6.3) and is a tensor algebra derivation, it is easy to show that L_X is an inner product derivation:

$$L_X(i(Y)\Omega) = i(L_X Y)\Omega + i(Y)L_X\Omega.$$

We also use the formula $L_X = di(X) + i(X)d$ (Theorem 4.4.1). So for closed 1–forms τ and θ, if $X = V\tau$ and $Y = V\theta$,

$$\begin{aligned}
[\tau, \theta] &= i([X, Y])\Omega \\
&= i(L_X Y)\Omega + i(Y)\,d\tau \\
&= i(L_X Y)\Omega + i(Y)\{di(X)\Omega + i(X)\,d\Omega\} \\
&= i(L_X Y)\Omega + i(Y)L_X\Omega \\
&= L_X i(Y)\Omega \\
&= di(X)\theta + i(X)\,d\theta \\
&= d\{i(X)\theta\}. \quad \blacksquare
\end{aligned}$$

The following theorem should be compared with the single point version, Theorem 2.23.1. The corresponding theorem for a 1–form ω, which should be thought of as a *primitive* for Ω, that is, $\Omega = d\omega$, is called Darboux's theorem and is stated below as Theorem 6.8.1. The proofs of these theorems make use of the converse of the Poincaré lemma (Theorem 4.5.1) and Frobenius' complete integrability theorem (Theorems 3.12.1 and 4.10.1).

Theorem 6.2.1. *Let Ω be a closed 2–form of rank $2k$ everywhere on the d-dimensional manifold M. Then, in a neighborhood of each point of M, coordinates p_i, q^i ($i = 1, \ldots, k$), and u^α ($\alpha = 1, \ldots, d - 2k$) exist such that*

$$\Omega = dp_i \wedge dq^i.$$

(Proof omitted.)

Since the fundamental form of a hamiltonian manifold (M, Ω) is closed and of rank d, it has local expressions

$$\Omega = dp_i \wedge dq^i, \qquad i = 1, \ldots, r = d/2.$$

We call such p_i, q^i *hamiltonian coordinates*. The jacobian matrix of two systems of hamiltonian coordinates, say, p_i, q^i and P_i, Q^i, is a symplectic matrix (see Theorem 2.23.2). Specifically, we have

$$\frac{\partial p_i}{\partial P_j} = \frac{\partial Q^j}{\partial q^i}, \qquad \frac{\partial p_i}{\partial Q^j} = -\frac{\partial P_j}{\partial q^i},$$
$$\frac{\partial q^i}{\partial P_j} = -\frac{\partial Q^j}{\partial p_i}, \qquad \frac{\partial q^i}{\partial Q^j} = \frac{\partial P_j}{\partial p_i}. \tag{6.2.2}$$

Problem 6.2.1. A manifold M admits a hamiltonian structure iff there is an atlas on M such that every pair of overlapping coordinate systems in the atlas satisfies equations (6.2.2).

Proposition 6.2.2. *In terms of hamiltonian coordinates p_i, q^i the operators Ω and V are given by*

$$\Omega\left(a_i \frac{\partial}{\partial p_i} + b^i \frac{\partial}{\partial q^i}\right) = -b^i \, dp_i + a_i \, dq^i, \tag{6.2.3}$$

$$V(f^i \, dp_i + g_i \, dq^i) = g_i \frac{\partial}{\partial p_i} - f^i \frac{\partial}{\partial q^i}. \tag{6.2.4}$$

In particular, for a function f on M,

$$V \, df = \frac{\partial f}{\partial q^i} \frac{\partial}{\partial p_i} - \frac{\partial f}{\partial p_i} \frac{\partial}{\partial q^i}. \tag{6.2.5}$$

[These follow immediately from the definition of the operator $i(X)$.]

If f and g are functions on M, then df and dg are closed 1-forms. The particular primitive for the bracket $[df, dg]$ found in the proof of Proposition 6.2.1 is denoted $\{f, g\}$ and is called the *Poisson bracket* of the functions f and g:

$$\{f, g\} = i(V \, df) \, dg = (V \, df)g. \tag{6.2.6}$$

In terms of hamiltonian coordinates p_i, q^i,

$$\{f, g\} = \frac{\partial f}{\partial q^i} \frac{\partial g}{\partial p_i} - \frac{\partial f}{\partial p_i} \frac{\partial g}{\partial q^i}. \tag{6.2.6}$$

The following proposition is left as an exercise.

Proposition 6.2.3. *Let f and g be functions on the hamiltonian manifold (M, Ω). Then the following are equivalent.*

(a) *f is constant along the integral curves of $V \, dg$.*
(b) *g is constant along the integral curves of $V \, df$.*
(c) *$\{f, g\} = 0$.*

Problem 6.2.2.   Verify:

$$\text{(a)} \; \{f, q^i\} = -\frac{\partial f}{\partial p_i}, \qquad \text{(a')} \; dq^i = \Omega \frac{\partial}{\partial p_i}.$$

$$\text{(b)} \; \{f, p_i\} = \frac{\partial f}{\partial q^i}, \qquad \text{(b')} \; dp_i = -\Omega \frac{\partial}{\partial q^i}.$$

Problem 6.2.3.   (a) The Poisson bracket operation is bilinear.
(b) Verify the identity

$$\{f, gh\} = g\{f, h\} + h\{f, g\}.$$

(c) Prove the *Jacobi identity*

$$\{f, \{g, h\}\} + \{g, \{h, f\}\} + \{h, \{f, g\}\} = 0.$$

A vector field X is said to be a *hamiltonian vector field* or an *infinitesimal automorphism of the hamiltonian structure* if it leaves the hamiltonian structure invariant, that is, if

$$L_X \Omega = 0.$$

Since Ω is closed,

$$L_X \Omega = di(X)\Omega + i(X)\, d\Omega$$
$$= di(X)\Omega.$$

Thus X is hamiltonian iff ΩX is closed. There are many closed 1–forms on a manifold (for example, the differential of any function), so hamiltonian structures are rich in automorphisms. This is just the opposite to the situation for riemannian structures, where the existence of isometries (automorphisms) and Killing fields is the exception rather than the rule.

Problem 6.2.4.   Show that the Lie bracket of two hamiltonian vector fields is hamiltonian.

Problem 6.2.5.   Let $N = P = S^2$ be the two-dimensional sphere of radius 1 in E^3, each provided with the inherited riemannian structure. Let $M = N \times P$ and let $q: M \to N$, $p: M \to P$ be the projections of the cartesian factorization of M. If the riemannian volume elements of N and P are α and β, show that $\Omega = q^*\alpha + p^*\beta$ is a hamiltonian structure on $M = S^2 \times S^2$.

6.3. Canonical Hamiltonian Structure on the Cotangent Bundle

The topological limitations for the existence of a hamiltonian structure on a manifold are quite severe, particularly in the compact case. However, there is one important class of hamiltonian manifolds—the cotangent bundles of other manifolds. By dualization with a riemannian metric we see that the tangent bundle of a manifold is diffeomorphic to the cotangent bundle, and therefore admits a hamiltonian structure also.

Theorem 6.3.1. *There is a canonical hamiltonian structure on the cotangent bundle T^*M of a manifold M.*

Proof. On the tangent bundle TT^*M of $T^*M = N$ we have two projections, the one into T^*M, the ordinary tangent bundle projection $\pi: TN \to N$, and the other into TM, the differential $p_*: TT^*M \to TM$ of the cotangent bundle projection $p: T^*M \to M$. When both are applied to the same vector $x \in TT^*M$ the two results interact to produce the value of a 1-form θ on x:

$$\langle x, \theta \rangle = \langle p_*x, \pi x \rangle.$$

That θ actually is a 1-form is clear, since $n = \pi x$ remains fixed as x runs through $(T^*M)_n$ and p_* is linear on each $(T^*M)_n$. Clearly $d\theta$ is closed. We shall show that it is of maximal rank. Local expressions for θ and $d\theta$ will be produced in the process.

Let $\{X_i\}$ be a local basis on M, $\{\omega^i\}$ the dual basis, and $p^*\omega^i = \tau^i$. Each X_i gives rise to a real-valued function p_i on T^*M, the evaluation on cotangents: If $n \in T^*M$, then

$$p_i n = \langle X_i(pn), n \rangle. \tag{6.3.1}$$

The expression in terms of $\{X_i(m)\}$ of an arbitrary vector $t \in M_m$ is given by the ω^i on t:

$$t = \langle t, \omega^i_m \rangle X_i(m).$$

For $x \in (T^*M)_n$, we have $p_*x \in M_m$, where $m = pn = p\pi x$, so

$$\begin{aligned} p_*x &= \langle p_*x, \omega^i_m \rangle X_i(m) \\ &= \langle x, p^*\omega^i_m \rangle X_i(m) \\ &= \langle x, \tau^i_n \rangle X_i(m). \end{aligned}$$

The local expression for θ is now easy to compute:

$$\begin{aligned} \langle x, \theta \rangle &= \langle p_*x, n \rangle \\ &= \langle \langle x, \tau^i_n \rangle X_i(m), n \rangle \\ &= \langle x, \tau^i_n \rangle \langle X_i(m), n \rangle \\ &= \langle x, \tau^i_n \rangle p_i n \\ &= \langle x, (p_i n)\tau^i_n \rangle, \end{aligned}$$

and thus

$$\theta = p_i \tau^i. \tag{6.3.2}$$

Taking the exterior derivative of θ gives

$$d\theta = dp_i \wedge \tau^i + p_i \, d\tau^i. \tag{6.3.3}$$

To show that $d\theta$ has maximal rank we consider the special case of a coordinate basis, $X_i = \partial/\partial x^i$. Then $\omega^i = dx^i$ and $d\tau^i = d(p^*dx^i) = p^*d^2x^i = 0$. Moreover, if $q^i = x^i \circ p$, then the p_i and q^i are the special coordinates on T^*M associated with the coordinates x^i on M (see Appendix 3A). Thus

$$d\theta = dp_i \wedge dq^i, \tag{6.3.4}$$

which is obviously of maximal rank. ∎

Note that the fundamental form $\Omega = d\theta$ of the hamiltonian structure on T^*M is exact. We call θ the *canonical 1–form* and Ω the *canonical 2–form* on T^*M.

For any C^∞ vector field X on M we define a C^∞ function P_X on T^*M, as we defined p_i for X_i in (6.3.1), by

$$P_X n = \langle X(pn), n \rangle.$$

We call P_X the *X-component of momentum*. Since dP_X is a closed 1–form on T^*M the vector field $V\,dP_X$ is an infinitesimal automorphism of the canonical hamiltonian structure on T^*M. In the following proposition we show how $V\,dP_X$ is obtained directly from the flow of X.

Proposition 6.3.1. *If $\{\mu_t\}$ is the flow of X, then $\{\mu_t^*\}$ is the flow of $V\,dP_X$. Moreover, $p_*V\,dP_X = -X \circ p$.*

Proof. Let $\theta_t = \mu_t^*: T^*M \to T^*M$, $M_m^* \to M_{\mu_{-t}m}^*$. Thus we have $p \circ \varphi_t = \mu_{-t} \circ p$. Now we show that the canonical 1–form θ is invariant under φ_t; that is, $\varphi_t^*\theta = \theta$. Indeed, for $y \in (T^*M)_n$,

$$\begin{aligned}
\langle y, (\varphi_t^*\theta)(n) \rangle &= \langle y, \varphi_t^*(\theta)\varphi_t n)) \rangle \\
&= \langle \varphi_{t*}y, \theta(\mu_t^*n) \rangle \\
&= \langle p_*\varphi_{t*}y, \mu_t^*n \rangle \\
&= \langle \mu_{t*}p_*\varphi_{t*}y, n \rangle \\
&= \langle (\mu_t \circ \mu_{-t} \circ p)_* y, n \rangle \\
&= \langle p_*y, n \rangle \\
&= \langle y, \theta(n) \rangle.
\end{aligned}$$

It is trivial to verify that $\{\varphi_t\}$ is a flow on T^*M, so there is a vector field W on T^*M whose flow is $\{\varphi_t\}$. What has been shown is that $L_W\theta = 0$. But $L_W\theta = di(W)\theta + i(W)\,d\theta = d\langle W, \theta \rangle + \Omega W$, so $W = -Vd\langle W, \theta \rangle$. However, for $n \in T^*M$, if γ is the integral curve of W starting at n, then $\gamma(t) = \varphi_t n$ and $p\gamma(t) = p\varphi_t n = \mu_{-t}pn$. Thus $p_*W(n) = -X(pn)$; that is, $p_*W = -X \circ p$. Finally,

$$\begin{aligned}
\langle W, \theta \rangle &= \langle p_*W, \pi W \rangle \\
&= -\langle X \circ p, \text{ the identity on } T^*M \rangle \\
&= -P_X. \quad \blacksquare
\end{aligned}$$

The vector field $-V\,dP_X$, which is p-related to X, is called the *canonical lift of X to T^*M*.

Problem 6.3.1. If x^i are coordinates on M, ∂_i the coordinate vector fields, p_i the ∂_i-component of momentum, $q^i = x^i \circ p$, and $X = f^i\,\partial_i$, then the canonical lift of X to T^*M is

$$-V\,dP_X = -p_i(\partial_j f^i \circ p)\frac{\partial}{\partial p_j} + f^i \circ p\,\frac{\partial}{\partial q^i}.$$

Problem 6.3.2. A tangent y to T^*M is called *vertical* if $p_*y = 0$.

(a) The vertical vectors in $(T^*M)_n$ are a d-dimensional subspace $W(n)$ of $(T^*M)_n$.

(b) The distribution W has as local basis $\{\partial/\partial p_i\}$.

(c) The map $\alpha_n \colon W(n) \to M^*_{pn}$ defined by $\alpha_n(c_i\,\partial/\partial p_i(n)) = c_i\,dx^i(pn)$ is a linear isomorphism independent of the choice of coordinates x^i.

(d) The vector field $V\theta$ is vertical and $\alpha_n(V\theta(n)) = n$. That is, $V\theta$ is the "displacement vector field" when tangent vectors to M^*_m are identified with elements of M^*_m.

Problem 6.3.3. We can define the *canonical lift of X to TM* as the vector field whose flow is the dual flow $\{\mu_{t*}\}$. Show that the canonical lift to TM is $X_* \colon TM \to TTM$.

6.4.　Geodesic Spray of a Semi-riemannian Manifold

The hamiltonian structure on T^*M enables us to obtain the geodesic spray (see Theorem 5.12.1) of a semi-riemannian connexion on M directly in terms of the energy function. We redefine the *energy function K on TM* by inserting a factor of $1/2$: For $v \in TM$,

$$Kv = \tfrac{1}{2}\langle v, v\rangle,$$

where $\langle\ ,\ \rangle$ is the semi-riemannian metric. The identification of tangents and cotangents due to $\langle\ ,\ \rangle$ is a bundle isomorphism $\mu \colon T^*M \to TM$. By means of μ we may transfer the energy function to a function on T^*M, $T = K \circ \mu$. Likewise, the geodesic spray G on TM is μ-related to a vector field J on T^*M, $J = \mu_*^{-1} \circ G \circ \mu$. The notation is justified by the following commutative diagram:

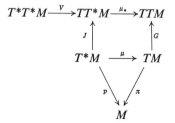

We retain the names "energy function" and "geodesic spray" for T and J.

Lemma 6.4.1.   *The geodesic spray J is characterized as follows. Let D be the semi-riemannian connexion and $E = p^*D$ the connexion over p induced by D. View μ as a vector field over p. Then*

(a') $p_* J = \mu$.
(b') $E_J \mu = 0$.

Proof. The equations (a') ánd (b') are immediate from (a) and (b) of Theorem 5.12.1 by chasing the diagram.

The following lemma is given by some simple computations which we omit.

Lemma 6.4.2.   *Let $\{F_i\}$ be a local orthonormal basis on M, $a_i = \langle F_i, F_i \rangle$, and $p_i = P_{F_i}$, the F_i-component of momentum. Then the local expressions for μ and T are*

$$\mu = \sum a_i p_i F_i \circ p, \tag{6.4.1}$$

$$T = \frac{1}{2} \sum a_i p_i^2. \tag{6.4.2}$$

Theorem 6.4.1.   *The geodesic spray on T^*M of a semi-riemannian connexion D on M is $J = -V\,dT$, where T is the energy function on T^*M and V is the inverse of the canonical hamiltonian operator Ω on T^*M.*

Proof. We shall use the notation of the above lemmas. Let $\{\omega^i\}$ be the dual basis of $\{F_i\}$, ω^i_j the connexion forms of D, $\tau^i = p^*\omega^i$, and $\tau^i_j = p^*\omega^i_j$. Then the τ^i_j are the connexion forms of $E = p^*D$ and they, as well as the ω^i_j, satisfy the skew-adjointness condition: $\tau^i_j = -a_i a_j \tau^j_i$ (no sum). The first structural equations pull back to T^*M to give

$$d\tau^i = -\tau^i_j \wedge \tau^j,$$

which may be substituted in (6.3.3) to obtain the local expression for the canonical 2–form

$$\Omega = (dp_i - p_j \tau^j_i) \wedge \tau^i.$$

Let $X = -V\,dT$. Then, by (6.4.2) and the definition of V,

$$dT = \sum a_i p_i \, dp_i$$
$$= -i(X)\Omega$$
$$= \tau^i(X)\,dp_i - \{Xp_i - p_j\tau^j_i(X)\}\tau^i - \tau^i(X)p_j\tau^j_i. \tag{6.4.3}$$

Since $\{dp_i, \tau^i\}$ is a local dual basis on T^*M and the τ^i_j are linear combinations of the τ^i, the coefficients of dp_i in (6.4.3) must match:

$$\tau^i(X) = a_i p_i. \tag{6.4.4}$$

It follows that $\tau^i(X)p_j\tau^j_i$ vanishes because

$$\tau^i(X)p_j\tau^j_i = \sum_{i,j} a_i p_i p_j \tau^j_i = -\sum_{i,j} a_i p_i p_j a_j a_i \tau^i_i$$

$$= -\sum_{i,j} a_j p_j p_i \tau^i_j.$$

Thus the remaining terms in (6.4.3) are zero; that is,

$$Xp_i - p_j \tau^j_i(X) = 0. \tag{6.4.5}$$

The formulas (6.4.4) and (6.4.5) are the local expressions for the fact that X satisfies (**a'**) and (**b'**):

(**a'**) $p_* X = \omega^i(p_* X)F_i \circ p = \tau^i(X)F_i \circ p = \sum a_i p_i F_i \circ p = \mu.$

(**b'**) $E_X \mu = E_X \sum a_i p_i F_i \circ p = \sum a_i(Xp_i)F_i \circ p + \sum a_j p_j E_X F_j \circ p$

$$= \sum_i \{a_i(Xp_i) + \sum_j a_j p_j \tau^i_j(X)\}F_i \circ p$$

$$= \sum_i a_i\{Xp_i - \sum_j p_j \tau^j_i(X)\}F_i \circ p$$

$$= 0.$$

Hence $X = J$ by Lemma 6.4.1. ∎

6.5. Phase Space

Let us consider the classical mechanics of e particles in R^3 with masses $m_1, \ldots, m_e$. Since no two particles can occupy the same position, the *configuration space* M of this system is a subset of R^{3e}. If $(x^{3i-2}, x^{3i-1}, x^{3i})$ are the coordinates of the ith particle, then the points of M are those points of R^{3e} for which

$$(x^{3i-2} - x^{3j-2})^2 + (x^{3i-1} - x^{3j-1})^2 + (x^{3i} - x^{3j})^2 \neq 0,$$

for all $i \neq j$. Thus M is an open submanifold of R^{3e} and has dimension $3e$.

If $(F_{3i-2}, F_{3i-1}, F_{3i})$ are the components of the force field on the ith particle, then the equations of motion are

$$F_{3i-\alpha} = m_i \frac{d^2 x^{3i-\alpha}}{dt^2} \quad \text{(no sum)},$$

where $i = 1, \ldots, e$, and $\alpha = 2, 1, 0$. Setting $k_1 = k_2 = k_3 = m_1$, $k_4 = k_5 = k_6 = m_2$, etc., the equations become

$$k_i \frac{d^2 x^i}{dt^2} = F_i \quad \text{(no sum)}, \tag{6.5.1}$$

$i = 1, \ldots, 3e$. The generalized force components F_i are given by $F_i = -\partial U/\partial x^i$ in the case where a *potential energy function* U exists.

Another feature of this system is that there is a *kinetic-energy function K*, the sum of the kinetic energies of the *e* particles. It is a function of the velocities of the particles and hence a function on the tangent bundle *TM* of the configuration space. For $v = v^i \, \partial_i(m) \in M_m$, the coordinate formula is

$$K = \frac{1}{2} \sum k_i (v^i)^2.$$

Since *K* is a quadratic form on each M_m, it may be polarized (see Section 2.21) to obtain a riemannian metric on *M* for which *K* is the energy function:

$$\langle \ , \ \rangle = \sum k_i (dx^i \otimes dx^i).$$

This riemannian metric is called the *kinetic-energy metric* on *M*. In the simple example under discussion this metric is affine, so the geodesics are the straight lines in R^d. If the force field vanishes, $F_i = 0$, then the solutions to (6.5.1) are exactly the geodesics, a fact which generalizes to more general systems.

Let us examine this example in light of the previous structures we have studied—riemannian and hamiltonian. The second-order differential equations of motion can be viewed as a vector field on the tangent bundle of the configuration space. However, from physical arguments we conclude that a force field should be a 1-form—not a vector field. Indeed, elementary evidence that a force field exists is usually the fact that work is done in moving along various curves. The nature of these work values associated with curves is precisely the same as the association of the value of an integral of a 1-form with a curve. Moreover, a force field is frequently given by the differentiation of a potential field *U*, which makes invariant sense only if the force is $-dU$. The amount of work done along a curve is independent of the mass moved, so the masses involved in (6.5.1) must be related to some other part of the structure. The change from a 1-form force to the vector field force apparent in (6.5.1) is due to the identification of tangents and cotangents by means of the kinetic-energy metric. The interaction of these two items, the force 1-form and the kinetic-energy metric, are a sufficient formulation of the structure of a classical mechanics problem. How, then, does the hamiltonian structure enter the picture? The answer seems to be one of convenience, improvement of insight, and better possibilities for generalization rather than necessity. The hamiltonian structure on T^*M is available, so we might as well use it. The abstract theorems of Sectiòn 6.2. translate quickly and naturally into significant theorems on conservation of momentum.

In summary, a possible mathematical model of a mechanical system consists of

(a) A configuration space *M*.

(b) A force field, a 1-form *F* on *M*.

(c) A kinetic-energy metric, which gives us a diffeomorphism $\mu: TM \to T^*M$ and allows us to view velocity fields of trajectories and all other features, as a part of T^*M (that is, we assign a momentum to a velocity via $\langle \ , \ \rangle$).

(d) The canonical hamiltonian structure on T^*M.

Let us carry out the transfer of all the structure of the above example to T^*M in terms of the local coordinates $p_i = P_{\partial_i}$ and $q^i = x^i \circ p$ on T^*M. Then the differential equations for the particle paths (trajectories), (6.5.1), are carried into first-order differential equations for the momentum path in T^*M,

$$\frac{dq^i}{dt} = \frac{1}{k_i} p_i, \qquad \frac{dp_i}{dt} = F_i \circ p. \qquad (6.5.2)$$

A flow is therefore defined on *phase space* T^*M with vector field

$$X = \sum_i \left(\frac{p_i}{k_i} \frac{\partial}{\partial q^i} + F_i \circ p \frac{\partial}{\partial p_i} \right). \qquad (6.5.3)$$

The integral curves of X are also called *trajectories*.

Using formula (6.2.3) for the operator Ω of the canonical hamiltonian structure on T^*M gives

$$\Omega X = - \sum \frac{p_i}{k_i} dp_i + \sum F_i \circ p \, dq^i$$
$$= -dT + p^*F, \qquad (6.5.4)$$

where $T = \frac{1}{2} \sum \frac{p_i^2}{k_i} = K \circ \mu$ is the *kinetic energy on* T^*M and $F = F_i \, dx^i$ is the force field on M. In the case where the force field is a potential field, say, $F = -dU$, then letting $V = U \circ p$ we get the *hamiltonian function* on T^*M; that is, the total energy of the system

$$H = T + V.$$

Then we can write (6.5.4) as

$$\Omega X = -dH.$$

It follows from Proposition 6.2.3 and the trivial fact $\{H, -H\} = 0$ that H is constant along the trajectories. This is called the *law of conservation of energy*. More generally, Proposition 6.2.3 shows a function f on T^*M is constant on trajectories iff $\{f, H\} = 0$; that is $(V \, df)H = 0$.

Suppose that the potential U depends only on the euclidean distances between particles. Then any euclidean motion of R^3 leaves U invariant. The extension of such a motion to TR^3 also leaves the kinetic energy of each particle invariant. We extend a euclidean motion to a diffeomorphism φ of R^{3e} by making it act the same on each of the e copies of R^3, and M is obviously an invariant subset of this extension. Finally, φ is extended to T^*M, that is, to $(\varphi^*)^{-1}$. The inverse is required to make the projection be φ, since φ^* pulls forms back rather than pushing them forward. The extension $(\varphi^*)^{-1}$ leaves T and V invariant, and hence leaves H invariant; that is, $H \circ \varphi^* = H$.

A parallel vector field $Y = a\partial_x + b\partial_y + c\partial_z$ on R^3, where a, b, and c are constant, has as its flow a 1-parameter group of translations. The extension to M is

$$Z = \sum_{i=1}^{e} a\, \partial/\partial x^{3i-2} + b\, \partial/\partial x^{3i-1} + c\, \partial/\partial x^{3i}.$$

The extension to T^*M is the canonical lift $-V\,dP_Z$ of Z (see Proposition 6.3.1). When U depends only on distances H is invariant under the flow of $-V\,dP_Z$; that is, H is constant along the integral curves of $-V\,dP_Z$. Again by Proposition 6.2.3, it follows that P_Z is constant along trajectories. This is the *law of conservation of linear momentum*. Similarly, the *law of conservation of angular momentum* is derived by taking Y to be the vector field on R^3 whose flow is a 1-parameter group of rotations about some axis.

There is no need to confine the above analysis to the case of e particles in R^3 or the case where the kinetic-energy metric is affine. More general mechanical systems are produced by introducing restraints on the positions and velocities of the particles. One example is the double pendulum which has been discussed in Section 1.2(c). The configuration space of a rigid object free to rotate around a fixed point is RP^3, the three-dimensional projective space.

A system in which all the restraints on the velocities are consequences of the restraints on the positions is called *holonomic*. In such a system every tangent to M, the configuration space, is the tangent to some trajectory, so the collection of possible velocities is all of TM. If the same particles are viewed as being in R^3, then the configuration space becomes a submanifold of R^{3e}, with dimension equal to the number of degrees of freedom. The kinetic-energy metric considered above restricts to M and still gives an identification of TM and T^*M, so the latter is the phase space in the holonomic case. The force field F is still a 1-form on M, the kinetic-energy function T is defined on T^*M, and the trajectory flow on T^*M is given as $X = V(-dT + p^*F)$, as above. If the force field vanishes, the trajectories in M are the geodesics of the kinetic-energy metric, by Theorem 6.4.1.

If the force field is a potential field $-dU$, then the hamiltonian $H = T + V$ is defined as above, and the equations of motion on T^*M are those given by the vector field $-V\,dH$. Thus if p_i, q^i are hamiltonian coordinates for the canonical structure on T^*M, the equations of motion are the *Hamilton-Jacobi equations*

$$\frac{dq^i}{dt} = \frac{\partial H}{\partial p_i}, \qquad \frac{dp_i}{dt} = -\frac{\partial H}{\partial q^i}. \tag{6.5.5}$$

One advantage of this analysis is that we know that arbitrary hamiltonian coordinates may be used. We do not require, for example, that they arise from coordinates x^i on M, that is, $p_i = P_{\partial_i}$ and $q^i = x^i \circ p$. Instead, we may try to find coordinates which simplify the expression for H. In particular, if we can include solutions of $\{f, H\} = 0$ (including H itself) among the coordinates, then these coordinates will not appear except in the specification of their initial (and hence perpetual) values.

A system for which there are restraints on the velocities which are not implicit in the restraints on the positions is called *nonholonomic* (or *anholonomic*). In the commonest type of nonholonomic system the velocity restraints are linear at each point of the configuration space and thus determine a distribution D on M. We call this a *linear nonholonomic system*. If the distribution D is completely integrable, then the maximal integral submanifolds slice M into a family of holonomic systems; that is, for each initial state there are additional positional restraints, giving a holonomic system which includes the trajectory of the initial state. In the genuinely nonholonomic system, the restraints determine a submanifold Q of TM, where M is the configuration space, and we define the *phase space* to be $P = \mu^{-1}Q$, a submanifold of T^*M. The force field F may fail to be consistent with the velocity restraints, so we must use the restriction of p^*F to P in the equations of motion (6.5.2). The first d of these equations, $dq^i/dt = p_i/k_i$, remain unchanged except for the restriction of the p_i to P, because they express the fact that the curve in T^*M is the velocity field of the curve in M transformed by μ^{-1}. However, the next step, corresponding to equation (6.5.4), breaks down because Ω becomes degenerate on P and is no longer a hamiltonian structure. To show this we note that locally a nonholonomic system is given by k restraints of the type

$$dp_i = \sum_{j=k+1}^{d} f^j\, dp_j, \qquad i = 1, \dots, k, \tag{6.5.6}$$

where p_i is the ∂_i-momentum component for some coordinate system x^i on M with $\partial_i = \partial/\partial x^i$. We leave as an exercise the proof of the fact that when (6.5.6) is substituted in $\Omega = dp_i \wedge dq^i$, the rank becomes $2(d - k)$. But $\dim P = 2d - k$.

An example of a linear nonholonomic system is given by a ball rolling on a surface without sliding. The configuration space is the same as in the case of a sliding ball and is thus five-dimensional. But there are two linear restraints on the velocities, so the phase space P is eight-dimensional. The velocity distribution is not integrable, since any configuration can be reached from any other by sufficient rolling.

For the remaining sections we assume that the systems under study are holonomic.

6.6.   State Space

In the above analysis we have ignored the possibility that the force may be time-dependent. This occurrence does not ruin the analysis since we may insert the time variable t as an extra parameter. The equations of motion (6.5.2) are still valid but we must bear in mind that $F_i \circ p$ is not defined on T^*M but on $S = T^*M \times R$. We also need another equation for the remaining variable t, namely, $dt/dt = 1$. Thus the vector field X, given by (6.5.3), must be replaced by

$$X = \sum_i \left(\frac{p_i}{k_i} \frac{\partial}{\partial q^i} + F_i \circ p \frac{\partial}{\partial p_i} \right) + \frac{\partial}{\partial t}. \tag{6.6.1}$$

We may regard the previous X as a family of vector fields on T^*M depending on a parameter t, in which case (6.5.4) still makes sense. We call S the *state space* of the system.

In the case where the force is the potential field of a time-dependent potential function $V = U \circ p$ on T^*M, the component $(\partial V/\partial t) \, dt$ of dH must be discarded. Thus we have

$$X = -V \, dH + \frac{\partial}{\partial t}, \tag{6.6.2}$$

where $V = \Omega^{-1}$. The condition that a function f on S be constant on trajectories, $Xf = 0$, may be written

$$\{f, H\} + \frac{\partial f}{\partial t} = 0. \tag{6.6.3}$$

A solution f to the partial differential equation (6.6.3) is called a *first integral* of the equations of motion. To simplify the coordinate expression for the equations of motion the obvious technique is to include numerous first integrals among the coordinate functions on S. However, H is no longer a first integral in the time-dependent case, since $\{H, H\} + \partial H/\partial t = \partial H/\partial t \neq 0$, so total energy is not conserved.

6.7.   Contact Coordinates

If $q: T^*M \times R \to T^*M$ is the cartesian product projection, that is, $q(n,t) = n$, then we get a 1–form $q^*\theta$ on S from the canonical 1–form θ on T^*M. We shall not distinguish $q^*\theta$ and θ notationally, thus viewing θ as a 1–form on S. We call $\omega = \theta - dt$ the *canonical contact form* on $T^*M \times R$. In terms of the special coordinates $p_i = P_{\theta_i}$ and $q^i = x^i \circ p$ we have

$$\omega = p_i \, dq^i - dt. \tag{6.7.1}$$

Coordinates P_i, Q^i, u on S are called *contact coordinates* if the expression for ω has the same appearance; that is,

$$\omega = P_i \, dQ^i - du. \tag{6.7.2}$$

If we take the exterior derivatives of (6.7.1) and (6.7.2) we obtain

$$\Omega = dp_i \wedge dq^i = dP_i \wedge dQ^i; \qquad (6.7.3)$$

that is, the expression for the 2–form Ω has the same appearance for any contact coordinate system. Moreover, the codistribution in T^*S spanned by the dp_i and the dq^i is the same as the codistribution E^* spanned by the dP_i and the dQ^i—the range of Ω viewed as a map $TS \rightarrow T^*S$ (see Theorem 2.23.1). The associated distribution E is thus one-dimensional and it is clearly spanned by $\partial/\partial t$ or $\partial/\partial u$. Thus $\partial/\partial t$ and $\partial/\partial u$ are linearly dependent; that is, $\partial/\partial u = f \partial/\partial t$. But $\langle \partial/\partial u, \omega \rangle = \langle \partial/\partial u, du \rangle = 1 = \langle f \partial/\partial t, \omega \rangle = f\langle \partial/\partial t, dt \rangle = f$, so $\partial/\partial u = \partial/\partial t$. In other words, the vector field $\partial/\partial t$ is determined uniquely by ω and is a coordinate vector field of any contact coordinate system. An immediate consequence is that

$$u = t + \text{a function of the } p_i \text{ and the } q^i.$$

Now we show that the trajectories are determined by ω, the 1–form $\tau = -dT + p^*F$, and the kinetic-energy function T. First we have that $i(\partial/\partial t)\Omega = 0$, so equation (6.5.4) is still valid with the new trajectory field X given by (6.6.1) on S; that is,

$$i(X)\Omega = \tau. \qquad (6.7.4)$$

This implies that τ belongs to the codistribution E^*. Second, we have

$$\langle X, \omega \rangle = 2T - 1, \qquad (6.7.5)$$

from (6.6.1) and (6.7.1). The desired result follows from (6.7.4), (6.7.5), and the following proposition.

Proposition 6.7.1. *Let τ be any 1–form belonging to the codistribution E^* annihilated by $\partial/\partial t$ and let f be any function on S. Then there is a unique vector field X such that :*

(a) $i(X)\Omega = \tau$.
(b) $\langle X, \omega \rangle = f$.

Outline of proof. Let $X = f_i \partial/\partial p_i + g^i \partial/\partial q^i + h \partial/\partial t$, $\tau = a^i dp_i + b_i dq^i$ and compute (a) and (b) in terms of coordinates. ∎

Suppose that P_i, Q^i are hamiltonian coordinates on T^*M. Then

$$d(P_i dQ^i - p_i dq^i) = \Omega - \Omega = 0,$$

so by the converse of the Poincaré lemma (Theorem 4.5.1) there is a function f such that

$$P_i dQ^i - p_i dq^i = df.$$

Letting $u = t + f$ we have $\omega = P_i dQ^i - du$; that is, P_i, Q^i, u are contact coordinates. These special contact coordinates, for which the P_i and the Q^i

are functions on T^*M, are called *homogeneous contact coordinates*. The fact that homogeneous contact coordinates always arise from hamiltonian coordinates as above is trivial to prove. Thus the contact coordinates are more general than hamiltonian coordinates, and consequently the freedom to operate with contact coordinates gives us greater simplifying power.

6.8. Contact Manifolds

A manifold M of dimension $d = 2r + 1$ is said to have a *contact structure*, and M is then called a *contact manifold*, if M has a distinguished 1–form ω such that

$$\omega \wedge (d\omega)^r \neq 0.$$

The form ω is then called a *contact form*. (The power of $d\omega$ is the iterated wedge product.)

As with the canonical contact structure on $T^*M \times R$ discussed above, we get a one-dimensional distribution E and the associated codistribution E^*, the latter being spanned by the range of $\Omega = d\omega$ considered as a map from tangents to cotangents, and E being the space annihilated by Ω; that is, a vector field Y belongs to E iff $i(Y)\Omega = 0$. A special basis for E is singled out by the further condition $\langle Y, \omega \rangle = 1$, and the $Y \in E$ which satisfies this is called the *contact vector field*. More generally, we have that Proposition 6.7.1 is valid for an arbitrary contact structure.

We define *contact coordinates* on a contact manifold to be coordinates $p_i, q^i, t, i = 1, \ldots, r$, such that the expression for ω is

$$\omega = p_i \, dq^i - dt.$$

That contact coordinates exist is a consequence of the following statement, known as *Darboux's theorem*, the proof of which is omitted. The purpose is to give a canonical simple coordinate expression for a 1–form whose algebraic relation to its exterior derivative is stable in a neighborhood.

Theorem 6.8.1.   *Let ω be a 1–form defined in a neighborhood of $m \in M$.*

(a) *If $\omega \wedge (d\omega)^k = 0$ and $(d\omega)^{k+1} = 0$ in a neighborhood of m, and $(d\omega_m)^k \neq 0$ and $\omega_m \neq 0$, then there are coordinates p_i, q^i $(i = 1, \ldots, k)$, and u^α $(\alpha = 1, \ldots, d - 2k)$ at m such that*

$$\omega = \sum_{i=1}^{k} p_i \, dq^i.$$

(b) *If $(d\omega)^{k+1} = 0$ in a neighborhood of m and $\omega_m \wedge (d\omega_m)^k \neq 0$, then there are coordinates p_i, q^i $(i = 1, \ldots, k)$, t, and u^α $(\alpha = 1, \ldots, d - 2k - 1)$ at m such that*

$$\omega = \sum_{i=1}^{k} p_i \, dq^i - dt.$$

It is case **(b)** which applies to a contact form, with $k = r$.

If M is a contact manifold with contact form ω, then the one-dimensional codistribution spanned by ω has as its associated distribution the $2r$-dimensional distribution D annihilated by ω; that is,

$$D(m) = \{x \in M_m \mid \langle x, \omega \rangle = 0\}.$$

We call D the *contact distribution*.

A diffeomorphism f of M onto M is said to be a *contact transformation* of M if $f^*\omega = h\omega$, where h is a nowhere-zero function on M. Equivalently, $f_* D = D$. It can be shown, using the techniques of Section 4.10, that the highest dimension of integral submanifolds of D is r. Moreover, if p_i, q^i, t are contact coordinates, then the coordinate slices $q^i = c^i$, $t = c$ are r-dimensional integral submanifolds. These facts allow us to state the following characterization of a contact transformation.

Theorem 6.8.2.   *A diffeomorphism of a contact manifold M maps every integral submanifold of D of highest dimension to another integral submanifold of D iff it is a contact transformation of M.*

Bibliography

1. Background material

Bartle, R. G., *The elements of real analysis*, Wiley, New York, 1964.
Kaplan, W., *Advanced calculus*, Addison-Wesley, Reading, Mass., 1952.

2. Set theory and topology

Gaal, S. A., *Point set topology*, Academic Press, New York, 1964.
Hocking, J., and Young, G., *Topology*, Addison-Wesley, Reading, Mass., 1961.
Kelley, J., *General topology*, Van Nostrand, Princeton, N.J., 1955.
Simmons, G., *Introduction to topology and modern analysis*, McGraw-Hill, New York, 1963.

3. Linear algebra and matrix theory

Greub, W., H., *Linear algebra*, 2nd ed. Academic Press, New York, 1963.
Hoffman, K., and Kunze, R., *Linear algebra*, Prentice-Hall, Englewood Cliffs, N,J., 1961.
Hohn, F. E., *Elementary matrix algebra*, Macmillan, New York, 1964.
Marcus, M., and Minc, H., *A survey of matrix theory and matrix inequalities*, Allyn & Bacon, Boston, 1964.

4. Differential equations

Birkhoff, Garrett, and Rota, Gian-Carlo, *Ordinary differential equations*, Ginn, Boston, 1962.
Coddington, E., and Levinson, N., *Theory of ordinary differential equations*, McGraw-Hill, New York, 1955.
Greenspan, D., *Theory and solution of ordinary differential equations*, Macmillan, New York, 1960.

5. Classical tensor calculus

Eisenhart, L. P., *Riemannian geometry*, Princeton University Press, Princeton, N.J., 1949.
Schouten, J., *Ricci calculus*, Springer, Berlin, 1954.
Sokolnikoff, I., *Tensor analysis*, Wiley, New York, 1964.

Spain, B., *Tensor calculus*, Interscience, New York, 1953.

Synge, J., and Schild, A., *Tensor calculus*, University of Toronto Press, Toronto, 1949.

6. Differential geometry

Auslander, L., and MacKenzie, R., *Introduction to differentiable manifolds*, McGraw-Hill, New York, 1963.

Flanders, H., *Differential forms*, Academic Press, New York, 1963.

Hicks, N., *Notes on differential geometry*, Van Nostrand, Princeton, N.J., 1965.

Laugwitz, D., *Differential and riemannian geometry*, Academic Press, New York, 1965.

O'Neill, B., *Elementary differential geometry*, Academic Press, New York, 1966.

Struik, D., *Differential geometry*, 2nd ed., Addison-Wesley, Reading, Mass., 1961.

Willmore, T. J., *An introduction to differential geometry*, Clarendon Press, Oxford, 1959.

Index